Astronomy: A Comprehensive Study

Astronomy: A Comprehensive Study

Edited by
Alejandro Cooke

Larsen & Keller
www.larsen-keller.com

Astronomy: A Comprehensive Study
Edited by Alejandro Cooke
ISBN: 978-1-63549-033-6 (Hardback)

☰ Larsen & Keller

Published by Larsen and Keller Education,
5 Penn Plaza,
19th Floor,
New York, NY 10001, USA

Cataloging-in-Publication Data

Astronomy : a comprehensive study / edited by Alejandro Cooke.
    p. cm.
Includes bibliographical references and index.
ISBN 978-1-63549-033-6
1. Astronomy. 2. Astronomy--Observations. 3. Astronomical instruments. I. Cooke, Alejandro.
QB43.3 .A88 2017
520--dc23

For more information regarding Larsen and Keller Education and its products, please visit the publisher's website www.larsen-keller.com

Table of Contents

Permissions

Index

Preface

This book provides comprehensive insights into the field of astronomy. It attempts to understand the multiple branches that fall under this discipline and explains how such concepts have practical applications. Astronomy refers to the study of the chemistry, evolution and physical properties of interstellar objects. Astronomy includes subjects like celestial navigation, calendar making, observational astronomy and astrophysics, etc. Most of the topics introduced in this text cover new techniques and the applications of astronomy. While understanding the long-term perspectives of the topics, the book makes an effort in highlighting their impact as a modern tool for the growth of the discipline. Through this book, we attempt to further enlighten the readers about the new concepts in this field.

A short introduction to every chapter is written below to provide an overview of the content of the book:

Chapter 1 - The study of astronomical objects and their processes along with the history of their evolution is known as astronomy. Astronomy is an ancient subject, which has been studied by civilizations such as the Mayans, the Greeks and the Chinese. This chapter will provide an integrated understanding of astronomy; **Chapter 2 -** Astronomical objects are physical entities that exist in the observable universe. Planets, minor planets, asteroids, galaxies, nebulas are some of the astronomical objects explained in this chapter. They provide us with comprehensive knowledge about our universe. This chapter is an overview of the subject matter incorporating all the major aspects of astronomical objects; **Chapter 3 -** Astronomy is an interdisciplinary subject. It spreads to other fields as well. Some of the branches of astronomy covered in this chapter are astrometry, astrophysics, galactic astronomy, infrared astronomy and neutrino astronomy. Astronomy is best understood in confluence with the major topics listed in the following chapter; **Chapter 4 -** Observational astronomy is concerned with recording data; it is the practice of observing objects by using telescopes and other astronomical tools. This chapter is a compilation of the various branches of astronomy such as observational astronomy, gravitational wave astronomy, visible light astronomy and radio astronomy; **Chapter 5 -** Techniques are an important component of any field of study. Techniques used in astronomy are polarimetry, photometry and astronomical spectroscopy. It can be used to accurately position astronomic devices. The following chapter elucidates the techniques used in astronomy in a critical manner providing key analysis to the subject matter; **Chapter 6 -** Telescope, zenith telescope, spectroscopy and astrolabe are the essential instruments used in astronomy. Telescope is a device used to perceive distant objects whereas a zenith telescope is particularly used to measure the distances of stars. The categories of the instruments are dealt with great details in this chapter; **Chapter 7 -** Some of the models of astronomy are theoretical astronomy, geocentric model, heliocentrism and cosmogony. The geocentric model of astronomy considers the Earth to be the center of the universe, under this model, the sun, the moon, the planets all circle around the Earth. The following content explains all the models of astronomy, providing the reader with a detailed understanding on the topic; **Chapter 8 -** Astronomy is one of the oldest sciences practiced and is an interdisciplinary subject. It spreads to other fields as well. Archaeoastronomy, astrochemistry along with cosmochemistry have been explained, this chapter will provide a glimpse of related fields of astronomy briefly.

Finally, I would like to thank my fellow scholars who gave constructive feedback and my family members who supported me at every step.

Editor

Introduction to Astronomy

The study of astronomical objects and their processes along with the history of their evolution is known as astronomy. Astronomy is an ancient subject, which has been studied by civilizations such as the Mayans, the Greeks and the Chinese. This chapter will provide an integrated understanding of astronomy.

Astronomy, a natural science, is the study of celestial objects (such as stars, galaxies, planets, moons, asteroids, comets and nebulae) and processes (such as supernovae explosions, gamma ray bursts, and cosmic microwave background radiation), the physics, chemistry, and evolution of such objects and processes, and more generally all phenomena that originate outside the atmosphere of Earth. A related but distinct subject, physical cosmology, is concerned with studying the Universe as a whole.

A star-forming region in the Large Magellanic Cloud, an irregular galaxy.

Astronomy is the oldest of the natural sciences. The early civilizations in recorded history, such as the Babylonians, Greeks, Indians, Egyptians, Nubians, Iranians, Chinese, and Maya performed methodical observations of the night sky. Historically, astronomy has included disciplines as diverse as astrometry, celestial navigation, observational astronomy and the making of calendars, but professional astronomy is nowadays often considered to be synonymous with astrophysics.

During the 20th century, the field of professional astronomy split into observational and theoretical branches. Observational astronomy is focused on acquiring data from observations of astronomical objects, which is then analyzed using basic principles of physics. Theoretical astronomy is oriented toward the development of computer or analytical models to describe astronomical objects and phenomena. The two fields complement each other, with theoretical astronomy seeking to explain the observational results and observations being used to confirm theoretical results.

A giant Hubble mosaic of the Crab Nebula, a supernova remnant

Astronomy is one of the few sciences where amateurs can still play an active role, especially in the discovery and observation of transient phenomena. Amateur astronomers have made and contributed to many important astronomical discoveries, such as finding new comets.

Etymology

Astronomy means "law of the stars" (or "culture of the stars" depending on the translation). Astronomy should not be confused with astrology, the belief system which claims that human affairs are correlated with the positions of celestial objects. Although the two ields share a common origin, they are now entirely distinct.

19th century Sydney Observatory, Australia (1873)

19th century Quito Astronomical Observatory is located 12 minutes south of the Equator in Quito, Ecuador.

Use of Terms "Astronomy" and "Astrophysics"

Generally, either the term "astronomy" or "astrophysics" may be used to refer to this subject. Based on strict dictionary definitions, "astronomy" refers to "the study of objects and matter outside the Earth's atmosphere and of their physical and chemical properties" and "astrophysics" refers to the branch of astronomy dealing with "the behavior, physical properties, and dynamic processes of celestial objects and phenomena". In some cases, as in the introduction of the introductory text-book *The Physical Universe* by Frank Shu, "astronomy" may be used to describe the qualitative study of the subject, whereas "astrophysics" is used to describe the physics-oriented version of the subject. However, since most modern astronomical research deals with subjects related to physics, modern astronomy could actually be called astrophysics. Few fields, such as astrometry, are purely astronomy rather than also astrophysics. Various departments in which scientists carry out research on this subject may use "astronomy" and "astrophysics," partly depending on whether the department is historically affiliated with a physics department, and many professional astronomers have physics rather than astronomy degrees. One of the leading scientific journals in the field is the European journal named *Astronomy and Astrophysics*. The leading American journals are *The Astrophysical Journal* and *The Astronomical Journal*.

History

A celestial map from the 17th century, by the Dutch cartographer Frederik de Wit.

In early times, astronomy only comprised the observation and predictions of the motions of objects visible to the naked eye. In some locations, early cultures assembled massive artifacts that possibly had some astronomical purpose. In addition to their ceremonial uses, these observatories could be employed to determine the seasons, an important factor in knowing when to plant crops, as well as in understanding the length of the year.

Before tools such as the telescope were invented, early study of the stars was conducted using the naked eye. As civilizations developed, most notably in Mesopotamia, Greece, India, China, Egypt, and Central America, astronomical observatories were assembled, and ideas on the nature of the Universe began to be explored. Most of early astronomy actually consisted of mapping the positions of the stars and planets, a science now referred to as astrometry. From these observations, early ideas about the motions of the planets were formed, and the nature of the Sun, Moon and the Earth in the Universe were explored philosophically. The Earth was believed to be the center of the Universe with the Sun, the Moon and the stars rotating around it. This is known as the geocentric model of the Universe, or the Ptolemaic system, named after Ptolemy.

A particularly important early development was the beginning of mathematical and scientific astronomy, which began among the Babylonians, who laid the foundations for the later astronomical traditions that developed in many other civilizations. The Babylonians discovered that lunar eclipses recurred in a repeating cycle known as a saros.

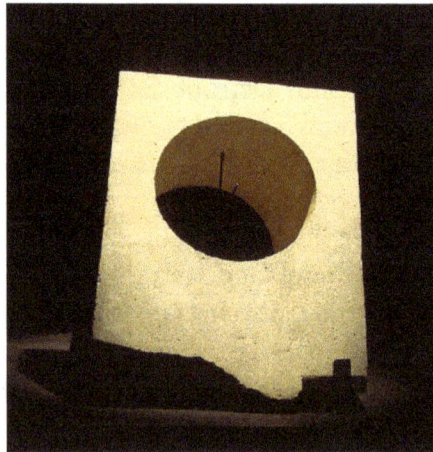

Greek equatorial sundial, Alexandria on the Oxus, present-day Afghanistan 3rd–2nd century BCE.

Following the Babylonians, significant advances in astronomy were made in ancient Greece and the Hellenistic world. Greek astronomy is characterized from the start by seeking a rational, physical explanation for celestial phenomena. In the 3rd century BC, Aristarchus of Samos estimated the size and distance of the Moon and Sun, and was the first to propose a heliocentric model of the solar system. In the 2nd century BC, Hipparchus discovered precession, calculated the size and distance of the Moon and invented the earliest known astronomical devices such as the astrolabe. Hipparchus also created a comprehensive catalog of 1020 stars, and most of the constellations of the northern hemisphere derive from Greek astronomy. The Antikythera mechanism (c. 150–80 BC) was an early analog computer designed to calculate the location of the Sun, Moon, and planets for a given date. Technological artifacts of similar complexity did not reappear until the 14th century, when mechanical astronomical clocks appeared in Europe.

During the Middle Ages, astronomy was mostly stagnant in medieval Europe, at least until the 13th century. However, astronomy flourished in the Islamic world and other parts of the world. This led to the emergence of the first astronomical observatories in the Muslim world by the early 9th century. In 964, the Andromeda Galaxy, the largest galaxy in the Local Group, was discovered by the Persian astronomer Azophi and first described in his *Book of Fixed Stars*. The SN 1006 supernova, the brightest apparent magnitude stellar event in recorded history, was observed by the Egyptian Arabic astronomer Ali ibn Ridwan and the Chinese astronomers in 1006. Some of the prominent Islamic (mostly Persian and Arab) astronomers who made significant contributions to the science include Al-Battani, Thebit, Azophi, Albumasar, Biruni, Arzachel, Al-Birjandi, and the astronomers of the Maragheh and Samarkand observatories. Astronomers during that time introduced many Arabic names now used for individual stars. It is also believed that the ruins at Great Zimbabwe and Timbuktu may have housed an astronomical observatory. Europeans had previously believed that there had been no astronomical observation in pre-colonial Middle Ages sub-Saharan Africa but modern discoveries show otherwise.

The Roman Catholic Church gave more financial and social support to the study of astronomy for over six centuries, from the recovery of ancient learning during the late Middle Ages into the Enlightenment, than any other, and, probably, all other, institutions. Among the Church's motives was finding the date for Easter.

Scientific Revolution

During the Renaissance, Nicolaus Copernicus proposed a heliocentric model of the solar system. His work was defended, expanded upon, and corrected by Galileo Galilei and Johannes Kepler. Galileo used telescopes to enhance his observations.

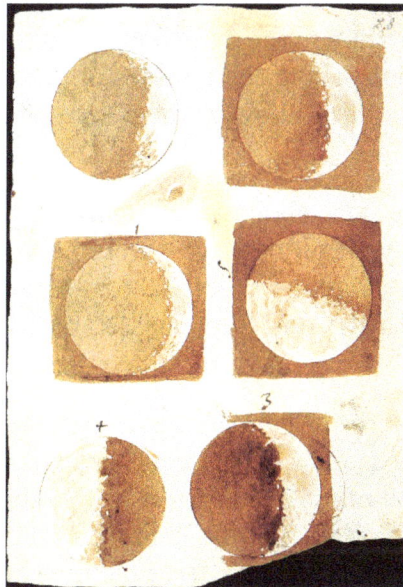

Galileo's sketches and observations of the Moon revealed that the surface was mountainous.

Kepler was the first to devise a system that described correctly the details of the motion of the planets with the Sun at the center. However, Kepler did not succeed in formulating a theory behind the laws he wrote down. It was left to Newton's invention of celestial dynamics and his law of gravitation to finally explain the motions of the planets. Newton also developed the reflecting telescope.

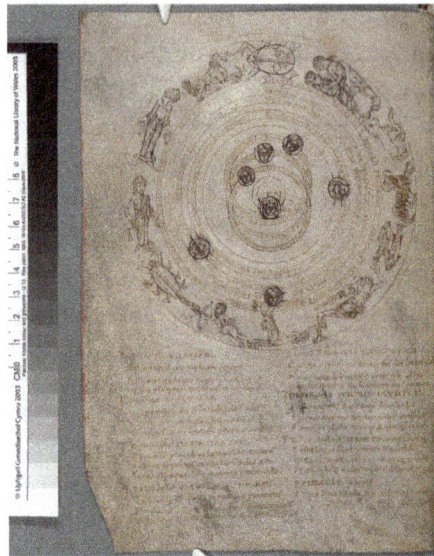

An astronomical chart from an early scientific manuscript. c.1000

The English astronomer John Flamsteed catalogued over 3000 stars. Further discoveries paralleled the improvements in the size and quality of the telescope. More extensive star catalogues were produced by Lacaille. The astronomer William Herschel made a detailed catalog of nebulosity and clusters, and in 1781 discovered the planet Uranus, the first new planet found. The distance to a star was first announced in 1838 when the parallax of 61 Cygni was measured by Friedrich Bessel.

During the 18–19th centuries, attention to the three body problem by Euler, Clairaut, and D'Alembert led to more accurate predictions about the motions of the Moon and planets. This work was further refined by Lagrange and Laplace, allowing the masses of the planets and moons to be estimated from their perturbations.

Significant advances in astronomy came about with the introduction of new technology, including the spectroscope and photography. Fraunhofer discovered about 600 bands in the spectrum of the Sun in 1814–15, which, in 1859, Kirchhoff ascribed to the presence of different elements. Stars were proven to be similar to the Earth's own Sun, but with a wide range of temperatures, masses, and sizes.

The existence of the Earth's galaxy, the Milky Way, as a separate group of stars, was only proved in the 20th century, along with the existence of "external" galaxies, and soon after, the expansion of the Universe, seen in the recession of most galaxies from us. Modern astronomy has also discovered many exotic objects such as quasars, pulsars, blazars, and radio galaxies, and has used these observations to develop physical theories which describe some of these objects in terms of equally exotic objects such as black holes and neutron stars. Physical cosmology made huge advances during the 20th century, with the model of the Big Bang heavily supported by the evidence provided by astronomy and physics, such as the cosmic microwave background radiation, Hubble's law, and cosmological abundances of elements. Space telescopes have enabled measurements in parts of the electromagnetic spectrum normally blocked or blurred by the atmosphere. Recently, in February 2016, it was revealed that the LIGO project had detected evidence of gravitational waves, in September 2015.

Observational Astronomy

In astronomy, the main source of information about celestial bodies and other objects is visible light or more generally electromagnetic radiation. Observational astronomy may be divided according to the observed region of the electromagnetic spectrum. Some parts of the spectrum can be observed from the Earth's surface, while other parts are only observable from either high altitudes or outside the Earth's atmosphere. Specific information on these subfields is given below.

Radio Astronomy

The Very Large Array in New Mexico, an example of a radio telescope

Radio astronomy studies radiation with wavelengths greater than approximately one millimeter. Radio astronomy is different from most other forms of observational astronomy in that the observed radio waves can be treated as waves rather than as discrete photons. Hence, it is relatively easier to measure both the amplitude and phase of radio waves, whereas this is not as easily done at shorter wavelengths.

Although some radio waves are produced by astronomical objects in the form of thermal emission, most of the radio emission that is observed from Earth is the result of synchrotron radiation, which is produced when electrons orbit magnetic fields. Additionally, a number of spectral lines produced by interstellar gas, notably the hydrogen spectral line at 21 cm, are observable at radio wavelengths.

A wide variety of objects are observable at radio wavelengths, including supernovae, interstellar gas, pulsars, and active galactic nuclei.

Infrared Astronomy

Infrared astronomy is founded on the detection and analysis of infrared radiation (wavelengths longer than red light). The infrared spectrum is useful for studying objects that are too cold to radiate visible light, such as planets, circumstellar disks or nebulae whose light is blocked by dust. Longer infrared wavelengths can penetrate clouds of dust that block visible light, allowing the observation of young stars in molecular clouds and the cores of galaxies. Observations from the Wide-field Infrared Survey Explorer (WISE) have been particularly effective at unveiling numerous Galactic protostars and their host star clusters. With the exception of wavelengths close to visible light, infrared radiation is heavily absorbed by the atmosphere, or masked, as the atmosphere itself produces significant infrared emission. Consequently, infrared observatories have to be located in high, dry places or in space. Some molecules radiate strongly in the infrared. This allows the study of the chemistry of space; more specifically it can detect water in comets.

ALMA Observatory is one of the highest observatory sites on Earth. Atacama, Chile.

Optical Astronomy

Historically, optical astronomy, also called visible light astronomy, is the oldest form of astronomy. Optical images of observations were originally drawn by hand. In the late 19th century and most of the 20th century, images were made using photographic equipment. Modern images are made using digital detectors, particularly detectors using charge-coupled devices (CCDs) and recorded on modern medium. Although visible light itself extends from approximately 4000 Å to 7000 Å (400 nm to 700 nm), that same equipment can be used to observe some near-ultraviolet and near-infrared radiation.

The Subaru Telescope (left) and Keck Observatory (center) on Mauna Kea, both examples of an observatory that operates at near-infrared and visible wavelengths. The NASA Infrared Telescope Facility (right) is an example of a telescope that operates only at near-infrared wavelengths.

Ultraviolet Astronomy

Ultraviolet astronomy refers to observations at ultraviolet wavelengths between approximately 100 and 3200 Å (10 to 320 nm). Light at these wavelengths is absorbed by the Earth's atmosphere, so observations at these wavelengths must be performed from the upper atmosphere or from space. Ultraviolet astronomy is best suited to the study of thermal radiation and spectral emission lines from hot blue stars (OB stars) that are very bright in this wave band. This includes the blue stars in other galaxies, which have been the targets of several ultraviolet surveys. Other objects commonly observed in ultraviolet light include planetary nebulae, supernova remnants, and active galactic nuclei. However, as ultraviolet light is easily absorbed by interstellar dust, an appropriate adjustment of ultraviolet measurements is necessary.

X-ray Astronomy

X-Ray jet made from a supermassive black hole found by NASA's Chandra X-ray Observatory, made visible by light from the early Universe.

X-ray astronomy is the study of astronomical objects at X-ray wavelengths. Typically, X-ray radiation is produced by synchrotron emission (the result of electrons orbiting magnetic field lines), thermal emission from thin gases above 10^7 (10 million) kelvins, and thermal emission from thick gases above 10^7 Kelvin. Since X-rays are absorbed by the Earth's atmosphere, all X-ray observations must be performed from high-altitude balloons, rockets, or spacecraft. Notable X-ray sources include X-ray binaries, pulsars, supernova remnants, elliptical galaxies, clusters of galaxies, and active galactic nuclei.

Gamma-ray Astronomy

Gamma ray astronomy is the study of astronomical objects at the shortest wavelengths of the electromagnetic spectrum. Gamma rays may be observed directly by satellites such as the Compton Gamma Ray Observatory or by specialized telescopes called atmospheric Cherenkov telescopes. The Cherenkov telescopes do not actually detect the gamma rays directly but instead detect the flashes of visible light produced when gamma rays are absorbed by the Earth's atmosphere.

Most gamma-ray emitting sources are actually gamma-ray bursts, objects which only produce gamma radiation for a few milliseconds to thousands of seconds before fading away. Only 10% of gamma-ray sources are non-transient sources. These steady gamma-ray emitters include pulsars, neutron stars, and black hole candidates such as active galactic nuclei.

Fields not based on the Electromagnetic Spectrum

In addition to electromagnetic radiation, a few other events originating from great distances may be observed from the Earth.

In neutrino astronomy, astronomers use heavily shielded underground facilities such as SAGE, GALLEX, and Kamioka II/III for the detection of neutrinos. The vast majority of the neutrinos streaming through the Earth originate from the Sun, but 24 neutrinos were also detected from supernova 1987A. Cosmic rays, which consist of very high energy particles that can decay or be absorbed when they enter the Earth's atmosphere, result in a cascade of particles which can be detected by current observatories. Additionally, some future neutrino detectors may also be sensitive to the particles produced when cosmic rays hit the Earth's atmosphere.

Gravitational-wave astronomy is an emerging new field of astronomy which aims to use gravitational-wave detectors to collect observational data about compact objects. A few observatories have been constructed, such as the *Laser Interferometer Gravitational Observatory* LIGO. LIGO made its first detection on 14 September 2015, observing gravitational waves from a binary black hole. A second gravitational wave was detected on 26 December 2015 and additional observations should continue but gravitational waves are extremely difficult to detect.

Combining observations made using electromagnetic radiation, neutrinos or gravitational waves with those made using a different means, which shall give complementary information, is known as multi-messenger astronomy.

Astrometry and Celestial Mechanics

Star cluster Pismis 24 with a nebula

One of the oldest fields in astronomy, and in all of science, is the measurement of the positions of celestial objects. Historically, accurate knowledge of the positions of the Sun, Moon, planets and stars has been essential in celestial navigation (the use of celestial objects to guide navigation) and in the making of calendars.

Careful measurement of the positions of the planets has led to a solid understanding of gravitational perturbations, and an ability to determine past and future positions of the planets with great accuracy, a field known as celestial mechanics. More recently the tracking of near-Earth objects will allow for predictions of close encounters, and potential collisions, with the Earth.

The measurement of stellar parallax of nearby stars provides a fundamental baseline in the cosmic distance ladder that is used to measure the scale of the Universe. Parallax measurements of nearby stars provide an absolute baseline for the properties of more distant stars, as their properties can be compared. Measurements of radial velocity and proper motion plot the movement of these systems through the Milky Way galaxy. Astrometric results are the basis used to calculate the distribution of dark matter in the galaxy.

During the 1990s, the measurement of the stellar wobble of nearby stars was used to detect large extrasolar planets orbiting nearby stars.

Theoretical Astronomy

Theoretical astronomers use several tools including analytical models (for example, polytropes to approximate the behaviors of a star) and computational numerical simulations. Each has some advantages. Analytical models of a process are generally better for giving insight into the heart of what is going on. Numerical models reveal the existence of phenomena and effects otherwise unobserved.

Theorists in astronomy endeavor to create theoretical models and from the results predict observational consequences of those models. The observation of a phenomenon predicted by a model allows astronomers to select between several alternate or conflicting models.

Theorists also try to generate or modify models to take into account new data. In the case of an inconsistency, the general tendency is to try to make minimal modifications to the model so that it produces results that fit the data. In some cases, a large amount of inconsistent data over time may lead to total abandonment of a model.

Topics studied by theoretical astronomers include: stellar dynamics and evolution; galaxy formation; large-scale structure of matter in the Universe; origin of cosmic rays; general relativity and physical cosmology, including string cosmology and astroparticle physics. Astrophysical relativity serves as a tool to gauge the properties of large scale structures for which gravitation plays a significant role in physical phenomena investigated and as the basis for black hole (*astro*)physics and the study of gravitational waves.

Some widely accepted and studied theories and models in astronomy, now included in the Lambda-CDM model are the Big Bang, Cosmic inflation, dark matter, and fundamental theories of physics.

Specific Subfields

Solar Astronomy

At a distance of about eight light-minutes, the most frequently studied star is the Sun, a typical main-sequence dwarf star of stellar class G2 V, and about 4.6 billion years (Gyr) old. The Sun is not considered a variable star, but it does undergo periodic changes in activity known as the sunspot cycle. This is an 11-year fluctuation in sunspot numbers. Sunspots are regions of lower-than- average temperatures that are associated with intense magnetic activity.

An ultraviolet image of the Sun's active photosphere as viewed by the TRACE space telescope. *NASA photo*

Solar observatory Lomnický štít (Slovakia) built in 1962.

The Sun has steadily increased in luminosity over the course of its life, increasing by 40% since it first became a main-sequence star. The Sun has also undergone periodic changes in luminosity that can have a significant impact on the Earth. The Maunder minimum, for example, is believed to have caused the Little Ice Age phenomenon during the Middle Ages.

The visible outer surface of the Sun is called the photosphere. Above this layer is a thin region known as the chromosphere. This is surrounded by a transition region of rapidly increasing temperatures, and finally by the super-heated corona.

At the center of the Sun is the core region, a volume of sufficient temperature and pressure for nuclear fusion to occur. Above the core is the radiation zone, where the plasma conveys the energy flux by means of radiation. Above that are the outer layers that form a convection zone where the gas material transports energy primarily through physical displacement of the gas. It is believed that this convection zone creates the magnetic activity that generates sunspots.

A solar wind of plasma particles constantly streams outward from the Sun until, at the outermost limit of the Solar System, it reaches the heliopause. This solar wind interacts with the magnetosphere of the Earth to create the Van Allen radiation belts about the Earth, as well as the aurora where the lines of the Earth's magnetic field descend into the atmosphere.

Planetary Science

Planetary science is the study of the assemblage of planets, moons, dwarf planets, comets, asteroids, and other bodies orbiting the Sun, as well as extrasolar planets. The Solar System has been relatively well-studied, initially through telescopes and then later by spacecraft. This has provided a good overall understanding of the formation and evolution of this planetary system, although many new discoveries are still being made.

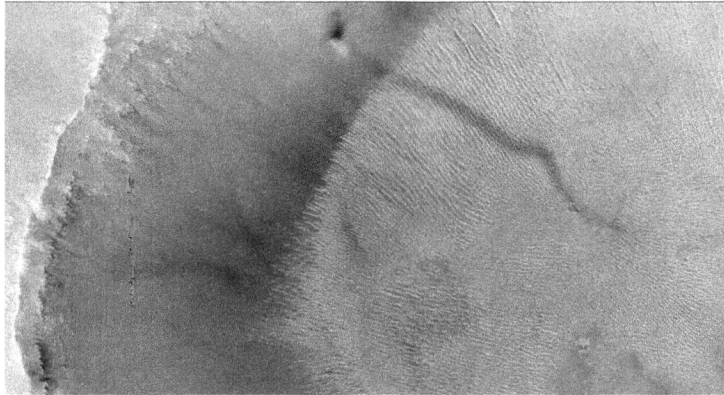

The black spot at the top is a dust devil climbing a crater wall on Mars. This moving, swirling column of Martian atmosphere (comparable to a terrestrial tornado) created the long, dark streak. *NASA image.*

The Solar System is subdivided into the inner planets, the asteroid belt, and the outer planets. The inner terrestrial planets consist of Mercury, Venus, Earth, and Mars. The outer gas giant planets are Jupiter, Saturn, Uranus, and Neptune. Beyond Neptune lies the Kuiper Belt, and finally the Oort Cloud, which may extend as far as a light-year.

The planets were formed in the protoplanetary disk that surrounded the early Sun. Through a process that included gravitational attraction, collision, and accretion, the disk formed clumps of matter that, with time, became protoplanets. The radiation pressure of the solar wind then expelled most of the unaccreted matter, and only those planets with sufficient mass retained their gaseous atmosphere. The planets continued to sweep up, or eject, the remaining matter during a period of intense bombardment, evidenced by the many impact craters on the Moon. During this period, some of the protoplanets may have collided, the leading hypothesis for how the Moon was formed.

Once a planet reaches sufficient mass, the materials of different densities segregate within, during planetary differentiation. This process can form a stony or metallic core, surrounded by a mantle and an outer surface. The core may include solid and liquid regions, and some planetary cores generate their own magnetic field, which can protect their atmospheres from solar wind stripping.

A planet or moon's interior heat is produced from the collisions that created the body, radioactive materials (*e.g.* uranium, thorium, and ^{26}Al), or tidal heating. Some planets and moons accumulate enough heat to drive geologic processes such as volcanism and tectonics. Those that accumulate or retain an atmosphere can also undergo surface erosion from wind or water. Smaller bodies, without tidal heating, cool more quickly; and their geological activity ceases with the exception of impact cratering.

Stellar Astronomy

The study of stars and stellar evolution is fundamental to our understanding of the Universe. The astrophysics of stars has been determined through observation and theoretical understanding; and from computer simulations of the interior. Star formation occurs in dense regions of dust and gas, known as giant molecular clouds. When destabilized, cloud fragments can collapse under the influence of gravity, to form a protostar. A sufficiently dense, and hot, core region will trigger nuclear fusion, thus creating a main-sequence star.

The Ant planetary nebula. Ejecting gas from the dying central star shows symmetrical patterns unlike the chaotic patterns of ordinary explosions.

Almost all elements heavier than hydrogen and helium were created inside the cores of stars.

The characteristics of the resulting star depend primarily upon its starting mass. The more massive the star, the greater its luminosity, and the more rapidly it expends the hydrogen fuel in its core. Over time, this hydrogen fuel is completely converted into helium, and the star begins to evolve. The fusion of helium requires a higher core temperature, so that the star both expands in size, and increases in core density. The resulting red giant enjoys a brief life span, before the helium fuel is in turn consumed. Very massive stars can also undergo a series of decreasing evolutionary phases, as they fuse increasingly heavier elements.

The final fate of the star depends on its mass, with stars of mass greater than about eight times the Sun becoming core collapse supernovae; while smaller stars form a white dwarf as it ejects matter that forms a planetary nebulae. The remnant of a supernova is a dense neutron star, or, if the stellar mass was at least three times that of the Sun, a black hole. Close binary stars can follow more complex evolutionary paths, such as mass transfer onto a white dwarf companion that can potentially cause a supernova. Planetary nebulae and supernovae are necessary for the distribution of metals to the interstellar medium; without them, all new stars (and their planetary systems) would be formed from hydrogen and helium alone.

Galactic Astronomy

Our solar system orbits within the Milky Way, a barred spiral galaxy that is a prominent member of the Local Group of galaxies. It is a rotating mass of gas, dust, stars and other objects, held together by mutual gravitational attraction. As the Earth is located within the dusty outer arms, there are large portions of the Milky Way that are obscured from view.

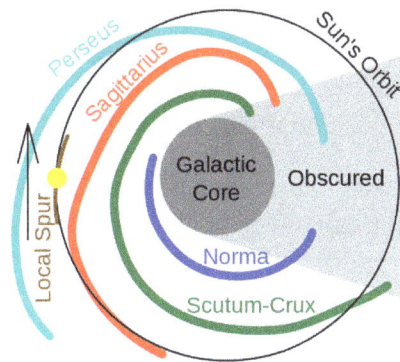

Observed structure of the Milky Way's spiral arms

In the center of the Milky Way is the core, a bar-shaped bulge with what is believed to be a super-massive black hole at the center. This is surrounded by four primary arms that spiral from the core. This is a region of active star formation that contains many younger, population I stars. The disk is surrounded by a spheroid halo of older, population II stars, as well as relatively dense concentrations of stars known as globular clusters.

Between the stars lies the interstellar medium, a region of sparse matter. In the densest regions, molecular clouds of molecular hydrogen and other elements create star-forming regions. These begin as a compact pre-stellar core or dark nebulae, which concentrate and collapse (in volumes determined by the Jeans length) to form compact protostars.

As the more massive stars appear, they transform the cloud into an H II region (ionized atomic hydrogen) of glowing gas and plasma. The stellar wind and supernova explosions from these stars eventually cause the cloud to disperse, often leaving behind one or more young open clusters of stars. These clusters gradually disperse, and the stars join the population of the Milky Way.

Kinematic studies of matter in the Milky Way and other galaxies have demonstrated that there is more mass than can be accounted for by visible matter. A dark matter halo appears to dominate the mass, although the nature of this dark matter remains undetermined.

Extragalactic Astronomy

This image shows several blue, loop-shaped objects that are multiple images of the same galaxy, duplicated by the gravitational lens effect of the cluster of yellow galaxies near the middle of the photograph. The lens is produced by the cluster's gravitational field that bends light to magnify and distort the image of a more distant object.

The study of objects outside our galaxy is a branch of astronomy concerned with the formation and evolution of Galaxies; their morphology (description) and classification; and the observation of active galaxies, and at a larger scale, the groups and clusters of galaxies. Finally, the latter is important for the understanding of the large-scale structure of the cosmos.

Most galaxies are organized into distinct shapes that allow for classification schemes. They are commonly divided into spiral, elliptical and Irregular galaxies.

As the name suggests, an elliptical galaxy has the cross-sectional shape of an ellipse. The stars move along random orbits with no preferred direction. These galaxies contain little or no interstellar dust; few star-forming regions; and generally older stars. Elliptical galaxies are more commonly found at the core of galactic clusters, and may have been formed through mergers of large galaxies.

A spiral galaxy is organized into a flat, rotating disk, usually with a prominent bulge or bar at the center, and trailing bright arms that spiral outward. The arms are dusty regions of star formation where massive young stars produce a blue tint. Spiral galaxies are typically surrounded by a halo of older stars. Both the Milky Way and our nearest galaxy neighbor, the Andromeda Galaxy, are spiral galaxies.

Irregular galaxies are chaotic in appearance, and are neither spiral nor elliptical. About a quarter of all galaxies are irregular, and the peculiar shapes of such galaxies may be the result of gravitational interaction.

An active galaxy is a formation that emits a significant amount of its energy from a source other than its stars, dust and gas. It is powered by a compact region at the core, thought to be a super-massive black hole that is emitting radiation from in-falling material.

A radio galaxy is an active galaxy that is very luminous in the radio portion of the spectrum, and is emitting immense plumes or lobes of gas. Active galaxies that emit shorter frequency, high-energy radiation include Seyfert galaxies, Quasars, and Blazars. Quasars are believed to be the most consistently luminous objects in the known universe.

The large-scale structure of the cosmos is represented by groups and clusters of galaxies. This structure is organized into a hierarchy of groupings, with the largest being the superclusters. The collective matter is formed into filaments and walls, leaving large voids between.

Cosmology

Hubble Extreme Deep Field.

Cosmology (from the Greek κόσμος (*kosmos*) "world, universe" and λόγος (*logos*) "word, study" or literally "logic") could be considered the study of the Universe as a whole.

Observations of the large-scale structure of the Universe, a branch known as physical cosmology, have provided a deep understanding of the formation and evolution of the cosmos. Fundamental to modern cosmology is the well-accepted theory of the big bang, wherein our Universe began at a single point in time, and thereafter expanded over the course of 13.8 billion years to its present condition. The concept of the big bang can be traced back to the discovery of the microwave background radiation in 1965.

In the course of this expansion, the Universe underwent several evolutionary stages. In the very early moments, it is theorized that the Universe experienced a very rapid cosmic inflation, which homogenized the starting conditions. Thereafter, nucleosynthesis produced the elemental abundance of the early Universe.

When the first neutral atoms formed from a sea of primordial ions, space became transparent to radiation, releasing the energy viewed today as the microwave background radiation. The expanding Universe then underwent a Dark Age due to the lack of stellar energy sources.

A hierarchical structure of matter began to form from minute variations in the mass density of space. Matter accumulated in the densest regions, forming clouds of gas and the earliest stars, the Population III stars. These massive stars triggered the reionization process and are believed to have created many of the heavy elements in the early Universe, which, through nuclear decay, create lighter elements, allowing the cycle of nucleosynthesis to continue longer.

Gravitational aggregations clustered into filaments, leaving voids in the gaps. Gradually, organizations of gas and dust merged to form the first primitive galaxies. Over time, these pulled in more matter, and were often organized into groups and clusters of galaxies, then into larger-scale superclusters.

Fundamental to the structure of the Universe is the existence of dark matter and dark energy. These are now thought to be its dominant components, forming 96% of the mass of the Universe. For this reason, much effort is expended in trying to understand the physics of these components.

Interdisciplinary Studies

Astronomy and astrophysics have developed significant interdisciplinary links with other major scientific fields. Archaeoastronomy is the study of ancient or traditional astronomies in their cultural context, utilizing archaeological and anthropological evidence. Astrobiology is the study of the advent and evolution of biological systems in the Universe, with particular emphasis on the possibility of non-terrestrial life. Astrostatistics is the application of statistics to astrophysics to the analysis of vast amount of observational astrophysical data.

The study of chemicals found in space, including their formation, interaction and destruction, is called astrochemistry. These substances are usually found in molecular clouds, although they may also appear in low temperature stars, brown dwarfs and planets. Cosmochemistry is the study of the chemicals found within the Solar System, including the origins of the elements and variations in the isotope ratios. Both of these fields represent an overlap of the disciplines of astronomy and chemistry. As "forensic astronomy", finally, methods from astronomy have been used to solve problems of law and history.

Amateur Astronomy

Amateur astronomers can build their own equipment, and can hold star parties and gatherings, such as Stellafane.

Astronomy is one of the sciences to which amateurs can contribute the most.

Collectively, amateur astronomers observe a variety of celestial objects and phenomena some-times with equipment that they build themselves. Common targets of amateur astronomers in-clude the Moon, planets, stars, comets, meteor showers, and a variety of deep-sky objects such as star clusters, galaxies, and nebulae. Astronomy clubs are located throughout the world and many have programs to help their members set up and complete observational programs including those to observe all the objects in the Messier (110 objects) or Herschel 400 catalogues of points of in-terest in the night sky. One branch of amateur astronomy, amateur astrophotography, involves the taking of photos of the night sky. Many amateurs like to specialize in the observation of particular objects, types of objects, or types of events which interest them.

Most amateurs work at visible wavelengths, but a small minority experiment with wavelengths outside the visible spectrum. This includes the use of infrared filters on conventional telescopes, and also the use of radio telescopes. The pioneer of amateur radio astronomy was Karl Jansky, who started observing the sky at radio wavelengths in the 1930s. A number of amateur astronomers use either homemade telescopes or use radio telescopes which were originally built for astronomy research but which are now available to amateurs (*e.g.* the One-Mile Telescope).

Amateur astronomers continue to make scientific contributions to the field of astronomy and it is one of the few scientific disciplines where amateurs can still make significant contributions. Amateurs can make occultation measurements that are used to refine the orbits of minor planets. They can also discover comets, and perform regular observations of variable stars. Improvements in digital technology have allowed amateurs to make impressive advances in the field of astropho-tography.

Unsolved Problems in Astronomy

Although the scientific discipline of astronomy has made tremendous strides in understanding the nature of the Universe and its contents, there remain some important unanswered questions. Answers to these may require the construction of new ground- and space-based instruments, and possibly new developments in theoretical and experimental physics.

- What is the origin of the stellar mass spectrum? That is, why do astronomers observe the same distribution of stellar masses – the initial mass function – apparently regardless of the initial conditions? A deeper understanding of the formation of stars and planets is needed.

- Is there other life in the Universe? Especially, is there other intelligent life? If so, what is the explanation for the Fermi paradox? The existence of life elsewhere has important scientific and philosophical implications. Is the Solar System normal or atypical?

- What caused the Universe to form? Is the premise of the Fine-tuned universe hypothesis correct? If so, could this be the result of cosmological natural selection? What caused the cosmic inflation that produced our homogeneous universe? Why is there a baryon asymmetry?

- What is the nature of dark matter and dark energy? These dominate the evolution and fate of the cosmos, yet their true nature remains unknown. What will be the ultimate fate of the universe?

- How did the first galaxies form? How did supermassive black holes form?

- What is creating the ultra-high-energy cosmic rays?

- Why is the abundance of lithium in the cosmos four times lower than predicted by the standard Big Bang model?

- What really happens beyond the event horizon?

References

- Unsöld, Albrecht; Baschek, Bodo (2001). The New Cosmos: An Introduction to Astronomy and Astrophysics. Translated by Brewer, W.D. Berlin, New York: Springer. ISBN 3-540-67877-8.

- DeWitt, Richard (2010). "The Ptolemaic System". Worldviews: An Introduction to the History and Philosophy of Science. Chichester, England: Wiley. p. 113. ISBN 1-4051-9563-0.

- Berry, Arthur (1961). A Short History of Astronomy From Earliest Times Through the 19th Century. New York: Dover Publications, Inc. ISBN 0-486-20210-0.

- McKissack, Pat; McKissack, Frederick (1995). The royal kingdoms of Ghana, Mali, and Songhay: life in medieval Africa. H. Holt. ISBN 978-0-8050-4259-7.

- Holbrook, Jarita C.; Medupe, R. Thebe; Urama, Johnson O. (2008). African Cultural Astronomy. Springer. ISBN 978-1-4020-6638-2.

- Belkora, Leila (2003). Minding the heavens: the story of our discovery of the Milky Way. CRC Press. pp. 1–14. ISBN 978-0-7503-0730-7.

- Beatty, J.K.; Petersen, C.C.; Chaikin, A., eds. (1999). The New Solar System. Cambridge press. p. 70edition = 4th. ISBN 0-521-64587-5.

- Smith, Michael David (2004). "Cloud formation, Evolution and Destruction". The Origin of Stars. Imperial College Press. pp. 53–86. ISBN 1-86094-501-5.

- Smith, Michael David (2004). "Massive stars". The Origin of Stars. Imperial College Press. pp. 185–199. ISBN 1-86094-501-5.

- "Planning for a bright tomorrow: Prospects for gravitational-wave astronomy with Advanced LIGO and Advanced Virgo". LIGO Scientific Collaboration. Retrieved 31 December 2015.

- Stenger, Richard "Star sheds light on African 'Stonehenge'". CNN. 5 December 2002.. CNN. 5 December 2002. Retrieved on 30 December 2011.

- Tammann, G. A.; Thielemann, F. K.; Trautmann, D. (2003). "Opening new windows in observing the Universe". Europhysics News. Retrieved 3 February 2010.

- Pogge, Richard W. (1997). "The Once & Future Sun". New Vistas in Astronomy. Archived from the original (lecture notes) on 27 May 2005. Retrieved 3 February 2010.

- "Edgar Wilson Award". IAU Central Bureau for Astronomical Telegrams. Archived from the original on 24 October 2010. Retrieved 24 October 2010.

2

Astronomical Objects: An Overview

Astronomical objects are physical entities that exist in the observable universe. Planets, minor planets, asteroids, galaxies, nebulas are some of the astronomical objects explained in this chapter. They provide us with comprehensive knowledge about our universe. This chapter is an overview of the subject matter incorporating all the major aspects of astronomical objects.

Astronomical Object

An astronomical object or celestial object is a naturally occurring physical entity, association, or structure that current science has demonstrated to exist in the observable universe. The term astronomical object is sometimes used interchangeably with astronomical body. Typically, an astronomical (celestial) body refers to a single, cohesive structure that is bound together by gravity (and sometimes by electromagnetism). Examples include the asteroids, moons, planets and the stars. Astronomical objects are gravitationally bound structures that are associated with a position in space, but may consist of multiple independent astronomical bodies or objects. These objects range from single planets to star clusters, nebulae or entire galaxies. A comet may be described as a body, in reference to the frozen nucleus of ice and dust, or as an object, when describing the nucleus with its diffuse coma and tail.

Galaxy and Larger

The universe can be viewed as having a hierarchical structure. At the largest scales, the fundamental component of assembly is the galaxy, which are assembled out of dwarf galaxies. The galaxies are organized into groups and clusters, often within larger superclusters, that are strung along great filaments between nearly empty voids, forming a web that spans the observable universe. Galaxies and dwarf galaxies have a variety of morphologies, with the shapes determined by their formation and evolutionary histories, including interaction with other galaxies. Depending on the category, a galaxy may have one or more distinct features, such as spiral arms, a halo and a nucleus. At the core, most galaxies have a supermassive black hole, which may result in an active galactic nucleus. Galaxies can also have satellites in the form of dwarf galaxies and globular clusters.

Within a Galaxy

The constituents of a galaxy are formed out of gaseous matter that assembles through gravitational self-attraction in a hierarchical manner. At this level, the resulting fundamental components are the stars, which are typically assembled in clusters from the various condensing nebulae. The great variety of stellar forms are determined almost entirely by the mass, composition and evolutionary state of these stars. Stars may be found in multi-star systems that orbit about each other in a hierarchical organization. A planetary system and various minor objects such as asteroids, comets and debris, can form in a hierarchical process of accretion from the protoplanetary disks that surrounds newly formed stars.

The various distinctive types of stars are shown by the Hertzsprung–Russell diagram (H–R diagram)—a plot of absolute stellar luminosity versus surface temperature. Each star follows an evolutionary track across this diagram. If this track takes the star through a region containing an intrinsic variable type, then its physical properties can cause it to become a variable star. An example of this is the instability strip, a region of the H-R diagram that includes Delta Scuti, RR Lyrae and Cepheid variables. Depending on the initial mass of the star and the presence or absence of a companion, a star may spend the last part of its life as a compact object; either a white dwarf, neutron star, or black hole.

Planet

A planet is an astronomical object orbiting a star or stellar remnant that

- is massive enough to be rounded by its own gravity,

- is not massive enough to cause thermonuclear fusion, and

- has cleared its neighbouring region of planetesimals.

The term *planet* is ancient, with ties to history, astrology, science, mythology, and religion. Several planets in the Solar System can be seen with the naked eye. These were regarded by many early cultures as divine, or as emissaries of deities. As scientific knowledge advanced, human perception of the planets changed, incorporating a number of disparate objects. In 2006, the International Astronomical Union (IAU) officially adopted a resolution defining planets within the Solar System. This definition is controversial because it excludes many objects of planetary mass based on where or what they orbit. Although eight of the planetary bodies discovered before 1950 remain "planets" under the modern definition, some celestial bodies, such as Ceres, Pallas, Juno and Vesta (each an object in the solar asteroid belt), and Pluto (the first trans-Neptunian object discovered), that were once considered planets by the scientific community, are no longer viewed as such.

The planets were thought by Ptolemy to orbit Earth in deferent and epicycle motions. Although the idea that the planets orbited the Sun had been suggested many times, it was not until the 17th century that this view was supported by evidence from the first telescopic astronomical observations, performed by Galileo Galilei. By careful analysis of the observation data, Johannes Kepler found the planets' orbits were not circular but elliptical. As observational tools improved, astronomers saw that, like Earth, the planets rotated around tilted axes, and some shared such features as ice caps and seasons. Since the dawn of the Space Age, close observation by space probes has found that Earth and the other planets share characteristics such as volcanism, hurricanes, tectonics, and even hydrology.

Planets are generally divided into two main types: large low-density giant planets, and smaller rocky terrestrials. Under IAU definitions, there are eight planets in the Solar System. In order of increasing distance from the Sun, they are the four terrestrials, Mercury, Venus, Earth, and Mars, then the four giant planets, Jupiter, Saturn, Uranus, and Neptune. Six of the planets are orbited by one or more natural satellites.

More than two thousand planets around other stars ("extrasolar planets" or "exoplanets") have been discovered in the Milky Way. As of 1 October 2016, 3,532 known extrasolar planets in 2,649 planetary systems (including 595 multiple planetary systems), ranging in size from just above the size of the Moon to gas giants about twice as large as Jupiter have been discovered, out of which more than 100 planets are the same size as Earth, nine of which are at the same relative distance from their star as Earth from the Sun, i.e. in the habitable zone. On December 20, 2011, the Kepler Space Telescope team reported the discovery of the first Earth-sized extrasolar planets, Kepler-20e and Kepler-20f, orbiting a Sun-like star, Kepler-20. A 2012 study, analyzing gravitational microlensing data, estimates an average of at least 1.6 bound planets for every star in the Milky Way. Around one in five Sun-like stars is thought to have an Earth-sized planet in its habitable zone.

History

Printed rendition of a geocentric cosmological model from *Cosmographia*, Antwerp, 1539

The word "planet" derives from the Ancient Greek astēr planētēs, or plánēs astēr, which means "wandering star," and originally referred to those objects in the night sky that moved relative to one another, as opposed to the "fixed stars", which maintained a constant relative position in the sky.

The idea of planets has evolved over its history, from the divine lights of antiquity to the earthly objects of the scientific age. The concept has expanded to include worlds not only in the Solar System, but in hundreds of other extrasolar systems. The ambiguities inherent in defining planets have led to much scientific controversy.

The five classical planets, being visible to the naked eye, have been known since ancient times and have had a significant impact on mythology, religious cosmology, and ancient astronomy. In ancient times, astronomers noted how certain lights moved across the sky in relation to the other stars. Ancient Greeks called these lights (*planētes asteres*, "wandering stars") or simply (*planētai*, "wanderers"), from which today's word "planet" was derived.

In ancient Greece, China, Babylon, and indeed all pre-modern civilizations, it was almost universally believed that Earth was the center of the Universe and that all the "planets" circled Earth. The reasons for this perception were that stars and planets appeared to revolve around Earth each day and the apparently common-sense perceptions that Earth was solid and stable and that it was not moving but at rest.

Babylon

The first civilization known to have a functional theory of the planets were the Babylonians, who lived in Mesopotamia in the first and second millennia BC. The oldest surviving planetary astronomical text is the Babylonian Venus tablet of Ammisaduqa, a 7th-century BC copy of a list of observations of the motions of the planet Venus, that probably dates as early as the second millennium BC. The MUL.APIN is a pair of cuneiform tablets dating from the 7th century BC that lays out the motions of the Sun, Moon and planets over the course of the year. The Babylonian astrologers also laid the foundations of what would eventually become Western astrology. The *Enuma anu enlil*, written during the Neo-Assyrian period in the 7th century BC, comprises a list of omens and their relationships with various celestial phenomena including the motions of the planets. Venus, Mercury and the outer planets Mars, Jupiter and Saturn were all identified by Babylonian astronomers. These would remain the only known planets until the invention of the telescope in early modern times.

Greco-Roman Astronomy

Ptolemy's 7 planetary spheres						
1 Moon ☾	2 Mercury ☿	3 Venus ♀	4 Sun ☉	5 Mars ♂	6 Jupiter ♃	7 Saturn ♄

The ancient Greeks initially did not attach as much significance to the planets as the Babylonians. The Pythagoreans, in the 6th and 5th centuries BC appear to have developed their own independent planetary theory, which consisted of the Earth, Sun, Moon, and planets revolving around a "Central Fire" at the center of the Universe. Pythagoras or Parmenides is said to have been the first to identify the evening star (Hesperos) and morning star (Phosphoros) as one and the same (Aphrodite, Greek corresponding to Latin Venus). In the 3rd century BC, Aristarchus of Samos proposed a heliocentric system, according to which Earth and the planets revolved around the Sun. The geocentric system remained dominant until the Scientific Revolution.

By the 1st century BC, during the Hellenistic period, the Greeks had begun to develop their own mathematical schemes for predicting the positions of the planets. These schemes, which were based on geometry rather than the arithmetic of the Babylonians, would eventually eclipse the Babylonians' theories in complexity and comprehensiveness, and account for most of the astronomical movements observed from Earth with the naked eye. These theories would reach their fullest expression in the *Almagest* written by Ptolemy in the 2nd century CE. So complete was the domination of Ptolemy's model that it superseded all previous works on astronomy and remained the definitive astronomical text in the Western world for 13 centuries. To the Greeks and Romans there were seven known planets, each presumed to be circling Earth according to the complex laws laid out by Ptolemy. They were, in increasing order from Earth (in Ptolemy's order): the Moon, Mercury, Venus, the Sun, Mars, Jupiter, and Saturn.

India

In 499 CE, the Indian astronomer Aryabhata propounded a planetary model that explicitly incorporated Earth's rotation about its axis, which he explains as the cause of what appears to be an apparent westward motion of the stars. He also believed that the orbits of planets are elliptical. Aryabhata's followers were particularly strong in South India, where his principles of the diurnal rotation of Earth, among others, were followed and a number of secondary works were based on them.

In 1500, Nilakantha Somayaji of the Kerala school of astronomy and mathematics, in his *Tantrasangraha*, revised Aryabhata's model. In his *Aryabhatiyabhasya*, a commentary on Aryabhata's *Aryabhatiya*, he developed a planetary model where Mercury, Venus, Mars, Jupiter and Saturn orbit the Sun, which in turn orbits Earth, similar to the Tychonic system later proposed by Tycho Brahe in the late 16th century. Most astronomers of the Kerala school who followed him accepted his planetary model.

Medieval Muslim Astronomy

In the 11th century, the transit of Venus was observed by Avicenna, who established that Venus was, at least sometimes, below the Sun. In the 12th century, Ibn Bajjah observed "two planets as black spots on the face of the Sun", which was later identified as a transit of Mercury and Venus by the Maragha astronomer Qotb al-Din Shirazi in the 13th century. Ibn Bajjah could not have observed a transit of Venus, because none occurred in his lifetime.

European Renaissance

Renaissance planets, c. 1543 to 1610 and c. 1680 to 1781					
1 Mercury ☿	2 Venus ♀	3 Earth ⊕	4 Mars ♂	5 Jupiter ♃	6 Saturn ♄

With the advent of the Scientific Revolution, use of the term "planet" changed from something that moved across the sky (in relation to the star field); to a body that orbited Earth (or that were believed to do so at the time); and by the 18th century to something that directly orbited the Sun when the heliocentric model of Copernicus, Galileo and Kepler gained sway.

Thus, Earth became included in the list of planets, whereas the Sun and Moon were excluded. At first, when the first satellites of Jupiter and Saturn were discovered in the 17th century, the terms "planet" and "satellite" were used interchangeably – although the latter would gradually become more prevalent in the following century. Until the mid-19th century, the number of "planets" rose rapidly because any newly discovered object directly orbiting the Sun was listed as a planet by the scientific community.

19th Century

Eleven planets, 1807–1845										
1 Mercury ☿	2 Venus ♀	3 Earth ⊕	4 Mars ♂	5 Vesta ⚶	6 Juno ⚵	7 Ceres ⚳	8 Pallas ⚴	9 Jupiter ♃	10 Saturn ♄	11 Uranus ⛢

In the 19th century astronomers began to realize that recently discovered bodies that had been classified as planets for almost half a century (such as Ceres, Pallas, and Vesta) were very different from the traditional ones. These bodies shared the same region of space between Mars and Jupiter (the asteroid belt), and had a much smaller mass; as a result they were reclassified as "asteroids". In the absence of any formal definition, a "planet" came to be understood as any "large" body that orbited the Sun. Because there was a dramatic size gap between the asteroids and the planets, and the spate of new discoveries seemed to have ended after the discovery of Neptune in 1846, there was no apparent need to have a formal definition.

20th Century

Planets 1854–1930, Solar planets 2006–present							
1 Mercury ☿	2 Venus ♀	3 Earth ⊕	4 Mars ♂	5 Jupiter ♃	6 Saturn ♄	7 Uranus ⛢	8 Neptune ♆

In the 20th century, Pluto was discovered. After initial observations led to the belief it was larger than Earth, the object was immediately accepted as the ninth planet. Further monitoring found the body was actually much smaller: in 1936, Raymond Lyttleton suggested that Pluto may be an escaped satellite of Neptune, and Fred Whipple suggested in 1964 that Pluto may be a comet. As it was still larger than all known asteroids and seemingly did not exist within a larger population, it kept its status until 2006.

(Solar) planets 1930–2006								
1 Mercury ☿	2 Venus ♀	3 Earth ⊕	4 Mars ♂	5 Jupiter ♃	6 Saturn ♄	7 Uranus ⛢	8 Neptune ♆	9 Pluto ♇

In 1992, astronomers Aleksander Wolszczan and Dale Frail announced the discovery of planets around a pulsar, PSR B1257+12. This discovery is generally considered to be the first definitive detection of a planetary system around another star. Then, on October 6, 1995, Michel Mayor and Didier Queloz of the Geneva Observatory announced the first definitive detection of an exoplanet orbiting an ordinary main-sequence star (51 Pegasi).

The discovery of extrasolar planets led to another ambiguity in defining a planet: the point at which a planet becomes a star. Many known extrasolar planets are many times the mass of Jupiter, approaching that of stellar objects known as brown dwarfs. Brown dwarfs are generally considered stars due to their ability to fuse deuterium, a heavier isotope of hydrogen. Although objects more massive than 75 times that of Jupiter fuse hydrogen, objects of only 13 Jupiter masses can fuse deuterium. Deuterium is quite rare, and most brown dwarfs would have ceased fusing deuterium long before their discovery, making them effectively indistinguishable from supermassive planets.

21st Century

With the discovery during the latter half of the 20th century of more objects within the Solar System and large objects around other stars, disputes arose over what should constitute a planet. There were particular disagreements over whether an object should be considered a planet if it was part of a distinct population such as a belt, or if it was large enough to generate energy by the thermonuclear fusion of deuterium.

A growing number of astronomers argued for Pluto to be declassified as a planet, because many similar objects approaching its size had been found in the same region of the Solar System (the Kuiper belt) during the 1990s and early 2000s. Pluto was found to be just one small body in a population of thousands.

Some of them, such as Quaoar, Sedna, and Eris, were heralded in the popular press as the tenth planet, failing to receive widespread scientific recognition. The announcement of Eris in 2005, an object then thought of as 27% more massive than Pluto, created the necessity and public desire for an official definition of a planet.

Acknowledging the problem, the IAU set about creating the definition of planet, and produced one in August 2006. The number of planets dropped to the eight significantly larger bodies that had cleared their orbit (Mercury, Venus, Earth, Mars, Jupiter, Saturn, Uranus, and Neptune), and a new class of dwarf planets was created, initially containing three objects (Ceres, Pluto and Eris).

Extrasolar Planets

There is no official definition of extrasolar planets. In 2003, the International Astronomical Union (IAU) Working Group on Extrasolar Planets issued a position statement, but this position statement was never proposed as an official IAU resolution and was never voted on by IAU members. The positions statement incorporates the following guidelines, mostly focused upon the boundary between planets and brown dwarfs:

1. Objects with true masses below the limiting mass for thermonuclear fusion of deuterium (currently calculated to be 13 times the mass of Jupiter for objects with the same isotopic abundance as the Sun) that orbit stars or stellar remnants are "planets" (no matter how they formed). The minimum mass and size required for an extrasolar object to be considered a planet should be the same as that used in the Solar System.

2. Substellar objects with true masses above the limiting mass for thermonuclear fusion of deuterium are "brown dwarfs", no matter how they formed or where they are located.

3. Free-floating objects in young star clusters with masses below the limiting mass for thermonuclear fusion of deuterium are not "planets", but are "sub-brown dwarfs" (or whatever name is most appropriate).

This working definition has since been widely used by astronomers when publishing discoveries of exoplanets in academic journals. Although temporary, it remains an effective working definition until a more permanent one is formally adopted. It does not address the dispute over the lower mass limit, and so it steered clear of the controversy regarding objects within the Solar System. This definition also makes no comment on the planetary status of objects orbiting brown dwarfs, such as 2M1207b.

One definition of a sub-brown dwarf is a planet-mass object that formed through cloud collapse rather than accretion. This formation distinction between a sub-brown dwarf and a planet is not universally agreed upon; astronomers are divided into two camps as whether to consider the formation process of a planet as part of its division in classification. One reason for the dissent is that often it may not be possible to determine the formation process. For example, a planet formed by accretion around a star may get ejected from the system to become free-floating, and likewise a sub-brown dwarf that formed on its own in a star cluster through cloud collapse may get captured into orbit around a star.

The 13 Jupiter-mass cutoff represents an average mass rather than a precise threshold value. Large objects will fuse most of their deuterium and smaller ones will fuse only a little, and the 13 M_J value is somewhere in between. In fact, calculations show that an object fuses 50% of its initial deuterium content when the total mass ranges between 12 and 14 M_J. The amount of deuterium fused depends not only on mass but also on the composition of the object, on the amount of helium and deuterium present. The Extrasolar Planets Encyclopaedia includes objects up to 25 Jupiter masses, saying, "The fact that there is no special feature around 13 M_J in the observed mass spectrum reinforces the choice to forget this mass limit." The Exoplanet Data Explorer includes objects up to 24 Jupiter masses with the advisory: "The 13 Jupiter-mass distinction by the IAU Working Group is physically unmotivated for planets with rocky cores, and observationally problematic due to the sin i ambiguity." The NASA Exoplanet Archive includes objects with a mass (or minimum mass) equal to or less than 30 Jupiter masses.

Another criterion for separating planets and brown dwarfs, rather than deuterium fusion, formation process or location, is whether the core pressure is dominated by coulomb pressure or electron degeneracy pressure.

2006 IAU Definition of Planet

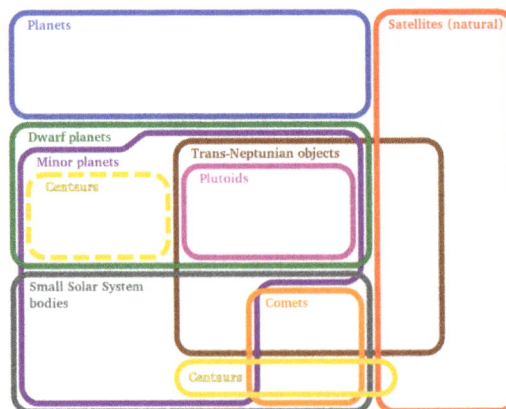

Euler diagram showing the types of bodies in the Solar System.

The matter of the lower limit was addressed during the 2006 meeting of the IAU's General Assembly. After much debate and one failed proposal, 232 members of the 10,000 member assembly, who nevertheless constituted a large majority of those remaining at the meeting, voted to pass a resolution. The 2006 resolution defines planets within the Solar System as follows:

A "planet" is a celestial body that (a) is in orbit around the Sun, (b) has sufficient mass for its self-gravity to overcome rigid body forces so that it assumes a hydrostatic equilibrium (nearly round) shape, and (c) has cleared the neighbourhood around its orbit.

The eight planets are: Mercury, Venus, Earth, Mars, Jupiter, Saturn, Uranus, and Neptune.

Under this definition, the Solar System is considered to have eight planets. Bodies that fulfill the first two conditions but not the third (such as Ceres, Pluto, and Eris) are classified as dwarf planets, provided they are not also natural satellites of other planets. Originally an IAU committee had proposed a definition that would have included a much larger number of planets as it did not include (c) as a criterion. After much discussion, it was decided via a vote that those bodies should instead be classified as dwarf planets.

This definition is based in theories of planetary formation, in which planetary embryos initially clear their orbital neighborhood of other smaller objects. As described by astronomer Steven Soter:

> "The end product of secondary disk accretion is a small number of relatively large bodies (planets) in either non-intersecting or resonant orbits, which prevent collisions between them. Minor planets and comets, including KBOs [Kuiper belt objects], differ from planets in that they can collide with each other and with planets."

The 2006 IAU definition presents some challenges for exoplanets because the language is specific to the Solar System and because the criteria of roundness and orbital zone clearance are not presently observable. Astronomer Jean-Luc Margot proposed a mathematical criterion that determines whether an object can clear its orbit during the lifetime of its host star, based on the mass of the planet, its semimajor axis, and the mass of its host star. This formula produces a value π that is greater than 1 for planets. The eight known planets and all known exoplanets have π values above 100, while Ceres, Pluto, and Eris have π values of 0.1 or less. Objects with π values of 1 or more are also expected to be approximately spherical, so that objects that fulfill the orbital zone clearance requirement automatically fulfill the roundness requirement.

Objects Formerly Considered Planets

The table below lists Solar System bodies once considered to be planets.

Body	Current classification	Notes
Sun	Star	Classified as classical planets (Ancient Greek πλαν□ται, wanderers) in classical antiquity and medieval Europe, in accordance with the now-disproved geocentric model.
Moon	Moon	

Io, Europa, Ganymede, and Callisto	Moons	The four largest moons of Jupiter, known as the Galilean moons after their discoverer Galileo Galilei. He referred to them as the "Medicean Planets" in honor of his patron, the Medici family. They were known as secondary planets.
Titan, Iapetus, Rhea, Tethys, and Dione	Moons	Five of Saturn's larger moons, discovered by Christiaan Huygens and Giovanni Domenico Cassini. As with Jupiter's major moons, they were known as secondary planets.
Pallas, Juno, and Vesta	Asteroids	Regarded as planets from their discoveries between 1801 and 1807 until they were reclassified as asteroids during the 1850s. Ceres was subsequently classified as a dwarf planet in 2006.
Ceres	Dwarf planet and asteroid	
Astraea, Hebe, Iris, Flora, Metis, Hygiea, Parthenope, Victoria, Egeria, Irene, Eunomia	Asteroids	More asteroids, discovered between 1845 and 1851. The rapidly expanding list of bodies between Mars and Jupiter prompted their reclassification as asteroids, which was widely accepted by 1854.
Pluto	Dwarf planet and Kuiper belt object	The first known trans-Neptunian object (i.e. minor planet with a semi-major axis beyond Neptune). Regarded as a planet from its discovery in 1930 until it was reclassified as a dwarf planet in 2006.

Beyond the scientific community, Pluto still holds cultural significance for many in the general public due to its historical classification as a planet from 1930 to 2006. A few astronomers, such as Alan Stern, consider dwarf planets and the larger moons to be planets, based on a purely geophysical definition of *planet*.

Mythology and Naming

The Greek gods of Olympus, after whom the Solar System's Roman names of the planets are derived

The names for the planets in the Western world are derived from the naming practices of the Romans, which ultimately derive from those of the Greeks and the Babylonians. In ancient Greece, the two great luminaries the Sun and the Moon were called *Helios* and *Selene*; the farthest planet (Saturn) was called *Phainon*, the shiner; followed by *Phaethon* (Jupiter), "bright"; the red planet (Mars) was known as *Pyroeis*, the "fiery"; the brightest (Venus) was known as *Phosphoros*, the light bringer; and the fleeting final planet (Mercury) was called *Stilbon*, the gleamer. The Greeks also made each planet sacred to one among their pantheon of gods, the Olympians: Helios and Selene were the names of both planets and gods; Phainon was sacred to Cronus, the Titan who fathered the Olympians; Phaethon was sacred to Zeus, Cronus's son who deposed him as king; Pyroeis was given to Ares, son of Zeus and god of war; Phosphoros was ruled by Aphrodite, the goddess of love; and Hermes, messenger of the gods and god of learning and wit, ruled over Stilbon.

The Greek practice of grafting of their gods' names onto the planets was almost certainly borrowed from the Babylonians. The Babylonians named Phosphoros after their goddess of love, *Ishtar*; Pyroeis after their god of war, *Nergal*, Stilbon after their god of wisdom Nabu, and Phaethon after their chief god, *Marduk*. There are too many concordances between Greek and Babylonian naming conventions for them to have arisen separately. The translation was not perfect. For instance, the Babylonian Nergal was a god of war, and thus the Greeks identified him with Ares. Unlike Ares, Nergal was also god of pestilence and the underworld.

Today, most people in the western world know the planets by names derived from the Olympian pantheon of gods. Although modern Greeks still use their ancient names for the planets, other European languages, because of the influence of the Roman Empire and, later, the Catholic Church, use the Roman (Latin) names rather than the Greek ones. The Romans, who, like the Greeks, were Indo-Europeans, shared with them a common pantheon under different names but lacked the rich narrative traditions that Greek poetic culture had given their gods. During the later period of the Roman Republic, Roman writers borrowed much of the Greek narratives and applied them to their own pantheon, to the point where they became virtually indistinguishable. When the Romans studied Greek astronomy, they gave the planets their own gods' names: *Mercurius* (for Hermes), *Venus* (Aphrodite), *Mars* (Ares), *Iuppiter* (Zeus) and *Saturnus* (Cronus). When subsequent planets were discovered in the 18th and 19th centuries, the naming practice was retained with *Neptūnus* (Poseidon). Uranus is unique in that it is named for a Greek deity rather than his Roman counterpart.

Some Romans, following a belief possibly originating in Mesopotamia but developed in Hellenistic Egypt, believed that the seven gods after whom the planets were named took hourly shifts in looking after affairs on Earth. The order of shifts went Saturn, Jupiter, Mars, Sun, Venus, Mercury, Moon (from the farthest to the closest planet). Therefore, the first day was started by Saturn (1st hour), second day by Sun (25th hour), followed by Moon (49th hour), Mars, Mercury, Jupiter and Venus. Because each day was named by the god that started it, this is also the order of the days of the week in the Roman calendar after the Nundinal cycle was rejected – and still preserved in many modern languages. In English, *Saturday, Sunday,* and *Monday* are straightforward translations of these Roman names. The other days were renamed after *Tiw* (Tuesday), *Wóden* (Wednesday), *Thunor* (Thursday), and *Frige* (Friday), the Anglo-Saxon gods considered similar or equivalent to Mars, Mercury, Jupiter, and Venus, respectively.

Earth is the only planet whose name in English is not derived from Greco-Roman mythology. Because it was only generally accepted as a planet in the 17th century, there is no tradition of naming it after a god. (The same is true, in English at least, of the Sun and the Moon, though they are no longer generally considered planets.) The name originates from the 8th century Anglo-Saxon word *erda*, which means ground or soil and was first used in writing as the name of the sphere of Earth perhaps around 1300. As with its equivalents in the other Germanic languages, it derives ultimately from the Proto-Germanic word *ertho*, "ground", as can be seen in the English *earth*, the German *Erde*, the Dutch *aarde*, and the Scandinavian *jord*. Many of the Romance languages retain the old Roman word *terra* (or some variation of it) that was used with the meaning of "dry land" as opposed to "sea". The non-Romance languages use their own native words. The Greeks retain their original name, *Γή (Ge)*.

Non-European cultures use other planetary-naming systems. India uses a system based on the Nava-graha, which incorporates the seven traditional planets (Surya for the Sun, Chandra for the Moon, and Budha, Shukra, Mangala, Bṛhaspati and Shani for Mercury, Venus, Mars, Jupiter and Saturn) and the ascending and descending lunar nodes Rahu and Ketu. China and the countries of eastern Asia historically subject to Chinese cultural influence (such as Japan, Korea and Vietnam) use a naming system based on the five Chinese elements: water (Mercury), metal (Venus), fire (Mars), wood (Jupiter) and earth (Saturn). In traditional Hebrew astronomy, the seven traditional planets have (for the most part) descriptive names - the Sun is המה *Hammah* or "the hot one," the Moon is הנבל *Levanah* or "the white one," Venus is הגונ בכוכ *Kokhav Nogah* or "the bright planet," Mercury is בכוכ *Kokhav* or "the planet" (given its lack of distinguishing features), Mars is מידאמ *Ma'adim* or "the red one," and Saturn is יאתבש *Shabbatai* or "the resting one" (in reference to its slow movement com-pared to the other visible planets). The odd one out is Jupiter, called קדצ *Tzedeq* or "justice." Steiglitz suggests that this may be a euphemism for the original name of לעב בכוכ *Kokhav Ba'al* or "Baal's plan-et," seen as idolatrous and euphemized in a similar manner to Ishbosheth from II Samuel

Formation

An artist's impression of protoplanetary disk

It is not known with certainty how planets are formed. The prevailing theory is that they are formed during the collapse of a nebula into a thin disk of gas and dust. A protostar forms at the core, surrounded by a rotating protoplanetary disk. Through accretion (a process of sticky collision) dust particles in the disk steadily accumulate mass to form ever-larger bodies. Local concentrations of mass known as planetesimals form, and these accelerate the accretion process by drawing in additional material by their gravitational attraction. These concentrations become ever denser until they collapse inward under gravity to form protoplanets. After a planet reaches a mass somewhat larger than Mars' mass, it begins to accumulate an extended atmosphere, greatly increasing the capture rate of the planetesimals by means of atmospheric drag. Depending on the accretion history of solids and gas, a giant planet, an ice giant, or a terrestrial planet may result.

Asteroid collision - building planets (artist concept).

When the protostar has grown such that it ignites to form a star, the surviving disk is removed from the inside outward by photoevaporation, the solar wind, Poynting–Robertson drag and other effects. Thereafter there still may be many protoplanets orbiting the star or each other, but over time many will collide, either to form a single larger planet or release material for other larger protoplanets or planets to absorb. Those objects that have become massive enough will capture most matter in their orbital neighbourhoods to become planets. Protoplanets that have avoided collisions may become natural satellites of planets through a process of gravitational capture, or remain in belts of other objects to become either dwarf planets or small bodies.

The energetic impacts of the smaller planetesimals (as well as radioactive decay) will heat up the growing planet, causing it to at least partially melt. The interior of the planet begins to differentiate by mass, developing a denser core. Smaller terrestrial planets lose most of their atmospheres because of this accretion, but the lost gases can be replaced by outgassing from the mantle and from the subsequent impact of comets. (Smaller planets will lose any atmosphere they gain through various escape mechanisms.)

With the discovery and observation of planetary systems around stars other than the Sun, it is becoming possible to elaborate, revise or even replace this account. The level of metallicity—an astronomical term describing the abundance of chemical elements with an atomic number greater than 2 (helium)—is now thought to determine the likelihood that a star will have planets. Hence, it is thought that a metal-rich population I star will likely have a more substantial planetary system than a metal-poor, population II star.

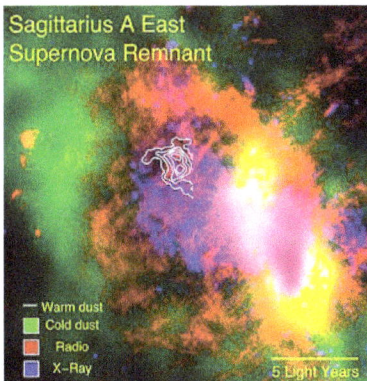

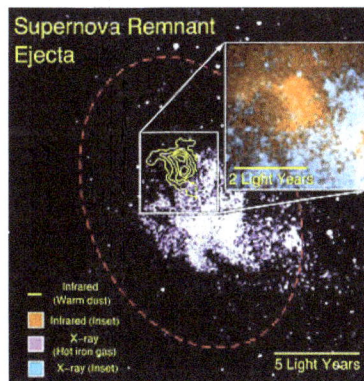

Supernova remnant ejecta producing planet-forming material.

Solar System

There are eight planets in the Solar System, which are in increasing distance from the Sun:

1. ☿ **Mercury**

2. ♀ **Venus**

3. ⊕ **Earth**

4. ♂ **Mars**

5. ♃ **Jupiter**

6. ♄ **Saturn**

7. ♅ **Uranus**

8. ♆ **Neptune**

Solar System – sizes but not distances are to scale

The Sun and the eight planets of the Solar System

The inner planets, Mercury, Venus, Earth, and Mars

The four giant planets Jupiter, Saturn, Uranus, and Neptune against the Sun and some sunspots

Jupiter is the largest, at 318 Earth masses, whereas Mercury is the smallest, at 0.055 Earth masses.

The planets of the Solar System can be divided into categories based on their composition:

- Terrestrials: Planets that are similar to Earth, with bodies largely composed of rock: Mercury, Venus, Earth and Mars. At 0.055 Earth masses, Mercury is the smallest terrestrial planet (and smallest planet) in the Solar System. Earth is the largest terrestrial planet.

- Giant planets (Jovians): Massive planets significantly more massive than the terrestrials: Jupiter, Saturn, Uranus, Neptune.

    o Gas giants, Jupiter and Saturn, are giant planets primarily composed of hydrogen and helium and are the most massive planets in the Solar System. Jupiter, at 318 Earth masses, is the largest planet in the Solar System, and Saturn is one third as massive, at 95 Earth masses.

    o Ice giants, Uranus and Neptune, are primarily composed of low-boiling-point materials such as water, methane, and ammonia, with thick atmospheres of hydrogen and helium. They have a significantly lower mass than the gas giants (only 14 and 17 Earth masses).

Planetary Attributes

Exoplanets

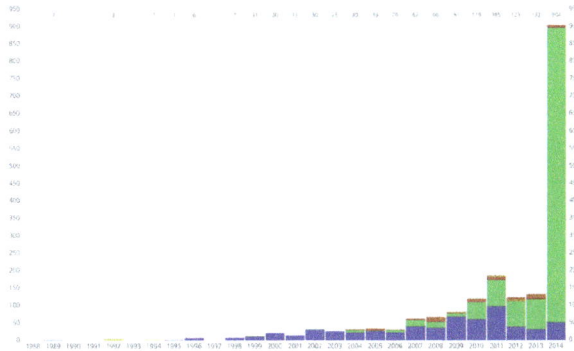

Exoplanets, by year of discovery, through September 2014.

An exoplanet (extrasolar planet) is a planet outside the Solar System. More than 2000 such planets have been discovered (3,532 planets in 2,649 planetary systems including 595 multiple planetary systems as of 1 October 2016).

In early 1992, radio astronomers Aleksander Wolszczan and Dale Frail announced the discovery of two planets orbiting the pulsar PSR 1257+12. This discovery was confirmed, and is generally considered to be the first definitive detection of exoplanets. These pulsar planets are believed to have formed from the unusual remnants of the supernova that produced the pulsar, in a second round of planet formation, or else to be the remaining rocky cores of giant planets that survived the supernova and then decayed into their current orbits.

The first confirmed discovery of an extrasolar planet orbiting an ordinary main-sequence star occurred on 6 October 1995, when Michel Mayor and Didier Queloz of the University of Geneva announced the detection of an exoplanet around 51 Pegasi. From then until the Kepler mission most known extrasolar planets were gas giants comparable in mass to Jupiter or larger as they were more easily detected. The catalog of Kepler candidate planets consists mostly of planets the size of Neptune and smaller, down to smaller than Mercury.

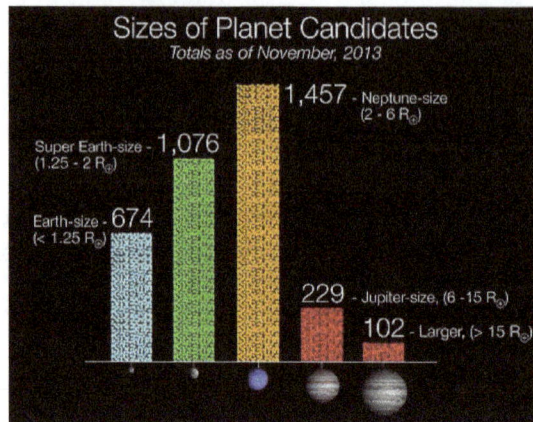

Sizes of *Kepler* Planet Candidates – based on 2,740 candidates orbiting 2,036 stars as of 4 November 2013 (NASA).

There are types of planets that do not exist in the Solar System: super-Earths and mini-Neptunes, which could be rocky like Earth or a mixture of volatiles and gas like Neptune—a radius of 1.75 times that of Earth is a possible dividing line between the two types of planet. There are hot Jupiters that orbit very close to their star and may evaporate to become chthonian planets, which are the leftover cores. Another possible type of planet is carbon planets, which form in systems with a higher proportion of carbon than in the Solar System.

A 2012 study, analyzing gravitational microlensing data, estimates an average of at least 1.6 bound planets for every star in the Milky Way.

On December 20, 2011, the Kepler Space Telescope team reported the discovery of the first Earth-size exoplanets, Kepler-20e and Kepler-20f, orbiting a Sun-like star, Kepler-20.

Around 1 in 5 Sun-like stars have an "Earth-sized" planet in the habitable zone, so the nearest would be expected to be within 12 light-years distance from Earth. The frequency of occurrence of such terrestrial planets is one of the variables in the Drake equation, which estimates the number of intelligent, communicating civilizations that exist in the Milky Way.

There are exoplanets that are much closer to their parent star than any planet in the Solar System is to the Sun, and there are also exoplanets that are much farther from their star. Mercury, the closest planet to the Sun at 0.4 AU, takes 88-days for an orbit, but the shortest known orbits for exoplanets take only a few hours, e.g. Kepler-70b. The Kepler-11 system has five of its planets in shorter orbits than Mercury's, all of them much more massive than Mercury. Neptune is 30 AU from the Sun and takes 165 years to orbit, but there are exoplanets that are hundreds of AU from their star and take more than a thousand years to orbit, e.g. 1RXS1609 b.

The next few space telescopes to study exoplanets are expected to be Gaia launched in December 2013, CHEOPS in 2017, TESS in 2017, and the James Webb Space Telescope in 2018.

Planetary-mass Objects

A planetary-mass object (PMO), planemo /ˈplænɨmoʊ/, or planetary body is a celestial object with a mass that falls within the range of the definition of a planet: massive enough to achieve hydrostatic equilibrium (to be rounded under its own gravity), but not enough to sustain core fusion like a star. By definition, all planets are *planetary-mass objects*, but the purpose of this term is to refer to objects that do not conform to typical expectations for a planet. These include dwarf planets, which are rounded by their own gravity but not massive enough to clear their own orbit, the larger moons, and free-floating planemos, which may have been ejected from a system (rogue planets) or formed through cloud-collapse rather than accretion (sometimes called sub-brown dwarfs).

Artist's impression of a super-Jupiter around the brown dwarf 2M1207.

Rogue Planets

Several computer simulations of stellar and planetary system formation have suggested that some objects of planetary mass would be ejected into interstellar space. Some scientists have argued that such objects found roaming in deep space should be classed as "planets", although others have suggested that they should be called low-mass brown dwarfs.

Sub-brown Dwarfs

Stars form via the gravitational collapse of gas clouds, but smaller objects can also form via cloud-collapse. Planetary-mass objects formed this way are sometimes called sub-brown dwarfs. Sub-brown dwarfs may be free-floating such as Cha 110913-773444 and OTS 44, or orbiting a larger object such as 2MASS J04414489+2301513.

Binary systems of sub-brown dwarfs are theoretically possible; Oph 162225-240515 was initially thought to be a binary system of a brown dwarf of 14 Jupiter masses and a sub-brown dwarf of 7 Jupiter masses, but further observations revised the estimated masses upwards to greater than 13 Jupiter masses, making them brown dwarfs according to the IAU working definitions.

Former Stars

In close binary star systems one of the stars can lose mass to a heavier companion. Accretion-powered pulsars may drive mass loss. The shrinking star can then become a planetary-mass object.

An example is a Jupiter-mass object orbiting the pulsar PSR J1719-1438. These shrunken white dwarfs may become a helium planet or carbon planet.

Satellite Planets and Belt Planets

Some large satellites are of similar size or larger than the planet Mercury, e.g. Jupiter's Galilean moons and Titan. Alan Stern has argued that location should not matter and that only geophysical attributes should be taken into account in the definition of a planet, and proposes the term *satellite planet* for a planet-sized satellite. Likewise, dwarf planets in the asteroid belt and Kuiper belt should be considered planets according to Stern.

Captured Planets

Free-floating planets in stellar clusters have similar velocities to the stars and so can be recaptured. They are typically captured into wide orbits between 100 and 10^5 AU. The capture efficiency decreases with increasing cluster volume, and for a given cluster size it increases with the host/primary mass. It is almost independent of the planetary mass. Single and multiple planets could be captured into arbitrary unaligned orbits, non-coplanar with each other or with the stellar host spin, or pre-existing planetary system.

Attributes

Although each planet has unique physical characteristics, a number of broad commonalities do exist among them. Some of these characteristics, such as rings or natural satellites, have only as yet been observed in planets in the Solar System, whereas others are also commonly observed in extrasolar planets.

Dynamic Characteristics

Orbit

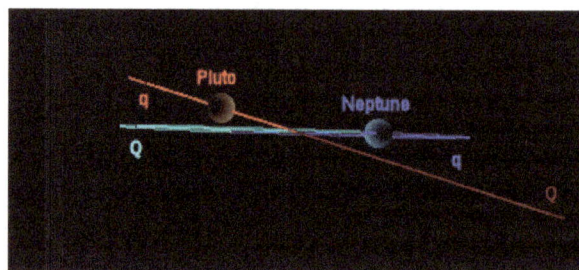

The orbit of the planet Neptune compared to that of Pluto. Note the elongation of Pluto's orbit in relation to Neptune's (eccentricity), as well as its large angle to the ecliptic (inclination).

According to current definitions, all planets must revolve around stars; thus, any potential "rogue planets" are excluded. In the Solar System, all the planets orbit the Sun in the same direction as the Sun rotates (counter-clockwise as seen from above the Sun's north pole). At least one extrasolar planet, WASP-17b, has been found to orbit in the opposite direction to its star's rotation. The period of one revolution of a planet's orbit is known as its sidereal period or *year*. A planet's year depends on its distance from its star; the farther a planet is from its star, not only the longer the distance it must travel, but also the slower its speed, because it is less affected by its star's gravity. No planet's orbit is perfectly circular, and hence the distance of each varies over the course of its year. The closest approach to its star is called its periastron (perihelion in the Solar System), whereas its farthest separation from the star is called its apastron (aphelion). As a planet approaches periastron, its speed increases as it trades gravitational potential energy for kinetic energy, just as a falling object on Earth accelerates as it falls; as the planet reaches apastron, its speed decreases, just as an object thrown upwards on Earth slows down as it reaches the apex of its trajectory.

Each planet's orbit is delineated by a set of elements:

- The *eccentricity* of an orbit describes how elongated a planet's orbit is. Planets with low eccentricities have more circular orbits, whereas planets with high eccentricities have more elliptical orbits. The planets in the Solar System have very low eccentricities, and thus nearly circular orbits. Comets and Kuiper belt objects (as well as several extrasolar planets) have very high eccentricities, and thus exceedingly elliptical orbits.

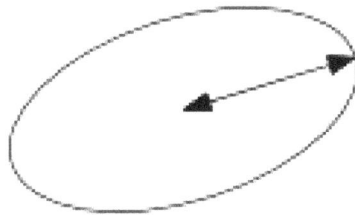

Illustration of the semi-major axis

The *semi-major axis* is the distance from a planet to the half-way point along the longest diameter of its elliptical orbit. This distance is not the same as its apastron, because no planet's orbit has its star at its exact centre.

- The *inclination* of a planet tells how far above or below an established reference plane its orbit lies. In the Solar System, the reference plane is the plane of Earth's orbit, called the ecliptic. For extrasolar planets, the plane, known as the *sky plane* or *plane of the sky*, is the plane perpendicular to the observer's line of sight from Earth. The eight planets of the Solar System all lie very close to the ecliptic; comets and Kuiper belt objects like Pluto are at far more extreme angles to it. The points at which a planet crosses above and below its reference plane are called its ascending and descending nodes. The longitude of the ascending node is the angle between the reference plane's 0 longitude and the planet's ascending node. The argument of periapsis (or perihelion in the Solar System) is the angle between a planet's ascending node and its closest approach to its star.

Axial Tilt

Planets also have varying degrees of axial tilt; they lie at an angle to the plane of their stars' equators. This causes the amount of light received by each hemisphere to vary over the course of its year; when the northern hemisphere points away from its star, the southern hemisphere points towards it, and vice versa. Each planet therefore has seasons, changes to the climate over the course of its year. The time at which each hemisphere points farthest or nearest from its star is known as its solstice. Each planet has two in the course of its orbit; when one hemisphere has its summer solstice, when its day is longest, the other has its winter solstice, when its day is shortest. The varying amount of light and heat received by each hemisphere creates annual changes in weather patterns for each half of the planet. Jupiter's axial tilt is very small, so its seasonal variation is minimal; Uranus, on the other hand, has an axial tilt so extreme it is virtually on its side, which means that its hemispheres are either perpetually in sunlight or perpetually in darkness around the time of its solstices. Among extrasolar planets, axial tilts are not known for certain, though most hot Jupiters are believed to have negligible to no axial tilt as a result of their proximity to their stars.

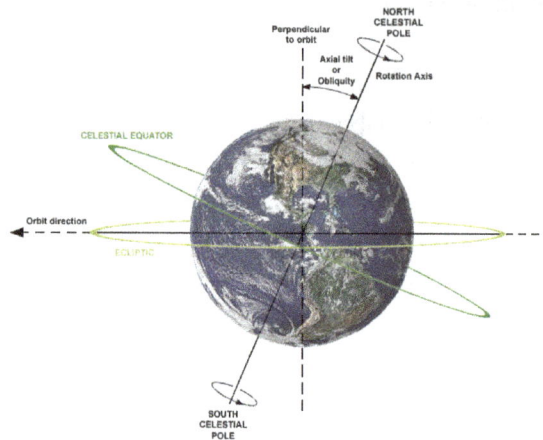

Earth's axial tilt is about 23.4°. It oscillates between 22.1° and 24.5° on a 41,000-year cycle and is currently decreasing.

Rotation

The planets rotate around invisible axes through their centres. A planet's rotation period is known as a stellar day. Most of the planets in the Solar System rotate in the same direction as they orbit the Sun, which is counter-clockwise as seen from above the Sun's north pole, the exceptions being Venus and Uranus, which rotate clockwise, though Uranus's extreme axial tilt means there are differing conventions on which of its poles is "north", and therefore whether it is rotating clockwise or anti-clockwise. Regardless of which convention is used, Uranus has a retrograde rotation relative to its orbit.

The rotation of a planet can be induced by several factors during formation. A net angular momentum can be induced by the individual angular momentum contributions of accreted objects. The accretion of gas by the giant planets can also contribute to the angular momentum. Finally, during the last stages of planet building, a stochastic process of protoplanetary accretion can randomly alter the spin axis of the planet. There is great variation in the length of day between the planets, with Venus taking 243 days to rotate, and the giant planets only a few hours. The rotational periods of extrasolar planets are not known. However, for "hot" Jupiters, their proximity to their stars means that they are tidally locked (i.e., their orbits are in sync with their rotations). This means, they always show one face to their stars, with one side in perpetual day, the other in perpetual night.

Orbital Clearing

The defining dynamic characteristic of a planet is that it has *cleared its neighborhood*. A planet that has cleared its neighborhood has accumulated enough mass to gather up or sweep away all the planetesimals in its orbit. In effect, it orbits its star in isolation, as opposed to sharing its orbit with a multitude of similar-sized objects. This characteristic was mandated as part of the IAU's official definition of a planet in August, 2006. This criterion excludes such planetary bodies as Pluto, Eris and Ceres from full-fledged planethood, making them instead dwarf planets. Although to date this criterion only applies to the Solar System, a number of young extrasolar systems have been found in which evidence suggests orbital clearing is taking place within their circumstellar discs.

Physical Characteristics

Mass

A planet's defining physical characteristic is that it is massive enough for the force of its own gravity to dominate over the electromagnetic forces binding its physical structure, leading to a state of hydrostatic equilibrium. This effectively means that all planets are spherical or spheroidal. Up to a certain mass, an object can be irregular in shape, but beyond that point, which varies depending on the chemical makeup of the object, gravity begins to pull an object towards its own centre of mass until the object collapses into a sphere.

Mass is also the prime attribute by which planets are distinguished from stars. The upper mass limit for planethood is roughly 13 times Jupiter's mass for objects with solar-type isotopic abundance, beyond which it achieves conditions suitable for nuclear fusion. Other than the Sun, no objects of such mass exist in the Solar System; but there are exoplanets of this size. The 13-Jupiter-mass limit is not universally agreed upon and the Extrasolar Planets Encyclopaedia includes objects up to 20 Jupiter masses, and the Exoplanet Data Explorer up to 24 Jupiter masses.

The smallest known planet is PSR B1257+12A, one of the first extrasolar planets discovered, which was found in 1992 in orbit around a pulsar. Its mass is roughly half that of the planet Mercury. The smallest known planet orbiting a main-sequence star other than the Sun is Kepler-37b, with a mass (and radius) slightly higher than that of the Moon.

Internal Differentiation

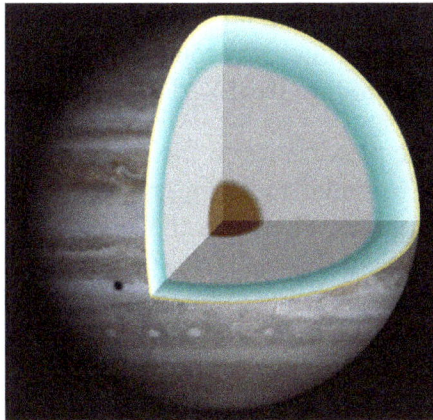

Illustration of the interior of Jupiter, with a rocky core overlaid by a deep layer of metallic hydrogen

Every planet began its existence in an entirely fluid state; in early formation, the denser, heavier materials sank to the centre, leaving the lighter materials near the surface. Each therefore has a differentiated interior consisting of a dense planetary core surrounded by a mantle that either is or was a fluid. The terrestrial planets are sealed within hard crusts, but in the giant planets the mantle simply blends into the upper cloud layers. The terrestrial planets have cores of elements such as iron and nickel, and mantles of silicates. Jupiter and Saturn are believed to have cores of rock and metal surrounded by mantles of metallic hydrogen. Uranus and Neptune, which are smaller, have rocky cores surrounded by mantles of water, ammonia, methane and other ices. The fluid action within these planets' cores creates a geodynamo that generates a magnetic field.

Atmosphere

All of the Solar System planets except Mercury have substantial atmospheres because their gravity is strong enough to keep gases close to the surface. The larger giant planets are massive enough to keep large amounts of the light gases hydrogen and helium, whereas the smaller planets lose these gases into space. The composition of Earth's atmosphere is different from the other planets because the various life processes that have transpired on the planet have introduced free molecular oxygen.

Earth's atmosphere

Planetary atmospheres are affected by the varying insolation or internal energy, leading to the formation of dynamic weather systems such as hurricanes, (on Earth), planet-wide dust storms (on Mars), a greater-than-Earth-sized anticyclone on Jupiter (called the Great Red Spot), and holes in the atmosphere (on Neptune). At least one extrasolar planet, HD 189733 b, has been claimed to have such a weather system, similar to the Great Red Spot but twice as large.

Hot Jupiters, due to their extreme proximities to their host stars, have been shown to be losing their atmospheres into space due to stellar radiation, much like the tails of comets. These planets may have vast differences in temperature between their day and night sides that produce supersonic winds, although the day and night sides of HD 189733 b appear to have very similar temperatures, indicating that that planet's atmosphere effectively redistributes the star's energy around the planet.

Magnetosphere

One important characteristic of the planets is their intrinsic magnetic moments, which in turn give rise to magnetospheres. The presence of a magnetic field indicates that the planet is still geologically alive. In other words, magnetized planets have flows of electrically conducting material in their interiors, which generate their magnetic fields. These fields significantly change the interaction of the planet and solar wind. A magnetized planet creates a cavity in the solar wind around itself called the magnetosphere, which the wind cannot penetrate. The magnetosphere can be much larger than the planet itself. In contrast, non-magnetized planets have only small magnetospheres induced by interaction of the ionosphere with the solar wind, which cannot effectively protect the planet.

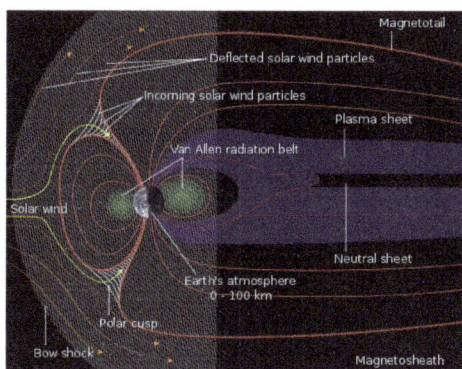

Earth's magnetosphere (diagram)

Of the eight planets in the Solar System, only Venus and Mars lack such a magnetic field. In addition, the moon of Jupiter Ganymede also has one. Of the magnetized planets the magnetic field of Mercury is the weakest, and is barely able to deflect the solar wind. Ganymede's magnetic field is several times larger, and Jupiter's is the strongest in the Solar System (so strong in fact that it poses a serious health risk to future manned missions to its moons). The magnetic fields of the other giant planets are roughly similar in strength to that of Earth, but their magnetic moments are significantly larger. The magnetic fields of Uranus and Neptune are strongly tilted relative the rotational axis and displaced from the centre of the planet.

In 2004, a team of astronomers in Hawaii observed an extrasolar planet around the star HD 179949, which appeared to be creating a sunspot on the surface of its parent star. The team hypothesized that the planet's magnetosphere was transferring energy onto the star's surface, increasing its already high 7,760 °C temperature by an additional 400 °C.

Secondary Characteristics

The rings of Saturn

Several planets or dwarf planets in the Solar System (such as Neptune and Pluto) have orbital periods that are in resonance with each other or with smaller bodies (this is also common in satellite systems). All except Mercury and Venus have natural satellites, often called "moons". Earth has one, Mars has two, and the giant planets have numerous moons in complex planetary-type systems. Many moons of the giant planets have features similar to those on the terrestrial planets and dwarf planets, and some have been studied as possible abodes of life (especially Europa).

The four giant planets are also orbited by planetary rings of varying size and complexity. The rings are composed primarily of dust or particulate matter, but can host tiny 'moonlets' whose gravity shapes and maintains their structure. Although the origins of planetary rings is not precisely known, they are believed to be the result of natural satellites that fell below their parent planet's Roche limit and were torn apart by tidal forces.

No secondary characteristics have been observed around extrasolar planets. The sub-brown dwarf Cha 110913-773444, which has been described as a rogue planet, is believed to be orbited by a tiny protoplanetary disc and the sub-brown dwarf OTS 44 was shown to be surrounded by a substantial protoplanetary disk of at least 10 Earth masses.

Minor Planet

A minor planet is an astronomical object in direct orbit around the Sun that is neither a planet nor exclusively classified as a comet. Minor planets can be dwarf planets, asteroids, trojans, centaurs, Kuiper belt objects, and other trans-Neptunian objects. As of 2016, the orbits of 709,706 minor planets were archived at the Minor Planet Center, 469,275 of which had received permanent numbers.

The first minor planet to be discovered was Ceres in 1801. The term *minor planet* has been used since the 19th century to describe these objects. The term planetoid has also been used, especially for larger (planetary) objects such as those the International Astronomical Union (IAU) has called dwarf planets since 2006. Historically, the terms *asteroid, minor planet*, and *planetoid* have been more or less synonymous. This terminology has become more complicated by the discovery of numerous minor planets beyond the orbit of Jupiter, especially trans-Neptunian objects that are generally not considered asteroids. Minor planets seen releasing gas may be dually classified as a comet.

Before 2006, the IAU had officially used the term *minor planet*. During its 2006 meeting, the IAU reclassified minor planets and comets into dwarf planets and small Solar System bodies (SSSB). Objects are called dwarf planets if their self-gravity is sufficient to achieve hydrostatic equilibrium and form an ellipsoidal shape. All other minor planets and comets are called *small Solar System bodies*. The IAU stated that the term *minor planet* may still be used, but the term *small Solar System body* will be preferred. However, for purposes of numbering and naming, the traditional distinction between minor planet and comet is still used.

Populations

Hundreds of thousands of minor planets have been discovered within the Solar System and thousands more are discovered each month. The Minor Planet Center has documented over 156 million observations and 717,768 minor planets, of which 474,120 have orbits known well enough to be assigned permanent official numbers. Of these, 20,215 have official names. As of August 2016, the lowest-numbered unnamed minor planet is (3708) 1974 FV1. As of August 2016, the highest-numbered named minor planet is 450931 Coculescu.

There are various broad minor-planet populations:

- Asteroids; traditionally, most have been bodies in the inner Solar System.

  o Near-Earth asteroids, those whose orbits take them inside the orbit of Mars. Further subclassification of these, based on orbital distance, is used:

    ▪ Apohele asteroids orbit inside of Earth's perihelion distance and thus are contained entirely within the orbit of Earth.

    ▪ Aten asteroids, those that have semi-major axes of less than Earth's and aphelion (furthest distance from the Sun) greater than 0.983 AU.

    ▪ Apollo asteroids are those asteroids with a semimajor axis greater than Earth's, while having a perihelion distance of 1.017 AU or less. Like Aten asteroids, Apollo asteroids are Earth-crossers.

    ▪ Amor asteroids are those near-Earth asteroids that approach the orbit of Earth from beyond, but do not cross it. Amor asteroids are further subdivided into four subgroups, depending on where their semimajor axis falls between Earth's orbit and the asteroid belt;

  o Earth trojans, asteroids sharing Earth's orbit and gravitationally locked to it. As of 2011, the only one known is 2010 TK$_7$.

  o Mars trojans, asteroids sharing Mars's orbit and gravitationally locked to it. As of 2007, eight such asteroids are known.

  o Asteroid belt, whose members follow roughly circular orbits between Mars and Jupiter. These are the original and best-known group of asteroids.

  o Jupiter trojans, asteroids sharing Jupiter's orbit and gravitationally locked to it. Numerically they are estimated to equal the main-belt asteroids.

- Distant minor planets; an umbrella term for minor planets in the outer Solar System.

  o Centaurs, bodies in the outer Solar System between Jupiter and Neptune. They have unstable orbits due to the gravitational influence of the giant planets, and therefore must have come from elsewhere, probably outside Neptune.

  o Neptune trojans, bodies sharing Neptune's orbit and gravitationally locked to it. Although only a handful are known, there is evidence that Neptune trojans are more numerous than either the asteroids in the asteroid belt or the Jupiter trojans.

  o Trans-Neptunian objects, bodies at or beyond the orbit of Neptune, the outermost planet.

    ▪ The Kuiper belt, objects inside an apparent population drop-off approximately 55 AU from the Sun.

    ▪ Classical Kuiper belt objects like Makemake, also known as cubewanos, are in primordial, relatively circular orbits that are not in resonance with Neptune.

    ▪ Resonant Kuiper belt objects

- Plutinos, bodies like Pluto that are in a 2:3 resonance with Neptune.

- Scattered disc objects like Eris, with aphelia outside the Kuiper belt. These are thought to have been scattered by Neptune.

- Resonant scattered disc objects.

- Detached objects such as Sedna, with both aphelia and perihelia outside the Kuiper belt.

- Sednoids, detached objects with perihelia greater than 75 AU (Sedna and 2012 VP113).

- The Oort cloud, a hypothetical population thought to be the source of long-period comets that may extend out to 50,000 AU from the Sun.

Naming Conventions

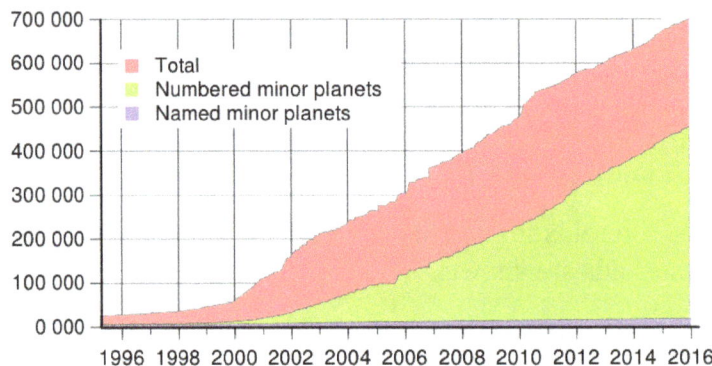

Out of a total of more than 700,000 discovered minor planets, 66% have been numbered (green) and 34% remain unnumberd (red). Only a small fraction of 20,071 minor planets (3%) have been named (blue).

All astronomical bodies in the Solar System need a distinct designation. The naming of minor planets runs through a three-step process. First, a provisional designation is given upon discovery—because the object still may turn out to be a false positive or become lost later on—called a *provisionally designated minor planet*. After the observation arc is accurate enough to predict its future location, a minor planet is formally designated and receives a number. It is then a *numbered minor planet*. Finally, in the third step, it may be named by its discoverers. However, only a small fraction of all minor planets have been named. The vast majority is either numbered or has still only a provisional designation. Example of the naming process:

- 1932 HA – provisional designation upon discovery on 24 April 1932

- (1862) 1932 HA – formal designation, receives an official number

- 1862 Apollo – named Minor planet, receives a name, the alphanumeric code is dropped

Provisional Designation

A newly discovered minor planet is given a provisional designation. For example, the provisional designation 2002 AT4 consists of the year of discovery (2002) and an alphanumeric code indicating the half-month of discovery and the sequence within that half-month. Once an asteroid's orbit has been confirmed, it is given a number, and later may also be given a name (e.g. 433 Eros). The formal naming convention uses parentheses around the number, but dropping the parentheses is quite common. Informally, it is common to drop the number altogether, or to drop it after the first mention when a name is repeated in running text.

Minor planets that have been given a number but not a name keep their provisional designation, e.g. (29075) 1950 DA. Because modern discovery techniques are finding vast numbers of new asteroids, they are increasingly being left unnamed. The earliest discovered to be left unnamed was for a long time (3360) 1981 VA, now 3360 Syrinx; as of September 2008, this distinction is held by (3708) 1974 FV1. On rare occasions, a small object's provisional designation may become used as a name in itself: the still unnamed (15760) 1992 QB1 gave its "name" to a group of objects that became known as Classical Kuiper belt objects ("cubewanos").

A few objects are cross-listed as both comets and asteroids, such as 4015 Wilson–Harrington, which is also listed as 107P/Wilson–Harrington.

Numbering

Minor planets are awarded an official number once their orbits are confirmed. With the increasing rapidity of discovery, these are now six-figure numbers. The switch from five figures to six figures arrived with the publication of the Minor Planet Circular (MPC) of October 19, 2005, which saw the highest numbered minor planet jump from 99947 to 118161.

Naming

The first few asteroids were named after figures from Greek and Roman mythology but as such names started to dwindle the names of famous people, literary characters, discoverer's wives, chil-dren, and even television characters were used.

Gender

The first asteroid to be given a non-mythological name was 20 Massalia, named after the Greek name for the city of Marseille. The first to be given an entirely non-Classical name was 45 Eugenia, named after Empress Eugénie de Montijo, the wife of Napoleon III. For some time only female (or feminized) names were used; Alexander von Humboldt was the first man to have an asteroid named after him, but his name was feminized to 54 Alexandra. This unspoken tradition lasted until 334 Chicago was named; even then, female names show up in the list for years after.

Eccentric

As the number of asteroids began to run into the hundreds, and eventually in the thousands, discoverers began to give them increasingly frivolous names. The first hints of this were 482 Petrina and 483 Seppina, named after the discoverer's pet dogs. However, there was little controversy about this until 1971, upon the naming of 2309 Mr. Spock (the name of the discoverer's cat). Although the IAU subsequently banned pet names as sources, eccentric asteroid names are still being proposed and accepted, such as 4321 Zero, 6042 Cheshirecat, 9007 James Bond, 13579 Allodd and 24680 Alleven, and 26858 Misterrogers.

Discoverer'S Name

A well-established rule is that, unlike comets, minor planets may not be named after their discoverer(s). One way to circumvent this rule has been for astronomers to exchange the courtesy of naming their discoveries after each other. An exception to this rule is 96747 Crespodasilva, which was named after its discoverer, Lucy d'Escoffier Crespo da Silva, because she died shortly after the discovery, at age 22.

Languages

Names were adapted to various languages from the beginning. 1 Ceres, *Ceres* being its Anglo-Latin name, was actually named *Cerere*, the Italian form of the name. German, French, Arabic and Hindi use forms similar to the English, whereas Russian uses a form, *Tserera*, similar to the Italian. In Greek the name was translated to (Demeter), the Greek equivalent of the Roman goddess Ceres. In the early years, before it started causing conflicts, asteroids named after Roman figures were generally translated in Greek; other examples are (Hera) for 3 Juno, (Hestia) for 4 Vesta, (Chloris) for 8 Flora, and (Pistis) for 37 Fides. In Chinese, the names are not given the Chinese forms of the deities they are named after, but rather typically have a syllable or two for the character of the deity or person, followed by 神 'god(dess)' or 女 'woman' if just one syllable, plus 星 'star/planet', so that most asteroid names are written with three Chinese characters. Thus Ceres is 谷神星 'grain goddess planet', Pallas is 智神星 'wisdom goddess planet', etc.

Physical Properties of Comets and Minor Planets

Commission 15 of the International Astronomical Union is dedicated to the Physical Study of Comets & Minor Planets.

Archival data on the physical properties of comets and minor planets are found in the PDS Asteroid/Dust Archive. This includes standard asteroid physical characteristics such as the properties of binary systems, occultation timings and diameters, masses, densities, rotation periods, surface temperatures, albedoes, spin vectors, taxonomy, and absolute magnitudes and slopes. In addition, European Asteroid Research Node (E.A.R.N.), an association of asteroid research groups, maintains a Data Base of Physical and Dynamical Properties of Near Earth Asteroids.

Most detailed information is available from Category: Asteroids visited by spacecraft and Category: Comets visited by spacecraft.

Asteroid

Asteroids are minor planets, especially those of the inner Solar System. The larger ones have also been called planetoids. These terms have historically been applied to any astronomical object orbiting the Sun that did not show the disc of a planet and was not observed to have the character-istics of an active comet. As minor planets in the outer Solar System were discovered and found to have volatile-based surfaces that resemble those of comets, they were often distinguished from asteroids of the asteroid belt. In this article, the term "asteroid" refers to the minor planets of the inner Solar System including those co-orbital with Jupiter.

253 Mathilde, a C-type asteroid measuring about 50 kilometres (30 mi) across, covered in craters half that size. Photograph taken in 1997 by the NEAR Shoemaker probe.

There are millions of asteroids, many thought to be the shattered remnants of planetesimals, bodies within the young Sun's solar nebula that never grew large enough to become planets. The large majority of known asteroids orbit in the asteroid belt between the orbits of Mars and Jupiter, or are co-orbital with Jupiter (the Jupiter trojans). However, other orbital families exist with significant populations, including the near-Earth objects. Individual asteroids are classified by their characteristic spectra, with the majority falling into three main groups: C-type, M-type, and S-type. These were named after and are generally identified with carbon-rich, metallic, and silicate (stony) compositions, respectively. The size of asteroids varies greatly, some reaching as much as 1000 km across.

Asteroids are differentiated from comets and meteoroids. In the case of comets, the difference is one of composition: while asteroids are mainly composed of mineral and rock, comets are composed of dust and ice. In addition, asteroids formed closer to the sun, preventing the development of the aforementioned cometary ice. The difference between asteroids and meteoroids is mainly one of size: meteoroids have a diameter of less than one meter, whereas asteroids have a diameter of greater than one meter. Finally, meteoroids can be composed of either cometary or asteroidal materials.

Only one asteroid, 4 Vesta, which has a relatively reflective surface, is normally visible to the naked eye, and this only in very dark skies when it is favorably positioned. Rarely, small asteroids passing close to Earth may be visible to the naked eye for a short time. As of March 2016, the Minor Planet Center had data on more than 1.3 million objects in the inner and outer Solar System, of which 750,000 had enough information to be given numbered designations.

Discovery

The first asteroid to be discovered, Ceres, was found in 1801 by Giuseppe Piazzi, and was originally considered to be a new planet. This was followed by the discovery of other similar bodies, which, with the equipment of the time, appeared to be points of light, like stars, showing little or no planetary disc, though readily distinguishable from stars due to their apparent motions. This prompted the astronomer Sir William Herschel to propose the term "asteroid", coined in Greek as *asteroeidēs*, meaning 'star-like, star-shaped', and derived from the Ancient Greek *astēr* 'star, planet'. In the early second half of the nineteenth century, the terms "asteroid" and "planet" (not always qualified as "minor") were still used interchangeably.

243 Ida and its moon Dactyl. Dactyl is the first satellite of an asteroid to be discovered.

Historical Methods

Asteroid discovery methods have dramatically improved over the past two centuries.

In the last years of the 18th century, Baron Franz Xaver von Zach organized a group of 24 astronomers to search the sky for the missing planet predicted at about 2.8 AU from the Sun by the Titius-Bode law, partly because of the discovery, by Sir William Herschel in 1781, of the planet Uranus at the distance predicted by the law. This task required that hand-drawn sky charts be prepared for all stars in the zodiacal band down to an agreed-upon limit of faintness. On subsequent nights, the sky would be charted again and any moving object would, hopefully, be spotted. The expected motion of the missing planet was about 30 seconds of arc per hour, readily discernible by observers.

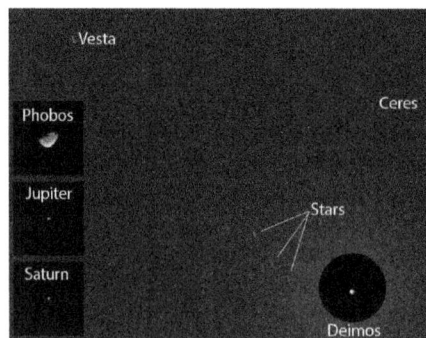

First asteroid image (Ceres and Vesta) from Mars – viewed by *Curiosity* (20 April 2014).

The first object, Ceres, was not discovered by a member of the group, but rather by accident in 1801 by Giuseppe Piazzi, director of the observatory of Palermo in Sicily. He discovered a new star-like object in Taurus and followed the displacement of this object during several nights. Later that year, Carl Friedrich Gauss used these observations to calculate the orbit of this unknown object, which was found to be between the planets Mars and Jupiter. Piazzi named it after Ceres, the Roman goddess of agriculture.

Three other asteroids (2 Pallas, 3 Juno, and 4 Vesta) were discovered over the next few years, with Vesta found in 1807. After eight more years of fruitless searches, most astronomers assumed that there were no more and abandoned any further searches.

However, Karl Ludwig Hencke persisted, and began searching for more asteroids in 1830. Fifteen years later, he found 5 Astraea, the first new asteroid in 38 years. He also found 6 Hebe less than two years later. After this, other astronomers joined in the search and at least one new asteroid was discovered every year after that (except the wartime year 1945). Notable asteroid hunters of this early era were J. R. Hind, Annibale de Gasparis, Robert Luther, H. M. S. Goldschmidt, Jean Chacornac, James Ferguson, Norman Robert Pogson, E. W. Tempel, J. C. Watson, C. H. F. Peters, A. Borrelly, J. Palisa, the Henry brothers and Auguste Charlois.

In 1891, Max Wolf pioneered the use of astrophotography to detect asteroids, which appeared as short streaks on long-exposure photographic plates. This dramatically increased the rate of detection compared with earlier visual methods: Wolf alone discovered 248 asteroids, beginning with 323 Brucia, whereas only slightly more than 300 had been discovered up to that point. It was known that there were many more, but most astronomers did not bother with them, calling them "vermin of the skies", a phrase variously attributed to Eduard Suess and Edmund Weiss. Even a century later, only a few thousand asteroids were identified, numbered and named.

Manual Methods of the 1900s and Modern Reporting

Until 1998, asteroids were discovered by a four-step process. First, a region of the sky was photographed by a wide-field telescope, or astrograph. Pairs of photographs were taken, typically one hour apart. Multiple pairs could be taken over a series of days. Second, the two films or plates of the same region were viewed under a stereoscope. Any body in orbit around the Sun would move slightly between the pair of films. Under the stereoscope, the image of the body would seem to float slightly above the background of stars. Third, once a moving body was identified, its location would be measured precisely using a digitizing microscope. The location would be measured relative to known star locations.

These first three steps do not constitute asteroid discovery: the observer has only found an apparition, which gets a provisional designation, made up of the year of discovery, a letter representing the half-month of discovery, and finally a letter and a number indicating the discovery's sequential number (example: 1998 FJ$_{74}$).

The last step of discovery is to send the locations and time of observations to the Minor Planet Center, where computer programs determine whether an apparition ties together earlier apparitions into a single orbit. If so, the object receives a catalogue number and the observer of the first apparition with a calculated orbit is declared the discoverer, and granted the honor of naming the object subject to the approval of the International Astronomical Union.

Computerized Methods

There is increasing interest in identifying asteroids whose orbits cross Earth's, and that could, given enough time, collide with Earth. The three most important groups of near-Earth asteroids are the Apollos, Amors, and Atens. Various asteroid deflection strategies have been proposed, as early as the 1960s.

2004 FH is the center dot being followed by the sequence; the object that flashes by during the clip is an artificial satellite.

The near-Earth asteroid 433 Eros had been discovered as long ago as 1898, and the 1930s brought a flurry of similar objects. In order of discovery, these were: 1221 Amor, 1862 Apollo, 2101 Adonis, and finally 69230 Hermes, which approached within 0.005 AU of Earth in 1937. Astronomers began to realize the possibilities of Earth impact.

Two events in later decades increased the alarm: the increasing acceptance of Walter Alvarez' hypothesis that an impact event resulted in the Cretaceous–Paleogene extinction, and the 1994 observation of Comet Shoemaker-Levy 9 crashing into Jupiter. The U.S. military also declassified the information that its military satellites, built to detect nuclear explosions, had detected hundreds of upper-atmosphere impacts by objects ranging from one to 10 metres across.

All these considerations helped spur the launch of highly efficient surveys that consist of Charge-Coupled Device (CCD) cameras and computers directly connected to telescopes. As of spring 2011, it was estimated that 89% to 96% of near-Earth asteroids one kilometer or larger in diameter had been discovered. A list of teams using such systems includes:

- The Lincoln Near-Earth Asteroid Research (LINEAR) team
- The Near-Earth Asteroid Tracking (NEAT) team
- Spacewatch
- The Lowell Observatory Near-Earth-Object Search (LONEOS) team
- The Catalina Sky Survey (CSS)
- The Campo Imperatore Near-Earth Object Survey (CINEOS) team
- The Japanese Spaceguard Association
- The Asiago-DLR Asteroid Survey (ADAS)

The LINEAR system alone has discovered 138,393 asteroids, as of 20 September 2013. Among all the surveys, 4711 near-Earth asteroids have been discovered including over 600 more than 1 km (0.6 mi) in diameter.

Terminology

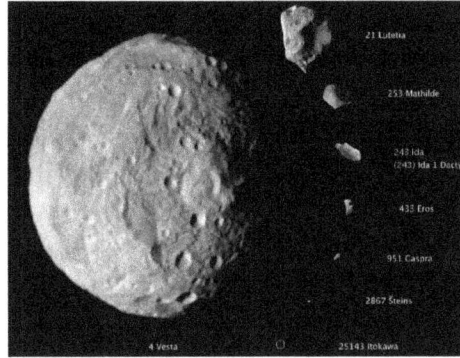

A composite image, to scale, of the asteroids that have been imaged at high resolution except Ceres. As of 2011 they are, from largest to smallest: 4 Vesta, 21 Lutetia, 253 Mathilde, 243 Ida and its moon Dactyl, 433 Eros, 951 Gaspra, 2867 Šteins, 25143 Itokawa.

The largest asteroid in the previous image, Vesta (left), with Ceres (center) and the Moon (right) shown to scale.

Traditionally, small bodies orbiting the Sun were classified as comets, asteroids, or meteoroids, with anything smaller than 10 meters across being called a meteoroid (such as in Beech and Steel's 1995 paper). The term "asteroid", from the Greek word for "star-like", never had a formal definition, with the broader term minor planet being preferred by the International Astronomical Union.

However, following the discovery of asteroids below 10 meters in size, Rubin and Grossman in a 2010 paper revised the previous definition of meteoroid to objects between 10 µm and 1 meter in size in order to maintain the distinction between asteroids and meteoroids. The smallest asteroids discovered (based on absolute magnitude H) are 2008 TS$_{26}$ with $H = 33.2$ and 2011 CQ1 with $H = 32.1$ both with an estimated size of about 1 meter.

In 2006, the term "small Solar System body" was also introduced to cover both most minor planets and comets. Other languages prefer "planetoid" (Greek for "planet-like"), and this term is occasionally used in English especially for larger minor planets such as the dwarf planets as well as an alternative for asteroids since they are not star-like. The word "planetesimal" has a similar meaning, but refers specifically to the small building blocks of the planets that existed when the Solar System was forming. The term "planetule" was coined by the geologist William Daniel Conybeare to describe minor planets, but is not in common use. The three largest objects in the asteroid belt, Ceres, Pallas, and Vesta, grew to the stage of protoplanets. Ceres is a dwarf planet, the only one in the inner Solar System.

When found, asteroids were seen as a class of objects distinct from comets, and there was no unified term for the two until "small Solar System body" was coined in 2006. The main difference between an asteroid and a comet is that a comet shows a coma due to sublimation of near surface ices by solar radiation. A few objects have ended up being dual-listed because they were first classified as minor planets but later showed evidence of cometary activity. Conversely, some (perhaps all) comets are eventually depleted of their surface volatile ices and become asteroid-like. A further distinction is that comets typically have more eccentric orbits than most asteroids; most "asteroids" with notably eccentric orbits are probably dormant or extinct comets.

For almost two centuries, from the discovery of Ceres in 1801 until the discovery of the first centaur, Chiron, in 1977, all known asteroids spent most of their time at or within the orbit of Jupiter, though a few such as Hidalgo ventured far beyond Jupiter for part of their orbit. When astronomers started finding more small bodies that permanently resided further out than Jupiter, now called centaurs, they numbered them among the traditional asteroids, though there was debate over whether they should be considered asteroids or as a new type of object. Then, when the first trans-Neptunian object (other than Pluto), 1992 QB$_1$, was discovered in 1992, and especially when large numbers of similar objects started turning up, new terms were invented to sidestep the issue: Kuiper-belt object, trans-Neptunian object, scattered-disc object, and so on. These inhabit the cold outer reaches of the Solar System where ices remain solid and comet-like bodies are not expected to exhibit much cometary activity; if centaurs or trans-Neptunian objects were to venture close to the Sun, their volatile ices would sublimate, and traditional approaches would classify them as comets and not asteroids.

The innermost of these are the Kuiper-belt objects, called "objects" partly to avoid the need to classify them as asteroids or comets. They are thought to be predominantly comet-like in composition, though some may be more akin to asteroids. Furthermore, most do not have the highly eccentric orbits associated with comets, and the ones so far discovered are larger than traditional comet nuclei. (The much more distant Oort cloud is hypothesized to be the main reservoir of dormant comets.) Other recent observations, such as the analysis of the cometary dust collected by the *Stardust* probe, are increasingly blurring the distinction between comets and asteroids, suggesting "a continuum between asteroids and comets" rather than a sharp dividing line.

The minor planets beyond Jupiter's orbit are sometimes also called "asteroids", especially in popular presentations. However, it is becoming increasingly common for the term "asteroid" to be restricted to minor planets of the inner Solar System. Therefore, this article will restrict itself for the most part to the classical asteroids: objects of the asteroid belt, Jupiter trojans, and near-Earth objects.

When the IAU introduced the class small Solar System bodies in 2006 to include most objects previously classified as minor planets and comets, they created the class of dwarf planets for the largest minor planets—those that have enough mass to have become ellipsoidal under their own gravity. According to the IAU, "the term 'minor planet' may still be used, but generally the term 'Small Solar System Body' will be preferred." Currently only the largest object in the asteroid belt, Ceres, at about 950 km (590 mi) across, has been placed in the dwarf planet category.

Formation

Artist's impression shows how an asteroid is torn apart by the strong gravity of a white dwarf.

It is thought that planetesimals in the asteroid belt evolved much like the rest of the solar nebula until Jupiter neared its current mass, at which point excitation from orbital resonances with Jupiter ejected over 99% of planetesimals in the belt. Simulations and a discontinuity in spin rate and spectral properties suggest that asteroids larger than approximately 120 km (75 mi) in diameter accreted during that early era, whereas smaller bodies are fragments from collisions between asteroids during or after the Jovian disruption. Ceres and Vesta grew large enough to melt and differentiate, with heavy metallic elements sinking to the core, leaving rocky minerals in the crust.

In the Nice model, many Kuiper-belt objects are captured in the outer asteroid belt, at distances greater than 2.6 AU. Most were later ejected by Jupiter, but those that remained may be the D-type asteroids, and possibly include Ceres.

Distribution within the Solar System

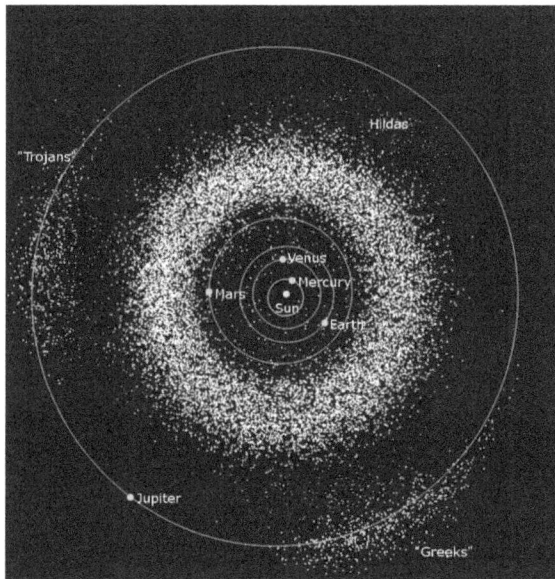

The asteroid belt (white) and Jupiter's trojan asteroids (green)

Various dynamical groups of asteroids have been discovered orbiting in the inner Solar System. Their orbits are perturbed by the gravity of other bodies in the Solar System and by the Yarkovsky effect. Significant populations include:

Asteroid Belt

The majority of known asteroids orbit within the asteroid belt between the orbits of Mars and Jupiter, generally in relatively low-eccentricity (i.e. not very elongated) orbits. This belt is now estimated to contain between 1.1 and 1.9 million asteroids larger than 1 km (0.6 mi) in diameter, and millions of smaller ones. These asteroids may be remnants of the protoplanetary disk, and in this region the accretion of planetesimals into planets during the formative period of the Solar System was prevented by large gravitational perturbations by Jupiter.

Trojans

Trojans are populations that share an orbit with a larger planet or moon, but do not collide with it because they orbit in one of the two Lagrangian points of stability, L4 and L5, which lie 60° ahead of and behind the larger body.

The most significant population of trojans are the Jupiter trojans. Although fewer Jupiter trojans have been discovered as of 2010, it is thought that they are as numerous as the asteroids in the asteroid belt.

A couple of trojans have also been found orbiting with Mars.

Near-Earth Asteroids

Near-Earth asteroids, or NEAs, are asteroids that have orbits that pass close to that of Earth. Asteroids that actually cross Earth's orbital path are known as *Earth-crossers*. As of June 2016, 14,464 near-Earth asteroids are known and the number over one kilometre in diameter is estimated to be 900–1,000.

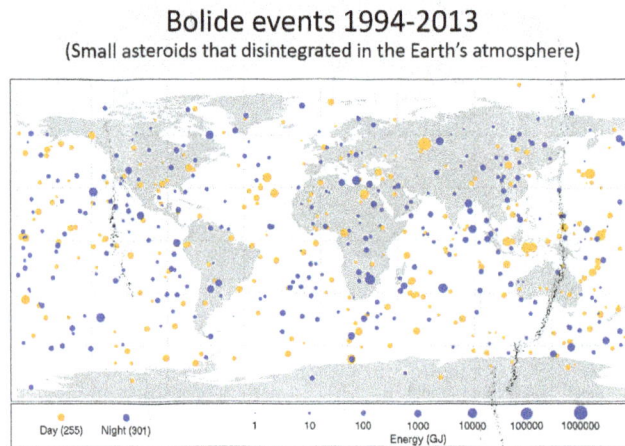

Frequency of bolides, small asteroids roughly 1 to 20 meters in diameter impacting Earth's atmosphere.

Characteristics

Size Distribution

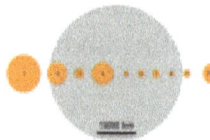

Sizes of the first ten asteroids to be discovered, compared to the Moon

Ceres as imaged by *Dawn* on 4 February 2015

Asteroids vary greatly in size, from almost 1000 km for the largest down to rocks just 1 meter across. The three largest are very much like miniature planets: they are roughly spherical, have at least partly differentiated interiors, and are thought to be surviving protoplanets. The vast majority, however, are much smaller and are irregularly shaped; they are thought to be either surviving planetesimals or fragments of larger bodies.

The dwarf planet Ceres is by far the largest asteroid, with a diameter of 975 km (610 mi). The next largest are 4 Vesta and 2 Pallas, both with diameters of just over 500 km (300 mi). Vesta is the only main-belt asteroid that can, on occasion, be visible to the naked eye. On some rare occasions, a near-Earth asteroid may briefly become visible without technical aid; see 99942 Apophis.

The mass of all the objects of the asteroid belt, lying between the orbits of Mars and Jupiter, is estimated to be about 2.8–3.2×10²¹ kg, or about 4% of the mass of the Moon. Of this, Ceres comprises 0.95×10²¹ kg, a third of the total. Adding in the next three most massive objects, Vesta (9%), Pallas (7%), and Hygiea (3%), brings this figure up to 51%; whereas the three after that, 511 Davida (1.2%), 704 Interamnia (1.0%), and 52 Europa (0.9%), only add another 3% to the total mass. The number of asteroids then increases rapidly as their individual masses decrease.

The number of asteroids decreases markedly with size. Although this generally follows a power law, there are 'bumps' at 5 km and 100 km, where more asteroids than expected from a logarithmic distribution are found.

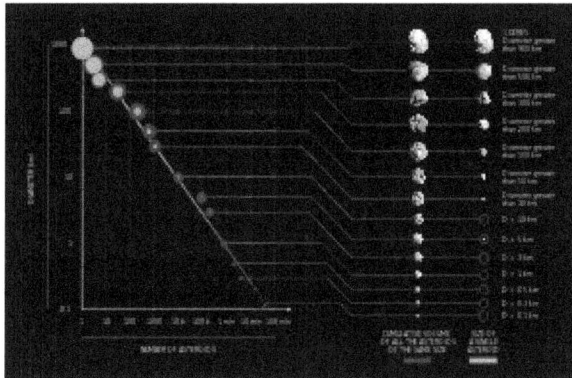

The asteroids of the Solar System, categorized by size and number

	Approximate number of asteroids (N) larger than a certain diameter (D)													
D	100 m	300 m	500 m	1 km	3 km	5 km	10 km	30 km	50 km	100 km	200 km	300 km	500 km	900 km
N	~25000000	4000000	2000000	750000	200000	90000	10000	1100	600	200	30	5	3	1

Largest Asteroids

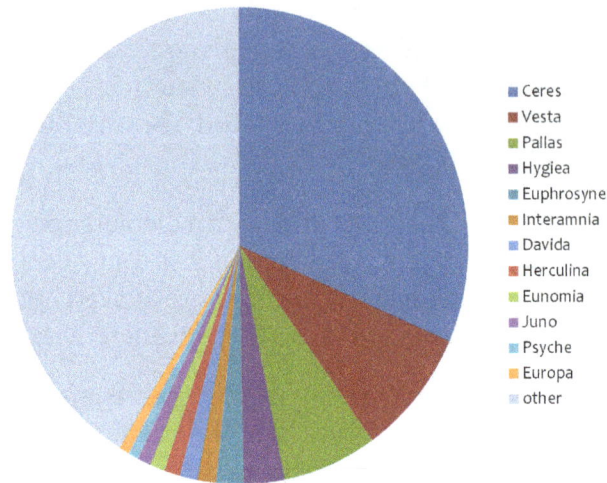

The relative masses of the twelve largest asteroids known, compared to the remaining mass of the asteroid belt.

1 ⬛ Ceres 4 ⬛ Vesta 2 ⬛ Pallas 10 ⬛ Hygiea 31 ⬛ Euphrosyne 704 ⬛ Interamnia 511 ⬛ Davida
532 ⬛ Herculina 15 ⬛ Eunomia 3 ⬛ Juno 16 ⬛ Psyche 52 ⬛ Europa ⬜ all others

Although their location in the asteroid belt excludes them from planet status, the three largest objects, Ceres, Vesta, and Pallas, are intact protoplanets that share many characteristics common to planets, and are atypical compared to the majority of "potato"-shaped asteroids.

Ceres is the only asteroid with a fully ellipsoidal shape and hence the only one that is a dwarf planet. It has a much higher absolute magnitude than the other asteroids, of around 3.32, and may possess a surface layer of ice. Like the planets, Ceres is differentiated: it has a crust, a mantle and a core. No meteorites from Ceres have been found on Earth.

Vesta, too, has a differentiated interior, though it formed inside the Solar System's frost line, and so is devoid of water; its composition is mainly of basaltic rock such as olivine. Aside from the large crater at its southern pole, Rheasilvia, Vesta also has an ellipsoidal shape. Vesta is the parent body of the Vestian family and other V-type asteroids, and is the source of the HED meteorites, which constitute 5% of all meteorites on Earth.

Pallas is unusual in that, like Uranus, it rotates on its side, with its axis of rotation tilted at high angles to its orbital plane. Its composition is similar to that of Ceres: high in carbon and silicon, and perhaps partially differentiated. Pallas is the parent body of the Palladian family of asteroids,

The fourth-most-massive asteroid, Hygiea, is the largest carbonaceous asteroid and, unlike the other largest asteroids, lies relatively close to the plane of the ecliptic. It is the largest member and presumed parent body of the Hygiean family of asteroids. Between them, the four largest asteroids constitute half the mass of the asteroid belt.

Attributes of largest asteroids												
Name	Orbital radius (AU)	Orbital period (years)	Inclination to ecliptic	Orbital eccentricity	Diameter (km)	Diameter (% of Moon)	Mass (×10^18 kg)	Mass (% of Ceres)	Density g/cm^3	Rotation period (hr)	Axial tilt	Surface temperature

Vesta	2.36	3.63	7.1°	0.089	573×557×446 (mean 525)	15%	260	28%	3.44 ± 0.12	5.34	29°	85–270 K
Ceres	2.77	4.60	10.6°	0.079	975×975×909 (mean 952)	28%	940	100%	2.12 ± 0.04	9.07	≈ 3°	167 K
Pallas	2.77	4.62	34.8°	0.231	580×555×500 (mean 545)	16%	210	22%	2.71 ± 0.11	7.81	≈ 80°	164 K
Hygiea	3.14	5.56	3.8°	0.117	530×407×370 (mean 430)	12%	87	9%	2.76 ± 1.2	27.6	≈ 60°	164 K

Rotation

Measurements of the rotation rates of large asteroids in the asteroid belt show that there is an upper limit. No asteroid with a diameter larger than 100 meters has a rotation period smaller than 2.2 hours. For asteroids rotating faster than approximately this rate, the inertia at the surface is greater than the gravitational force, so any loose surface material would be flung out. However, a solid object should be able to rotate much more rapidly. This suggests that most asteroids with a diameter over 100 meters are rubble piles formed through accumulation of debris after collisions between asteroids.

Composition

Cratered terrain on 4 Vesta

The physical composition of asteroids is varied and in most cases poorly understood. Ceres appears to be composed of a rocky core covered by an icy mantle, where Vesta is thought to have a nickeliron core, olivine mantle, and basaltic crust. 10 Hygiea, however, which appears to have a uniformly primitive composition of carbonaceous chondrite, is thought to be the largest undif-ferentiated asteroid. Most of the smaller asteroids are thought to be piles of rubble held together loosely by gravity, though the largest are probably solid. Some asteroids have moons or are coorbiting binaries: Rubble piles, moons, binaries, and scattered asteroid families are thought to be the results of collisions that disrupted a parent asteroid.

Asteroids contain traces of amino acids and other organic compounds, and some speculate that asteroid impacts may have seeded the early Earth with the chemicals necessary to initiate life, or may have even brought life itself to Earth. In August 2011, a report, based on NASA studies with meteorites found on Earth, was published suggesting DNA and RNA components (adenine, guanine and related organic molecules) may have been formed on asteroids and comets in outer space.

Composition is calculated from three primary sources: albedo, surface spectrum, and density. The last can only be determined accurately by observing the orbits of moons the asteroid might have. So far, every asteroid with moons has turned out to be a rubble pile, a loose conglomeration of rock and metal that may be half empty space by volume. The investigated asteroids are as large as 280 km in diameter, and include 121 Hermione (268×186×183 km), and 87 Sylvia (384×262×232 km). Only half a dozen asteroids are larger than 87 Sylvia, though none of them have moons; however, some smaller asteroids are thought to be more massive, suggesting they may not have been disrupted, and indeed 511 Davida, the same size as Sylvia to within measurement error, is estimated to be two and a half times as massive, though this is highly uncertain. The fact that such large asteroids as Sylvia can be rubble piles, presumably due to disruptive impacts, has important consequences for the formation of the Solar System: Computer simulations of collisions involving solid bodies show them destroying each other as often as merging, but colliding rubble piles are more likely to merge. This means that the cores of the planets could have formed relatively quickly.

On 7 October 2009, the presence of water ice was confirmed on the surface of 24 Themis using NASA's Infrared Telescope Facility. The surface of the asteroid appears completely covered in ice. As this ice layer is sublimated, it may be getting replenished by a reservoir of ice under the surface. Organic compounds were also detected on the surface. Scientists hypothesize that some of the first water brought to Earth was delivered by asteroid impacts after the collision that produced the Moon. The presence of ice on 24 Themis supports this theory.

In October 2013, water was detected on an extrasolar body for the first time, on an asteroid orbiting the white dwarf GD 61. On 22 January 2014, European Space Agency (ESA) scientists reported the detection, for the first definitive time, of water vapor on Ceres, the largest object in the asteroid belt. The detection was made by using the far-infrared abilities of the Herschel Space Observatory. The finding is unexpected because comets, not asteroids, are typically considered to "sprout jets and plumes". According to one of the scientists, "The lines are becoming more and more blurred between comets and asteroids." In May 2016, significant asteroid data arising from the Wide-field Infrared Survey Explorer and NEOWISE missions have been questioned, but the criticism has yet to undergo peer review.

Surface Features

Most asteroids outside the "big four" (Ceres, Pallas, Vesta, and Hygiea) are likely to be broadly similar in appearance, if irregular in shape. 50-km (31-mi) 253 Mathilde is a rubble pile saturated with craters with diameters the size of the asteroid's radius, and Earth-based observations of 300-km (186-mi) 511 Davida, one of the largest asteroids after the big four, reveal a similarly angular profile, suggesting it is also saturated with radius-size craters. Medium-sized asteroids such as Mathilde and 243 Ida that have been observed up close also reveal a deep regolith covering the surface. Of the big four, Pallas and Hygiea are practically unknown. Vesta has compression fractures encircling a radius-size crater at its south pole but is otherwise a spheroid. Ceres seems quite different in the glimpses Hubble has provided, with surface features that are unlikely to be due to simple craters and impact basins, but details will be expanded with the *Dawn* spacecraft, which entered Ceres orbit on 6 March 2015.

Color

Asteroids become darker and redder with age due to space weathering. However evidence suggests most of the color change occurs rapidly, in the first hundred thousands years, limiting the usefulness of spectral measurement for determining the age of asteroids.

Classification

Asteroids are commonly classified according to two criteria: the characteristics of their orbits, and features of their reflectance spectrum.

Orbital Classification

Many asteroids have been placed in groups and families based on their orbital characteristics. Apart from the broadest divisions, it is customary to name a group of asteroids after the first member of that group to be discovered. Groups are relatively loose dynamical associations, whereas families are tighter and result from the catastrophic break-up of a large parent asteroid sometime in the past. Families have only been recognized within the asteroid belt. They were first recognized by Kiyotsugu Hirayama in 1918 and are often called Hirayama families in his honor.

About 30–35% of the bodies in the asteroid belt belong to dynamical families each thought to have a common origin in a past collision between asteroids. A family has also been associated with the plutoid dwarf planet Haumea.

Quasi-satellites and Horseshoe Objects

Some asteroids have unusual horseshoe orbits that are co-orbital with Earth or some other planet. Examples are 3753 Cruithne and 2002 AA29. The first instance of this type of orbital arrangement was discovered between Saturn's moons Epimetheus and Janus.

Sometimes these horseshoe objects temporarily become quasi-satellites for a few decades or a few hundred years, before returning to their earlier status. Both Earth and Venus are known to have quasi-satellites.

Such objects, if associated with Earth or Venus or even hypothetically Mercury, are a special class of Aten asteroids. However, such objects could be associated with outer planets as well.

Spectral Classification

This picture of 433 Eros shows the view looking from one end of the asteroid across the gouge on its underside and toward the opposite end. Features as small as 35 m (115 ft) across can be seen.

In 1975, an asteroid taxonomic system based on color, albedo, and spectral shape was developed by Clark R. Chapman, David Morrison, and Ben Zellner. These properties are thought to correspond to the composition of the asteroid's surface material. The original classification system had three categories: C-types for dark carbonaceous objects (75% of known asteroids), S-types for stony (silicaceous) objects (17% of known asteroids) and U for those that did not fit into either C or S. This classification has since been expanded to include many other asteroid types. The number of types continues to grow as more asteroids are studied.

The two most widely used taxonomies now used are the Tholen classification and SMASS classification. The former was proposed in 1984 by David J. Tholen, and was based on data collected from an eight-color asteroid survey performed in the 1980s. This resulted in 14 asteroid categories. In 2002, the Small Main-Belt Asteroid Spectroscopic Survey resulted in a modified version of the Tholen taxonomy with 24 different types. Both systems have three broad categories of C, S, and X asteroids, where X consists of mostly metallic asteroids, such as the M-type. There are also several smaller classes.

The proportion of known asteroids falling into the various spectral types does not necessarily reflect the proportion of all asteroids that are of that type; some types are easier to detect than others, biasing the totals.

Problems

Originally, spectral designations were based on inferences of an asteroid's composition. However, the correspondence between spectral class and composition is not always very good, and a variety of classifications are in use. This has led to significant confusion. Although asteroids of different spectral classifications are likely to be composed of different materials, there are no assurances that asteroids within the same taxonomic class are composed of similar materials.

Naming

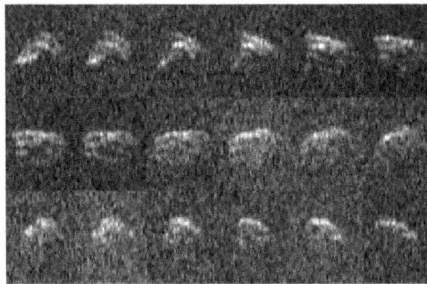

2013 EC, shown here in radar images, has a provisional designation

A newly discovered asteroid is given a provisional designation (such as 2002 AT4) consisting of the year of discovery and an alphanumeric code indicating the half-month of discovery and the sequence within that half-month. Once an asteroid's orbit has been confirmed, it is given a number, and later may also be given a name (e.g. 433 Eros). The formal naming convention uses parentheses around the number (e.g. (433) Eros), but dropping the parentheses is quite common. Informally, it is common to drop the number altogether, or to drop it after the first mention when a name is repeated in running text. In addition, names can be proposed by the asteroid's discoverer, within guidelines established by the International Astronomical Union.

Symbols

The first asteroids to be discovered were assigned iconic symbols like the ones traditionally used to designate the planets. By 1855 there were two dozen asteroid symbols, which often occurred in multiple variants.

Asteroid	Symbol		Year
1 Ceres		Ceres' scythe, reversed to double as the letter *C*	1801
2 Pallas		Athena's (Pallas') spear	1801
3 Juno		A star mounted on a scepter, for Juno, the Queen of Heaven	1804
4 Vesta		The altar and sacred fire of Vesta	1807
5 Astraea		A scale, or an inverted anchor, symbols of justice	1845
6 Hebe		Hebe's cup	1847
7 Iris		A rainbow (*iris*) and a star	1847
8 Flora		A flower (*flora*), specifically the Rose of England	1847
9 Metis		The eye of wisdom and a star	1848
10 Hygiea		Hygiea's serpent and a star, or the Rod of Asclepius	1849
11 Parthenope		A harp, or a fish and a star; symbols of the sirens	1850
12 Victoria		The laurels of victory and a star	1850
13 Egeria		A shield, symbol of Egeria's protection, and a star	1850
14 Irene		A dove carrying an olive branch (symbol of *irene* 'peace') with a star on its head, or an olive branch, a flag of truce, and a star	1851
15 Eunomia		A heart, symbol of good order (*eunomia*), and a star	1851
16 Psyche		A butterfly's wing, symbol of the soul (*psyche*), and a star	1852
17 Thetis		A dolphin, symbol of Thetis, and a star	1852
18 Melpomene		The dagger of Melpomene, and a star	1852
19 Fortuna		The wheel of fortune and a star	1852
26 Proserpina		Proserpina's pomegranate	1853
28 Bellona		Bellona's whip and lance	1854
29 Amphitrite		The shell of Amphitrite and a star	1854

| 35 Leukothea | ⌐☞ | A lighthouse beacon, symbol of Leucothea | 1855 |
| 37 Fides | † | The cross of faith (*fides*) | 1855 |

In 1851, after the fifteenth asteroid (Eunomia) had been discovered, Johann Franz Encke made a major change in the upcoming 1854 edition of the *Berliner Astronomisches Jahrbuch* (BAJ, *Berlin Astronomical Yearbook*). He introduced a disk (circle), a traditional symbol for a star, as the generic symbol for an asteroid. The circle was then numbered in order of discovery to indicate a specific asteroid (although he assigned ① to the fifth, Astraea, while continuing to designate the first four only with their existing iconic symbols). The numbered-circle convention was quickly adopted by astronomers, and the next asteroid to be discovered (16 Psyche, in 1852) was the first to be designated in that way at the time of its discovery. However, Psyche was given an iconic symbol as well, as were a few other asteroids discovered over the next few years. 20 Massalia was the first asteroid that was not assigned an iconic symbol, and no iconic symbols were created after the 1855 discovery of 37 Fides. That year Astraea's number was increased to ⑤, but the first four asteroids, Ceres to Vesta, were not listed by their numbers until the 1867 edition. The circle was soon abbreviated to a pair of parentheses, which were easier to typeset and sometimes omitted altogether over the next few decades, leading to the modern convention.

Exploration

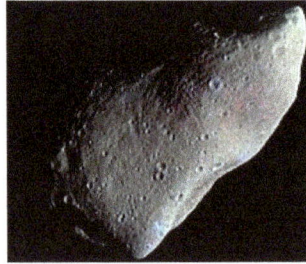

951 Gaspra is the first asteroid to be imaged in close-up (enhanced color).

Vesta imaged by the Dawn spacecraft

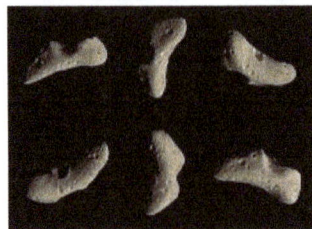

Several views of 433 Eros in natural colour

Until the age of space travel, objects in the asteroid belt were merely pinpricks of light in even the largest telescopes and their shapes and terrain remained a mystery. The best modern ground-based telescopes and the Earth-orbiting Hubble Space Telescope can resolve a small amount of detail on the surfaces of the largest asteroids, but even these mostly remain little more than fuzzy blobs. Limited information about the shapes and compositions of asteroids can be inferred from their light curves (their variation in brightness as they rotate) and their spectral properties, and asteroid sizes can be estimated by timing the lengths of star occulations (when an asteroid passes directly in front of a star). Radar imaging can yield good information about asteroid shapes and orbital and rotational parameters, especially for near-Earth asteroids. In terms of delta-v and propellant requirements, NEOs are more easily accessible than the Moon.

The first close-up photographs of asteroid-like objects were taken in 1971, when the *Mariner 9* probe imaged Phobos and Deimos, the two small moons of Mars, which are probably captured asteroids. These images revealed the irregular, potato-like shapes of most asteroids, as did later images from the Voyager probes of the small moons of the gas giants.

The first true asteroid to be photographed in close-up was 951 Gaspra in 1991, followed in 1993 by 243 Ida and its moon Dactyl, all of which were imaged by the *Galileo* probe en route to Jupiter.

The first dedicated asteroid probe was *NEAR Shoemaker*, which photographed 253 Mathilde in 1997, before entering into orbit around 433 Eros, finally landing on its surface in 2001.

Other asteroids briefly visited by spacecraft en route to other destinations include 9969 Braille (by *Deep Space 1* in 1999), and 5535 Annefrank (by *Stardust* in 2002).

In September 2005, the Japanese *Hayabusa* probe started studying 25143 Itokawa in detail and was plagued with difficulties, but returned samples of its surface to Earth on 13 June 2010.

The European *Rosetta* probe (launched in 2004) flew by 2867 Šteins in 2008 and 21 Lutetia, the third-largest asteroid visited to date, in 2010.

In September 2007, NASA launched the *Dawn* spacecraft, which orbited 4 Vesta from July 2011 to September 2012, and has been orbiting the dwarf planet 1 Ceres since 2015. 4 Vesta is the second-largest asteroid visited to date.

On 13 December 2012, China's lunar orbiter *Chang'e 2* flew within 2 miles (3.2 km) of the asteroid 4179 Toutatis on an extended mission.

Planned and Future Missions

The Japan Aerospace Exploration Agency (JAXA) launched the *Hayabusa 2* probe in December 2014, and plans to return samples from 162173 Ryugu in December 2020.

In May 2011, NASA selected the OSIRIS-REx sample return mission to asteroid 101955 Bennu; it is expected to launch in September 2016.

In early 2013, NASA announced the planning stages of a mission to capture a near-Earth asteroid and move it into lunar orbit where it could possibly be visited by astronauts and later impacted into the Moon. On 19 June 2014, NASA reported that asteroid 2011 MD was a prime candidate for capture by a robotic mission, perhaps in the early 2020s.

It has been suggested that asteroids might be used as a source of materials that may be rare or exhausted on Earth (asteroid mining), or materials for constructing space habitats . Materials that are heavy and expensive to launch from Earth may someday be mined from asteroids and used for space manufacturing and construction.

In the U.S. Discovery program the *Psyche* spacecraft proposal to 16 Psyche and *Lucy* spacecraft to Jupiter trojans made it to the semifinalist stage of mission selection.

Fiction

Asteroids and the asteroid belt are a staple of science fiction stories. Asteroids play several potential roles in science fiction: as places human beings might colonize, resources for extracting minerals, hazards encountered by spacecraft traveling between two other points, and as a threat to life on Earth or other inhabited planets, dwarf planets and natural satellites by potential impact.

Natural Satellite

A natural satellite or moon is a astronomical object that orbits a planet or minor planet.

Nineteen natural satellites are large enough to be round, and one, Saturn's moon, Titan, has a substantial atmosphere.

In the Solar System there are 173 known natural satellites which orbit within 6 planetary satellite systems. In addition, several other objects are known to have satellites, including three IAU-listed dwarf planets: Pluto, Haumea, and Eris. As of January 2012, over 200 minor-planet moons have been discovered. There are 76 known objects in the asteroid belt with satellites (five with two each), four Jupiter trojans, 39 near-Earth objects (two with two satellites each), and 14 Mars-crossers. There are also 84 known natural satellites of trans-Neptunian objects. Some 150 additional small bodies have been observed within the rings of Saturn, but only a few were tracked long enough to establish orbits. Planets around other stars are likely to have satellites as well, and although numerous candidates have been detected to date, none have yet been confirmed.

Of the inner planets, Mercury and Venus have no natural satellites; Earth has one large natural satellite, known as the Moon; and Mars has two tiny natural satellites, Phobos and Deimos. The giant planets have extensive systems of natural satellites, including half a dozen comparable in size to Earth's Moon: the four Galilean moons, Saturn's Titan, and Neptune's Triton. Saturn has an additional six mid-sized natural satellites massive enough to have achieved hydrostatic equilibrium, and Uranus has five. It has been suggested that some satellites may potentially harbour life.

The Earth–Moon system is unique in that the ratio of the mass of the Moon to the mass of Earth is much greater than that of any other natural-satellite–planet ratio in the Solar System (although there are minor-planet systems with even greater ratios, notably the Pluto–Charon system).

Among the identified dwarf planets, Ceres has no known natural satellites. Pluto has the relatively large natural satellite Charon and four smaller natural satellites; Styx, Nix, Kerberos, and Hydra. Haumea has two natural satellites, and Eris and Makemake have one. The Pluto–Charon system is unusual in that the center of mass lies in open space between the two, a characteristic sometimes associated with a double-planet system.

Origin and Orbital Characteristics

Two moons: Saturn's natural satellite Dione occults Enceladus

The natural satellites orbiting relatively close to the planet on prograde, uninclined circular orbits (*regular* satellites) are generally thought to have been formed out of the same collapsing region of the protoplanetary disk that created its primary. In contrast, irregular satellites (generally orbit-ing on distant, inclined, eccentric and/or retrograde orbits) are thought to be captured asteroids possibly further fragmented by collisions. Most of the major natural satellites of the Solar System have regular orbits, while most of the small natural satellites have irregular orbits. The Moon and possibly Charon are exceptions among large bodies in that they are thought to have originated by the collision of two large proto-planetary objects. The material that would have been placed in orbit around the central body is predicted to have reaccreted to form one or more orbiting natural satellites. As opposed to planetary-sized bodies, asteroid moons are thought to commonly form by this process. Triton is another exception; although large and in a close, circular orbit, its motion is retrograde and it is thought to be a captured dwarf planet.

Tidal Locking

Most regular moons (natural satellites following relatively close and prograde orbits with small orbital inclination and eccentricity) in the Solar System are tidally locked to their respective primaries, meaning that the same side of the natural satellite always faces its planet. The only known exception is Saturn's natural satellite Hyperion, which rotates chaotically because of the gravitational influence of Titan.

In contrast, the outer natural satellites of the giants planet (irregular satellites) are too far away to have become locked. For example, Jupiter's Himalia, Saturn's Phoebe, and Neptune's Nereid have rotation periods in the range of ten hours, whereas their orbital periods are hundreds of days.

Satellites of Satellites

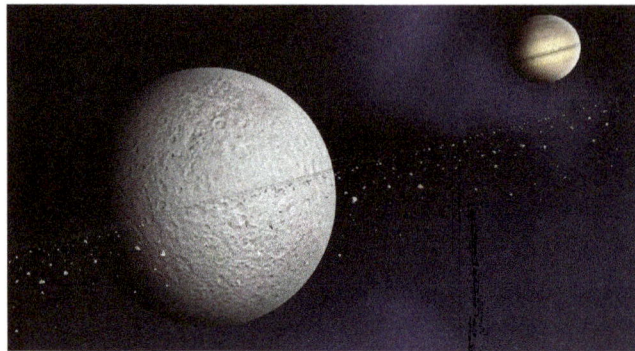

Artist impression of Rhea's proposed rings
No "moons of moons" (natural satellites that orbit a natural satellite of a planet) are currently known as of 2016. In most cases, the tidal effects of the planet would make such a system unstable.

However, calculations performed after the recent detection of a possible ring system around Saturn's moon Rhea indicate that satellites orbiting Rhea could have stable orbits. Furthermore, the suspected rings are thought to be narrow, a phenomenon normally associated with shepherd moons. However, targeted images taken by the *Cassini* spacecraft failed to detect rings around Rhea.

It has also been proposed that Saturn's moon Iapetus had a satellite in the past; this is one of several hypotheses that have been put forward to account for its equatorial ridge.

Trojan Satellites

Two natural satellites are known to have small companions at their L_4 and L_5 Lagrangian points, sixty degrees ahead and behind the body in its orbit. These companions are called trojan moons, as their orbits are analogous to the Trojan asteroids of Jupiter. The trojan moons are Telesto and Calypso, which are the leading and following companions, respectively, of Tethys; and Helene and Polydeuces, the leading and following companions of Dione.

Asteroid Satellites

The discovery of 243 Ida's natural satellite Dactyl in the early 1990s confirmed that some asteroids have natural satellites; indeed, 87 Sylvia has two. Some, such as 90 Antiope, are double asteroids with two comparably sized components.

Shape

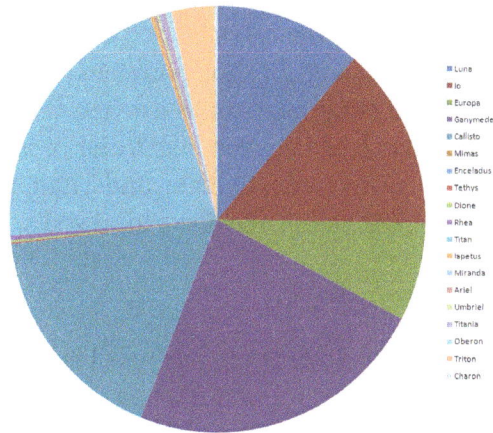

The relative masses of the natural satellites of the Solar System. Mimas, Enceladus, and Miranda are too small to be visible at this scale. All the irregularly shaped natural satellites, even added together, would also be too small to be visible.

Neptune's moon Proteus is the largest irregularly shaped natural satellite. All other known natural satellites that are at least the size of Uranus's Miranda have lapsed into rounded ellipsoids under hydrostatic equilibrium, i.e. are "round/rounded satellites". The larger natural satellites, being tidally locked, tend toward ovoid (egg-like) shapes: squat at their poles and with longer equatorial axes in the direction of their primaries (their planets) than in the direction of their motion. Saturn's moon Mimas, for example, has a major axis 9% greater than its polar axis and 5% greater than its other equatorial axis. Methone, another of Saturn's moons, is only around 3 km in diameter and visibly egg-shaped. The effect is smaller on the largest natural satellites, where their own gravity is greater relative to the effects of tidal distortion, especially those that orbit less massive planets or, as in the case of the Moon, at greater distances.

Name	Satellite of	Difference in axes	
		km	% of mean diameter
Mimas	Saturn	33.4 (20.4 / 13.0)	8.4 (5.1 / 3.3)
Enceladus	Saturn	16.6	3.3
Miranda	Uranus	14.2	3.0
Tethys	Saturn	25.8	2.4
Io	Jupiter	29.4	0.8
Moon (*Luna*)	Earth	4.3	0.1

Geological Activity

Of the nineteen known natural satellites in the Solar System that are massive enough to have lapsed into hydrostatic equilibrium, several remain geologically active today. Io is the most volcanically active body in the Solar System, while Europa, Enceladus, Titan and Triton display evidence of ongoing tectonic activity and cryovolcanism. In the first three cases, the geological activity is powered by the tidal heating resulting from having eccentric orbits close to their giant-planet primaries. (This mechanism would have also operated on Triton in the past, before its orbit was circularized.) Many other natural satellites, such as Earth's Moon, Ganymede, Tethys and Miranda, show evidence of past geological activity, resulting from energy sources such as the decay of their primordial radioisotopes, greater past orbital eccentricities (due in some cases to past orbital resonances), or the differentiation or freezing of their interiors. Enceladus and Triton both have active features resembling geysers, although in the case of Triton solar heating appears to provide the energy. Titan and Triton have significant atmospheres; Titan also has hydrocarbon lakes, and presumably methane rain. Four of the largest natural satellites, Europa, Ganymede, Callisto, and Titan, are thought to have subsurface oceans of liquid water, while smaller Enceladus may have localized subsurface liquid water.

Natural Satellites of the Solar System

The seven largest natural satellites in the Solar System (those bigger than 2,500 km across) are Jupiter's Galilean moons (Ganymede, Callisto, Io, and Europa), Saturn's moon Titan, Earth's moon, and Neptune's captured natural satellite Triton. Triton, the smallest of these, has more mass than all smaller natural satellites together. Similarly in the next size group of nine natural satellites, between 1,000 km and 1,600 km across, Titania, Oberon, Rhea, Iapetus, Charon, Ariel, Umbriel, Dione, and Tethys, the smallest, Tethys, has more mass than all smaller natural satellites together. As well as the natural satellites of the various planets, there are also over 80 known natural satellites of the dwarf planets, minor planets and other small Solar System bodies. Some studies estimate that up to 15% of all trans-Neptunian objects could have satellites.

The following is a comparative table classifying the natural satellites in the Solar System by diameter. The column on the right includes some notable planets, dwarf planets, asteroids, and trans-Neptunian objects for comparison. The natural satellites of the planets are named after mythological figures. These are predominantly Greek, except for the Uranian natural satellites, which are named after Shakespearean characters. The nineteen bodies massive enough to have achieved hydrostatic equilibrium are in bold in the table below. Minor planets and satellites suspected but not proven to have achieved a hydrostatic equilibrium are italicized in the table below.

Terminology

Montage comparing the relative sizes of selected natural satellites with the terrestrial planets and the dwarf planet Pluto

The first known natural satellite was the Moon, but it was considered a "planet" until Copernicus' introduction of heliocentrism in 1543. Until the discovery of the Galilean satellites in 1610, however, there was no opportunity for referring to such objects as a class. Galileo chose to refer to his discoveries as *Planetæ* ("planets"), but later discoverers chose other terms to distinguish them from the objects they orbited.

Christiaan Huygens, the discoverer of Titan, was the first to use the term *moon* for such objects, calling Titan *Luna Saturni* or *Luna Saturnia* – "Saturn's moon" or "The Saturnian moon", because it stood in the same relation to Saturn as the Moon did to Earth.

The first to use of the term *satellite* to describe orbiting bodies was the German astronomer Johannes Kepler in his pamphlet *Narratio de Observatis a se quatuor Iouis satellitibus erronibus* ("Narration About Four Satellites of Jupiter Observed") in 1610. He derived the term from the Latin word *satelles*, meaning "guard", "attendant", or "companion", because the *satellites* accompanied their primary planet in their journey through the heavens.

As additional natural satellites of Saturn were discovered the term "moon" was abandoned. Giovanni Domenico Cassini sometimes referred to his discoveries as *planètes* in French, but more often as *satellites*.

The term *satellite* thus became the normal one for referring to an object orbiting a planet, as it avoided the ambiguity of "moon". In 1957, however, the launching of the artificial object Sputnik created a need for new terminology. The terms *man-made satellite* or *artificial moon* were very quickly abandoned in favor of the simpler *satellite*, and as a consequence, the term has become linked primarily with artificial objects flown in space – including, sometimes, even those not in orbit around a planet.

Because of this shift in meaning, the term *moon*, which had continued to be used in a generic sense in works of popular science and in fiction, has regained respectability and is now used interchangeably with *natural satellite*, even in scientific articles. When it is necessary to avoid both the ambiguity of confusion with Earth's natural satellite the Moon and the natural satellites of the other planets on the one hand, and artificial satellites on the other, the term *natural satellite* (using "natural" in a sense opposed to "artificial") is used. To further avoid ambiguity, the convention is to capitalize the word Moon when referring to Earth's natural satellite, but not when referring to other natural satellites.

Many authors define "satellite" or "natural satellite" as orbiting some planet or minor planet, synonymous with "moon" -- by such a definition all natural satellites are moons, but Earth and other planets are not satellites. A few recent authors define "moon" as "a satellite of a planet or minor planet", and "planet" as "a satellite of a star" -- such authors consider Earth as a "natural satellite of the sun".

The Definition of a Moon

There is no established lower limit on what is considered a "moon". Every natural celestial body with an identified orbit around a planet of the Solar System, some as small as a kilometer across, has been considered a moon, though objects a tenth that size within Saturn's rings, which have not been directly observed, have been called *moonlets*. Small asteroid moons (natural satellites of asteroids), such as Dactyl, have also been called moonlets.

Size comparison of Earth and the Moon

The upper limit is also vague. Two orbiting bodies are sometimes described as a double body rather than primary and satellite. Asteroids such as 90 Antiope are considered double asteroids, but they have not forced a clear definition of what constitutes a moon. Some authors consider the Pluto–Charon system to be a double (dwarf) planet. The most common dividing line on what is considered a moon rests upon whether the barycentre is below the surface of the larger body, though this is somewhat arbitrary, because it depends on distance as well as relative mass.

Galaxy

A galaxy is a gravitationally bound system of stars, stellar remnants, interstellar gas, dust, and dark matter. The word galaxy is derived from the Greek *galaxias*, literally «milky», a reference to the Milky Way. Galaxies range in size from dwarfs with just a few billion (10^9) stars to giants with one hundred trillion (10^{14}) stars, each orbiting its galaxy's center of mass. Galaxies are categorized according to their visual morphology as elliptical, spiral and irregular. Many galaxies are thought to have black holes at their active centers. The Milky Way's central black hole, known as Sagittarius A*, has a mass four million times greater than the Sun. As of March 2016, GN-z11 is the oldest and most distant observed galaxy with a comoving distance of 32 billion light-years from Earth, and observed as it existed just 400 million years after the Big Bang. Previously, as of July 2015, EGSY8p7 was the most distant known galaxy, estimated to have a light travel distance of 13.2 billion light-years away.

NGC 4414, a typical spiral galaxy in the constellation Coma Berenices, is about 55,000 light-years in diameter and approximately 60 million light-years away from Earth

Approximately 170 billion (1.7×10^{11}) to 200 billion (2.0×10^{11}) galaxies exist in the observable universe. Most of the galaxies are 1,000 to 100,000 parsecs in diameter and usually separated by distances on the order of millions of parsecs (or megaparsecs). The space between galaxies is filled with a tenuous gas having an average density of less than one atom per cubic meter. The majority of galaxies are gravitationally organized into associations known as galaxy groups, clusters, and superclusters. At the largest scale, these associations are generally arranged into sheets and filaments surrounded by immense voids.

Etymology

The origin of the word *galaxy* derives from the Greek term for the Milky Way, *galaxias* ("milky one"), or *kyklos galaktikos* ("milky circle") due to its appearance as a "milky" band of light in the sky. In Greek mythology, Zeus places his son born by a mortal woman, the infant Heracles, on Hera's breast while she is asleep so that the baby will drink her divine milk and will thus become immortal. Hera wakes up while breastfeeding and then realizes she is nursing an unknown baby: she pushes the baby away, some of her milk spills and, produces the faint band of light known as the Milky Way.

In the astronomical literature, the capitalized word "Galaxy" is often used to refer to our galaxy, the Milky Way, to distinguish it from the other galaxies in our universe. The English term *Milky Way* can be traced back to a story by Chaucer c. 1380:

> "See yonder, lo, the Galaxyë
> Which men clepeth *the Milky Wey*,
> For hit is whyt."

> — *Geoffrey Chaucer, The House of Fame*

Certain astronomical objects known as *spiral nebula* such as M31 would later be recognized as conglomerations of stars when the true distance to these objects began to be discovered, and they would be deemed *island universes*. However, the word *Universe* was later understood to mean the entirety of existence, so this expression fell into disuse and the objects instead became known as galaxies.

Nomenclature

Tens of thousands of galaxies have been catalogued, but only a few have well-established names, such as the Andromeda Galaxy, the Magellanic Clouds, the Whirlpool Galaxy, and the Sombrero Galaxy. Astronomers work with numbers from certain catalogues, such as the Messier catalogue, the NGC (New General Catalogue), the IC (Index Catalogue), the CGCG (Catalogue of Galaxies and of Clusters of Galaxies), the MCG (Morphological Catalogue of Galaxies) and UGC (Uppsala General Catalogue of Galaxies). All of the well-known galaxies appear in one or more of these catalogues but each time under a different number. For example, Messier 109 is a spiral galaxy having the number 109 in the catalogue of Messier, but also codes NGC3992, UGC6937, CGCG 269-023, MCG +09-20-044, and PGC 37617.

Because it is customary in science to assign names to most of the studied objects, even to the smallest ones, the Belgian astrophysicist Gerard Bodifee and the classicist Michel Berger started a new catalogue (CNG-Catalogue of Named Galaxies) in which a thousand well-known galaxies are given meaningful, descriptive names in Latin (or Latinized Greek) in accordance with the binomial nomenclature that one uses in other sciences such as biology, anatomy, paleontology and in other fields of astronomy such as the geography of Mars. One of the arguments to do so is that these impressive objects deserve better than uninspired codes. For instance, Bodifee and Berger propose the informal, descriptive name *Callimorphus Ursae Majoris* for the well-formed barred galaxy Messier 109 in Ursa Major.

Observation History

The realization that we live in a galaxy, and that ours is one among many, parallels major discoveries that were made about the Milky Way and other nebulae in the night sky.

Milky Way

The Greek philosopher Democritus (450–370 BCE) proposed that the bright band on the night sky known as the Milky Way might consist of distant stars. Aristotle (384–322 BCE), however, believed the Milky Way to be caused by "the ignition of the fiery exhalation of some stars that were large, numerous and close together" and that the "ignition takes place in the upper part of the atmosphere, in the region of the World that is continuous with the heavenly motions." The Neoplatonist philosopher Olympiodorus the Younger (c. 495–570 CE) was critical of this view, arguing that if the Milky Way is sublunary (situated between Earth and the Moon) it should appear different at different times and places on Earth, and that it should have parallax, which it does not. In his view, the Milky Way is celestial.

According to Mohani Mohamed, the Arabian astronomer Alhazen (965–1037) made the first attempt at observing and measuring the Milky Way's parallax, and he thus "determined that because the Milky Way had no parallax, it must be remote from the Earth, not belonging to the atmosphere." The Persian astronomer al-Bīrūnī (973–1048) proposed the Milky Way galaxy to be "a collection of countless fragments of the nature of nebulous stars." The Andalusian astronomer Ibn Bajjah ("Avempace", d. 1138) proposed that the Milky Way is made up of many stars that almost touch one another and appear to be a continuous image due to the effect of refraction from sublunary material, citing his observation of the conjunction of Jupiter and Mars as evidence of this occurring when two objects are near. In the 14th century, the Syrian-born Ibn Qayyim proposed the Milky Way galaxy to be "a myriad of tiny stars packed together in the sphere of the fixed stars."

The shape of the Milky Way as estimated from star counts by William Herschel in 1785; the Solar System was assumed to be near the center.

Actual proof of the Milky Way consisting of many stars came in 1610 when the Italian astronomer Galileo Galilei used a telescope to study the Milky Way and discovered that it is composed of a huge number of faint stars. In 1750 the English astronomer Thomas Wright, in his *An original theory or new hypothesis of the Universe*, speculated (correctly) that the galaxy might be a rotating body of a huge number of stars held together by gravitational forces, akin to the Solar System but on a much larger scale. The resulting disk of stars can be seen as a band on the sky from our perspective inside the disk. In a treatise in 1755, Immanuel Kant elaborated on Wright's idea about the structure of the Milky Way.

The first project to describe the shape of the Milky Way and the position of the Sun was undertaken by William Herschel in 1785 by counting the number of stars in different regions of the sky. He produced a diagram of the shape of the galaxy with the Solar System close to the center. Using a refined approach, Kapteyn in 1920 arrived at the picture of a small (diameter about 15 kiloparsecs) ellipsoid galaxy with the Sun close to the center. A different method by Harlow Shapley based on the cataloguing of globular clusters led to a radically different picture: a flat disk with diameter approximately 70 kiloparsecs and the Sun far from the center. Both analyses failed to take into account the absorption of light by interstellar dust present in the galactic plane, but after Robert Julius Trumpler quantified this effect in 1930 by studying open clusters, the present picture of our host galaxy, the Milky Way, emerged.

A fish-eye mosaic of the Milky Way arching at a high inclination across the night sky, shot from a dark-sky location in Chile

Distinction from other Nebulae

A few galaxies outside the Milky Way are visible in the night sky to the unaided eye. In the 10th century, the Persian astronomer Al-Sufi made the earliest recorded identification of the Andromeda Galaxy, describing it as a "small cloud". In 964, Al-Sufi identified the Large Magellanic Cloud in his *Book of Fixed Stars*; it was not seen by Europeans until Magellan's voyage in the 16th century. The Andromeda Galaxy was independently noted by Simon Marius in 1612.

In 1750, Thomas Wright speculated (correctly) that the Milky Way is a flattened disk of stars, and that some of the nebulae visible in the night sky might be separate Milky Ways. In 1755, Immanuel Kant used the term "island Universe" to describe these distant nebulae.

Photograph of the "Great Andromeda Nebula" from 1899, later identified as the Andromeda Galaxy

Toward the end of the 18th century, Charles Messier compiled a catalog containing the 109 brightest celestial objects having nebulous appearance. Subsequently, William Herschel assembled a catalog of 5,000 nebulae. In 1845, Lord Rosse constructed a new telescope and was able to distinguish between elliptical and spiral nebulae. He also managed to make out individual point sources in some of these nebulae, lending credence to Kant's earlier conjecture.

In 1912, Vesto Slipher made spectrographic studies of the brightest spiral nebulae to determine their composition. Slipher discovered that the spiral nebulae have high Doppler shifts, indicating that they are moving at a rate exceeding the velocity of the stars he had measured. He found that the majority of these nebulae are moving away from us.

In 1917, Heber Curtis observed nova S Andromedae within the "Great Andromeda Nebula" (as the Andromeda Galaxy, Messier object M31, was then known). Searching the photographic record, he found 11 more novae. Curtis noticed that these novae were, on average, 10 magnitudes fainter than those that occurred within our galaxy. As a result, he was able to come up with a distance estimate of 150,000 parsecs. He became a proponent of the so-called "island universes" hypothesis, which holds that spiral nebulae are actually independent galaxies.

In 1920 a debate took place between Harlow Shapley and Heber Curtis (the Great Debate), concerning the nature of the Milky Way, spiral nebulae, and the dimensions of the Universe. To support his claim that the Great Andromeda Nebula is an external galaxy, Curtis noted the appearance of dark lanes resembling the dust clouds in the Milky Way, as well as the significant Doppler shift.

In 1922, the Estonian astronomer Ernst Öpik gave a distance determination that supported the theory that the Andromeda Nebula is indeed a distant extra-galactic object. Using the new 100 inch Mt. Wilson telescope, Edwin Hubble was able to resolve the outer parts of some spiral nebulae as collections of individual stars and identified some Cepheid variables, thus allowing him to estimate the distance to the nebulae: they were far too distant to be part of the Milky Way. In 1936 Hubble produced a classification of galactic morphology that is used to this day.

Modern Research

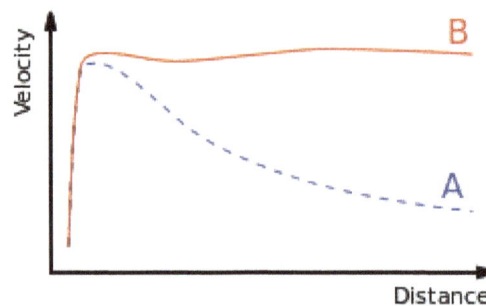

Rotation curve of a typical spiral galaxy: predicted based on the visible matter (A) and observed (B). The distance is from the galactic core.

In 1944, Hendrik van de Hulst predicted that microwave radiation with wavelength of 21 cm would be detectable from interstellar atomic hydrogen gas; and in 1951 it was observed. This radiation is not affected by dust absorption, and so its Doppler shift can be used to map the motion of the gas in our galaxy. These observations led to the hypothesis of a rotating bar structure in the center of our galaxy. With improved radio telescopes, hydrogen gas could also be traced in other galaxies. In the 1970s, Vera Rubin uncovered a discrepancy between observed galactic rotation speed and that predicted by the visible mass of stars and gas. Today, the galaxy rotation problem is thought to be explained by the presence of large quantities of unseen dark matter. A concept known as the universal rotation curve of spirals, moreover, shows that the problem is ubiquitous in these objects.

Beginning in the 1990s, the Hubble Space Telescope yielded improved observations. Among other things, Hubble data helped establish that the missing dark matter in our galaxy cannot solely consist of inherently faint and small stars. The Hubble Deep Field, an extremely long exposure of a relatively empty part of the sky, provided evidence that there are about 125 billion (1.25×10^{11}) galaxies in the observable universe. Improved technology in detecting the spectra invisible to humans (radio telescopes, infrared cameras, and x-ray telescopes) allow detection of other galaxies that are not detected by Hubble. Particularly, galaxy surveys in the Zone of Avoidance (the region of the sky blocked by the Milky Way) have revealed a number of new galaxies.

Types and Morphology

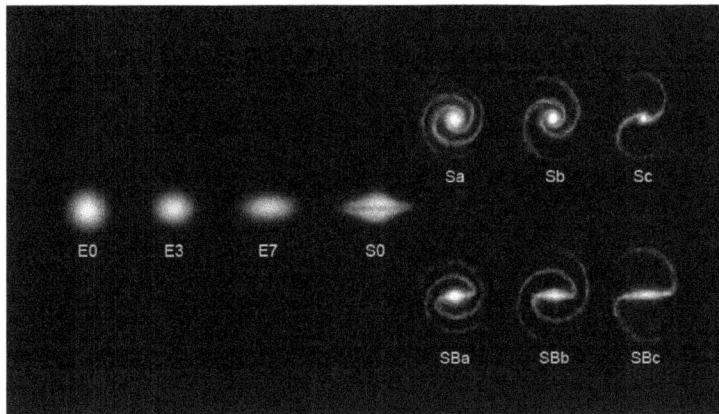

Types of galaxies according to the Hubble classification scheme: an *E* indicates a type of elliptical galaxy; an *S* is a spiral; and *SB* is a barred-spiral galaxy.

Galaxies come in three main types: ellipticals, spirals, and irregulars. A slightly more extensive description of galaxy types based on their appearance is given by the Hubble sequence. Since the Hubble sequence is entirely based upon visual morphological type(shape), it may miss certain important characteristics of galaxies such as star formation rate in starburst galaxies and activity in the cores of active galaxies.

Ellipticals

The Hubble classification system rates elliptical galaxies on the basis of their ellipticity, ranging from E0, being nearly spherical, up to E7, which is highly elongated. These galaxies have an ellipsoidal profile, giving them an elliptical appearance regardless of the viewing angle. Their appearance shows little structure and they typically have relatively little interstellar matter. Consequently, these galaxies also have a low portion of open clusters and a reduced rate of new star formation. Instead they are dominated by generally older, more evolved stars that are orbiting the common center of gravity in random directions. The stars contain low abundances of heavy elements because star formation ceases after the initial burst. In this sense they have some similarity to the much smaller globular clusters.

The largest galaxies are giant ellipticals. Many elliptical galaxies are believed to form due to the interaction of galaxies, resulting in a collision and merger. They can grow to enormous sizes (compared to spiral galaxies, for example), and giant elliptical galaxies are often found near the core of large galaxy clusters.

Starburst galaxies are the result of a galactic collision that can result in the formation of an elliptical galaxy.

Shell Galaxy

NGC 3923 Elliptical Shell Galaxy-Hubble Space Telescope photograph

A shell galaxy is a type of elliptical galaxy where the stars in the galaxy's halo are arranged in concentric shells. About one-tenth of elliptical galaxies have a shell-like structure, which has never been observed in spiral galaxies. The shell-like structures are thought to develop when a larger galaxy absorbs a smaller companion galaxy. As the two galaxy centers approach, the centers start to oscillate around a center point, the oscillation creates gravitational ripples forming the shells of stars, similar to ripples spreading on water. For example, galaxy NGC 3923 has over twenty shells.

Spirals

The Pinwheel Galaxy, NGC 5457.

Spiral galaxies resemble spiraling pinwheels. Though the stars and other visible material contained in such a galaxy lie mostly on a plane, the majority of mass in spiral galaxies exists in a roughly spherical halo of dark matter that extends beyond the visible component, as demonstrated by the universal rotation curve concept.

Spiral galaxies consist of a rotating disk of stars and interstellar medium, along with a central bulge of generally older stars. Extending outward from the bulge are relatively bright arms. In the Hubble classification scheme, spiral galaxies are listed as type *S*, followed by a letter (*a*, *b*, or *c*) that indicates the degree of tightness of the spiral arms and the size of the central bulge. An *Sa* galaxy has tightly wound, poorly defined arms and possesses a relatively large core region. At the other extreme, an *Sc* galaxy has open, well-defined arms and a small core region. A galaxy with poorly defined arms is sometimes referred to as a flocculent spiral galaxy; in contrast to the grand design spiral galaxy that has prominent and well-defined spiral arms.

It appears the reason that some spiral galaxies are fat and bulging while some are flat discs is because of how fast they rotate.

NGC 1300, an example of a barred spiral galaxy.

In spiral galaxies, the spiral arms do have the shape of approximate logarithmic spirals, a pattern that can be theoretically shown to result from a disturbance in a uniformly rotating mass of stars. Like the stars, the spiral arms rotate around the center, but they do so with constant angular velocity. The spiral arms are thought to be areas of high-density matter, or "density waves". As stars move through an arm, the space velocity of each stellar system is modified by the gravitational force of the higher density. (The velocity returns to normal after the stars depart on the other side of the arm.) This effect is akin to a "wave" of slowdowns moving along a highway full of moving cars. The arms are visible because the high density facilitates star formation, and therefore they harbor many bright and young stars.

Hoag's Object, an example of a ring galaxy.

Barred Spiral Galaxy

A majority of spiral galaxies, including our own Milky Way galaxy, have a linear, bar-shaped band of stars that extends outward to either side of the core, then merges into the spiral arm structure. In the Hubble classification scheme, these are designated by an *SB*, followed by a lower-case letter (*a*, *b* or *c*) that indicates the form of the spiral arms (in the same manner as the categorization of normal spiral galaxies). Bars are thought to be temporary structures that can occur as a result of a density wave radiating outward from the core, or else due to a tidal interaction with another galaxy. Many barred spiral galaxies are active, possibly as a result of gas being channeled into the core along the arms.

Our own galaxy, the Milky Way, is a large disk-shaped barred-spiral galaxy about 30 kiloparsecs in diameter and a kiloparsec thick. It contains about two hundred billion (2×10^{11}) stars and has a total mass of about six hundred billion (6×10^{11}) times the mass of the Sun.

Super Luminous Spiral

Can have a diameter of 437,000 light-years. The Milky Way is about 100,000 light-years in diameter. They can have a mass of 340 billion solar masses and generate large amount of ultraviolet and mid-infrared light. They create new stars 30 times faster than the Milky Way.

Other morphologies

- Peculiar galaxies are galactic formations that develop unusual properties due to tidal interactions with other galaxies.

    o A ring galaxy has a ring-like structure of stars and interstellar medium surrounding a bare core. A ring galaxy is thought to occur when a smaller galaxy passes through the core of a spiral galaxy. Such an event may have affected the Andromeda Galaxy, as it displays a multi-ring-like structure when viewed in infrared radiation.

- A lenticular galaxy is an intermediate form that has properties of both elliptical and spiral galaxies. These are categorized as Hubble type S0, and they possess ill-defined spiral arms with an elliptical halo of stars (barred lenticular galaxies receive Hubble classification SB0.)

- Irregular galaxies are galaxies that can not be readily classified into an elliptical or spiral morphology.

    o An Irr-I galaxy has some structure but does not align cleanly with the Hubble classification scheme.

    o Irr-II galaxies do not possess any structure that resembles a Hubble classification, and may have been disrupted. Nearby examples of (dwarf) irregular galaxies include the Magellanic Clouds.

- An ultra diffuse galaxy (UDG) is an extremely-low-density galaxy. The galaxy may be the same size as the Milky Way but has a visible star count of only 1% of the Milky Way. The lack of luminosity is because there is a lack of star-forming gas in the galaxy which results in old stellar populations.

Dwarfs

Despite the prominence of large elliptical and spiral galaxies, most galaxies in the Universe are dwarf galaxies. These galaxies are relatively small when compared with other galactic formations, being about one hundredth the size of the Milky Way, containing only a few billion stars. Ultra-compact dwarf galaxies have recently been discovered that are only 100 parsecs across.

Many dwarf galaxies may orbit a single larger galaxy; the Milky Way has at least a dozen such satellites, with an estimated 300–500 yet to be discovered. Dwarf galaxies may also be classified as elliptical, spiral, or irregular. Since small dwarf ellipticals bear little resemblance to large ellipticals, they are often called dwarf spheroidal galaxies instead.

A study of 27 Milky Way neighbors found that in all dwarf galaxies, the central mass is approximately 10 million solar masses, regardless of whether the galaxy has thousands or millions of stars. This has led to the suggestion that galaxies are largely formed by dark matter, and that the minimum size may indicate a form of warm dark matter incapable of gravitational coalescence on a smaller scale.

Other Types of Galaxies

Interacting

Interactions between galaxies are relatively frequent, and they can play an important role in galactic evolution. Near misses between galaxies result in warping distortions due to tidal interactions, and may cause some exchange of gas and dust. Collisions occur when two galaxies pass directly through each other and have sufficient relative momentum not to merge. The stars of interacting galaxies will usually not collide, but the gas and dust within the two forms will interact, sometimes triggering star formation. A collision can severely distort the shape of the galaxies, forming bars, rings or tail-like structures.

The Antennae Galaxies are undergoing a collision that will result in their eventual merger.

At the extreme of interactions are galactic mergers. In this case the relative momentum of the two galaxies is insufficient to allow the galaxies to pass through each other. Instead, they gradually merge to form a single, larger galaxy. Mergers can result in significant changes to morphology, as compared to the original galaxies. If one of the merging galaxies is much more massive than the other merging galaxies then the result is known as cannibalism. The more massive larger galaxy will remain relatively undisturbed by the merger, while the smaller galaxy is torn apart. The Milky Way galaxy is currently in the process of cannibalizing the Sagittarius Dwarf Elliptical Galaxy and the Canis Major Dwarf Galaxy.

Starburst

M82, a starburst galaxy that has ten times the star formation of a "normal" galaxy.

Stars are created within galaxies from a reserve of cold gas that forms into giant molecular clouds. Some galaxies have been observed to form stars at an exceptional rate, which is known as a starburst. If they continue to do so, then they would consume their reserve of gas in a time span less than the lifespan of the galaxy. Hence starburst activity usually lasts for only about ten million years, a relatively brief period in the history of a galaxy. Starburst galaxies were more common during the early history of the Universe, and, at present, still contribute an estimated 15% to the total star production rate.

Starburst galaxies are characterized by dusty concentrations of gas and the appearance of newly formed stars, including massive stars that ionize the surrounding clouds to create H II regions. These massive stars produce supernova explosions, resulting in expanding remnants that interact powerfully with the surrounding gas. These outbursts trigger a chain reaction of star building that spreads throughout the gaseous region. Only when the available gas is nearly consumed or dispersed does the starburst activity end.

Starbursts are often associated with merging or interacting galaxies. The prototype example of such a starburst-forming interaction is M82, which experienced a close encounter with the larger M81. Irregular galaxies often exhibit spaced knots of starburst activity.

Active Galaxy

A jet of particles is being emitted from the core of the elliptical radio galaxy M87.

A portion of the observable galaxies are classified as an active galaxy if the galaxy contains an active galactic nucleus. A significant portion of the total energy output from the galaxy is emitted by the active galactic nucleus, instead of the stars, dust and interstellar medium of the galaxy .

The standard model for an active galactic nucleus is based upon an accretion disc that forms around a supermassive black hole (SMBH) at the core region of the galaxy. The radiation from an active galactic nucleus results from the gravitational energy of matter as it falls toward the black hole from the disc. In about 10% of these galaxies, a diametrically opposed pair of energetic jets ejects particles from the galaxy core at velocities close to the speed of light. The mechanism for producing these jets is not well understood.

- Seyfert galaxies or quasars, are classified depending on the luminosity, are active galaxies that emit high-energy radiation in the form of x-rays.

Blazars

Blazars are believed to be an active galaxy with a relativistic jet that is pointed in the direction of Earth. A radio galaxy emits radio frequencies from relativistic jets. A unified model of these types of active galaxies explains their differences based on the viewing angle of the observer.

LINERS

Possibly related to active galactic nuclei (as well as starburst regions) are low-ionization nuclear emission-line regions (LINERs). The emission from LINER-type galaxies is dominated by weakly ionized elements. The excitation sources for the weakly ionized lines include post-AGB stars, AGN, and shocks. Approximately one-third of nearby galaxies are classified as containing LINER nuclei.

Seyfert Galaxy

Seyfert galaxies are one of the two largest groups of active galaxies, along with quasars. They have quasar-like nuclei (very luminous, distant and bright sources of electromagnetic radiation) with very high surface brightnesses but unlike quasars, their host galaxies are clearly detectable. Seyfert galaxies account for about 10% of all galaxies. Seen in visible light, most Seyfert galaxies look like normal spiral galaxies, but when studied under other wavelengths, the luminosity of their cores is equivalent to the luminosity of whole galaxies the size of the Milky Way.

Quasar

Quasars or quasi-stellar radio sources are the most energetic and distant members of a class of objects called active galactic nuclei (AGN). Quasars are extremely luminous and were first identified as being high redshift sources of electromagnetic energy, including radio waves and visible light, that appeared to be similar to stars, rather than extended sources similar to galaxies. Their luminosity can be 100 times greater than that of the Milky Way.

Luminous Infrared Galaxy

Luminous Infrared Galaxies or (LIRG's) are galaxies with luminosities, the measurement of brightness, above 10^{11} L$\odot$. LIRG's are more abundant than starburst galaxies, Seyfert galaxies and quasi-stellar objects at comparable total luminosity. Infrared galaxies emit more energy in the infrared than at all other wavelengths combined. An LIRG's luminosity is 100 billion times that of our Sun.

Properties

Galaxies have galactic magnetic fields.

Formation and Evolution

Galactic formation and evolution is an active area of research in astrophysics.

Formation

Artist's impression of a protocluster forming in the early Universe.

Current cosmological models of the early Universe are based on the Big Bang theory. About 300,000 years after this event, atoms of hydrogen and helium began to form, in an event called recombination. Nearly all the hydrogen was neutral (non-ionized) and readily absorbed light, and no stars had yet formed. As a result, this period has been called the "dark ages". It was from density fluctuations (or anisotropic irregularities) in this primordial matter that larger structures began to appear. As a result, masses of baryonic matter started to condense within cold dark matter halos. These primordial structures would eventually become the galaxies we see today.

Artist's impression of a young galaxy accreting material.

Early Galaxies

Evidence for the early appearance of galaxies was found in 2006, when it was discovered that the galaxy IOK-1 has an unusually high redshift of 6.96, corresponding to just 750 million years after the Big Bang and making it the most distant and primordial galaxy yet seen. While some scientists have claimed other objects (such as Abell 1835 IR1916) have higher redshifts (and therefore are seen in an earlier stage of the Universe's evolution), IOK-1's age and composition have been more reliably established. In December 2012, astronomers reported that the UDFj-39546284 is the most distant object known and has a redshift value of 11.9. The object, is estimated to have existed around "380 million years" after the Big Bang (which was about 13.8 billion years ago), is about 13.42 billion light travel distance years away. The existence of such early protogalaxies suggests that they must have grown in the so-called "dark ages". As of May 5, 2015, the galaxy EGS-zs8-1 is the most distant and earliest galaxy measured, forming 670 million years after the Big Bang. The light from EGS-zs8-1 has taken 13 billion years to reach Earth, and is now 30 billion light-years away, because of the expansion of the universe during 13 billion years.

Early Galaxy Formation

Different components of near-infrared background light detected by the Hubble Space Telescope in deep-sky surveys.

The detailed process by which early galaxies formed is an open question in astrophysics. Theories can be divided into two categories: top-down and bottom-up. In top-down theories (such as the Eggen–Lynden-Bell–Sandage [ELS] model), protogalaxies form in a large-scale simultaneous collapse lasting about one hundred million years. In bottom-up theories (such as the Searle-Zinn [SZ] model), small structures such as globular clusters form first, and then a number of such bodies accrete to form a larger galaxy.

Once protogalaxies began to form and contract, the first halo stars (called Population III stars) appeared within them. These were composed almost entirely of hydrogen and helium, and may have been massive. If so, these huge stars would have quickly consumed their supply of fuel and became supernovae, releasing heavy elements into the interstellar medium. This first generation of stars re-ionized the surrounding neutral hydrogen, creating expanding bubbles of space through which light could readily travel.

In June 2015, astronomers reported evidence for Population III stars in the Cosmos Redshift 7 galaxy at $z = 6.60$. Such stars are likely to have existed in the very early universe (i.e., at high redshift), and may have started the production of chemical elements heavier than hydrogen that are needed for the later formation of planets and life as we know it.

Evolution

Within a billion years of a galaxy's formation, key structures begin to appear. Globular clusters, the central supermassive black hole, and a galactic bulge of metal-poor Population II stars form. The creation of a supermassive black hole appears to play a key role in actively regulating the growth of galaxies by limiting the total amount of additional matter added. During this early epoch, galaxies undergo a major burst of star formation.

During the following two billion years, the accumulated matter settles into a galactic disc. A galaxy will continue to absorb infalling material from high-velocity clouds and dwarf galaxies throughout its life. This matter is mostly hydrogen and helium. The cycle of stellar birth and death slowly increases the abundance of heavy elements, eventually allowing the formation of planets.

Hubble eXtreme Deep Field (XDF)		
XDF view field compared to the angular size of the Moon. Several thousand galaxies, each consisting of billions of stars, are in this small view.	*XDF* (2012) view: Each light speck is a galaxy, some of which are as old as 13.2 billion years – the observable universe is estimated to contain 200 billion galaxies.	*XDF* image shows (from left) fully mature galaxies, nearly mature galaxies (from 5 to 9 billion years ago), protogalaxies, blazing with young stars (beyond 9 billion years).

The evolution of galaxies can be significantly affected by interactions and collisions. Mergers of galaxies were common during the early epoch, and the majority of galaxies were peculiar in morphology. Given the distances between the stars, the great majority of stellar systems in colliding galaxies will be unaffected. However, gravitational stripping of the interstellar gas and dust that makes up the spiral arms produces a long train of stars known as tidal tails. Examples of these formations can be seen in NGC 4676 or the Antennae Galaxies.

The Milky Way galaxy and the nearby Andromeda Galaxy are moving toward each other at about 130 km/s, and—depending upon the lateral movements—the two might collide in about five to six billion years. Although the Milky Way has never collided with a galaxy as large as Andromeda before, evidence of past collisions of the Milky Way with smaller dwarf galaxies is increasing.

Such large-scale interactions are rare. As time passes, mergers of two systems of equal size become less common. Most bright galaxies have remained fundamentally unchanged for the last few billion years, and the net rate of star formation probably also peaked approximately ten billion years ago.

Future Trends

Spiral galaxies, like the Milky Way, produce new generations of stars as long as they have dense molecular clouds of interstellar hydrogen in their spiral arms. Elliptical galaxies are largely devoid of this gas, and so form few new stars. The supply of star-forming material is finite; once stars have converted the available supply of hydrogen into heavier elements, new star formation will come to an end.

The current era of star formation is expected to continue for up to one hundred billion years, and then the "stellar age" will wind down after about ten trillion to one hundred trillion years (10^{13}–

10^{14} years), as the smallest, longest-lived stars in our universe, tiny red dwarfs, begin to fade. At the end of the stellar age, galaxies will be composed of compact objects: brown dwarfs, white dwarfs that are cooling or cold ("black dwarfs"), neutron stars, and black holes. Eventually, as a result of gravitational relaxation, all stars will either fall into central supermassive black holes or be flung into intergalactic space as a result of collisions.

Larger-scale Structures

Deep sky surveys show that galaxies are often found in groups and clusters. Solitary galaxies that have not significantly interacted with another galaxy of comparable mass during the past billion years are relatively scarce. Only about 5% of the galaxies surveyed have been found to be truly isolated; however, these isolated formations may have interacted and even merged with other galaxies in the past, and may still be orbited by smaller, satellite galaxies. Isolated galaxies can produce stars at a higher rate than normal, as their gas is not being stripped by other nearby galaxies.

On the largest scale, the Universe is continually expanding, resulting in an average increase in the separation between individual galaxies. Associations of galaxies can overcome this expansion on a local scale through their mutual gravitational attraction. These associations formed early in the Universe, as clumps of dark matter pulled their respective galaxies together. Nearby groups later merged to form larger-scale clusters. This on-going merger process (as well as an influx of infalling gas) heats the inter-galactic gas within a cluster to very high temperatures, reaching 30–100 megakelvins. About 70–80% of the mass in a cluster is in the form of dark matter, with 10–30% consisting of this heated gas and the remaining few percent of the matter in the form of galaxies.

Seyfert's Sextet is an example of a compact galaxy group.

Most galaxies in the Universe are gravitationally bound to a number of other galaxies. These form a fractal-like hierarchical distribution of clustered structures, with the smallest such associations being termed groups. A group of galaxies is the most common type of galactic cluster, and these formations contain a majority of the galaxies (as well as most of the baryonic mass) in the Universe. To remain gravitationally bound to such a group, each member galaxy must have a sufficiently low velocity to prevent it from escaping. If there is insufficient kinetic energy, however, the group may evolve into a smaller number of galaxies through mergers.

Clusters of galaxies consist of hundreds to thousands of galaxies bound together by gravity. Clusters of galaxies are often dominated by a single giant elliptical galaxy, known as the brightest cluster galaxy, which, over time, tidally destroys its satellite galaxies and adds their mass to its own.

Superclusters contain tens of thousands of galaxies, which are found in clusters, groups and sometimes individually. At the supercluster scale, galaxies are arranged into sheets and filaments surrounding vast empty voids. Above this scale, the Universe appears to be the same in all directions (isotropic and homogeneous).

The Milky Way galaxy is a member of an association named the Local Group, a relatively small group of galaxies that has a diameter of approximately one megaparsec. The Milky Way and the Andromeda Galaxy are the two brightest galaxies within the group; many of the other member galaxies are dwarf companions of these two galaxies. The Local Group itself is a part of a cloud-like structure within the Virgo Supercluster, a large, extended structure of groups and clusters of galaxies centered on the Virgo Cluster. And the Virgo Supercluster itself is a part of the Pisces-Cetus Supercluster Complex, a giant galaxy filament.

Multi-wavelength Observation

This ultraviolet image of Andromeda shows blue regions containing young, massive stars.

The peak radiation of most stars lies in the visible spectrum, so the observation of the stars that form galaxies has been a major component of optical astronomy. It is also a favorable portion of the spectrum for observing ionized H II regions, and for examining the distribution of dusty arms.

The dust present in the interstellar medium is opaque to visual light. It is more transparent to far-infrared, which can be used to observe the interior regions of giant molecular clouds and galactic cores in great detail. Infrared is also used to observe distant, red-shifted galaxies that were formed much earlier in the history of the Universe. Water vapor and carbon dioxide absorb a number of useful portions of the infrared spectrum, so high-altitude or space-based telescopes are used for infrared astronomy.

The first non-visual study of galaxies, particularly active galaxies, was made using radio frequencies. The Earth's atmosphere is nearly transparent to radio between 5 MHz and 30 GHz. (The ionosphere blocks signals below this range.) Large radio interferometers have been used to map the active jets emitted from active nuclei. Radio telescopes can also be used to observe neutral hydrogen (via 21 cm radiation), including, potentially, the non-ionized matter in the early Universe that later collapsed to form galaxies.

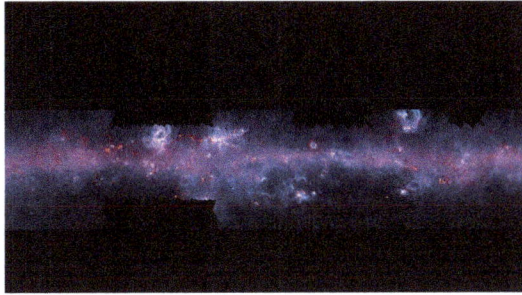

The southern plane of the Milky Way from submillimeter wavelengths.

Ultraviolet and X-ray telescopes can observe highly energetic galactic phenomena. Ultraviolet flares are sometimes observed when a star in a distant galaxy is torn apart from the tidal forces of a nearby black hole. The distribution of hot gas in galactic clusters can be mapped by X-rays. The existence of supermassive black holes at the cores of galaxies was confirmed through X-ray astronomy.

Star

A star-forming region in the Large Magellanic Cloud.

False-color imagery of the Sun, a G-type main-sequence star, the closest to Earth

A **star** is a luminous sphere of plasma held together by its own gravity. The nearest star to Earth is the Sun. Many other stars are visible to the naked eye from Earth during the night, appearing as a multitude of fixed luminous points in the sky due to their immense distance from Earth. Historically, the most prominent stars were grouped into constellations and asterisms, the brightest of which gained proper names. Astronomers have assembled star catalogues that identify the known stars and provide standardized stellar designations. However, most of the stars in the Universe, including all stars outside our galaxy, the Milky Way, are invisible to the naked eye from Earth. Indeed, most are invisible from Earth even through the most powerful telescopes.

For at least a portion of its life, a star shines due to thermonuclear fusion of hydrogen into helium in its core, releasing energy that traverses the star's interior and then radiates into outer space. Almost all naturally occurring elements heavier than helium are created by stellar nucleosynthesis during the star's lifetime, and for some stars by supernova nucleosynthesis when it explodes. Near the end of its life, a star can also contain degenerate matter. Astronomers can determine the mass, age, metallicity (chemical composition), and many other properties of a star by observing its motion through space, its luminosity, and spectrum respectively. The total mass of a star is the main factor that determines its evolution and eventual fate. Other characteristics of a star, including diameter and temperature, change over its life, while the star's environment affects its rotation and movement. A plot of the temperature of many stars against their luminosities produces a plot known as a Hertzsprung–Russell diagram (H–R diagram). Plotting a particular star on that diagram allows the age and evolutionary state of that star to be determined.

A star's life begins with the gravitational collapse of a gaseous nebula of material composed primarily of hydrogen, along with helium and trace amounts of heavier elements. When the stellar core is sufficiently dense, hydrogen becomes steadily converted into helium through nuclear fusion, releasing energy in the process. The remainder of the star's interior carries energy away from the core through a combination of radiative and convective heat transfer processes. The star's internal pressure prevents it from collapsing further under its own gravity. A star with mass greater than 0.4 times the Sun's will expand to become a red giant when the hydrogen fuel in its core is exhausted. In some cases, it will fuse heavier elements at the core or in shells around the core. As the star expands it throws a part of its mass, enriched with those heavier elements, into the interstellar environment, to be recycled later as new stars. Meanwhile, the core becomes a stellar remnant: a white dwarf, a neutron star, or if it is sufficiently massive a black hole.

Binary and multi-star systems consist of two or more stars that are gravitationally bound and generally move around each other in stable orbits. When two such stars have a relatively close orbit, their gravitational interaction can have a significant impact on their evolution. Stars can form part of a much larger gravitationally bound structure, such as a star cluster or a galaxy.

Observation History

People have seen patterns in the stars since ancient times. This 1690 depiction of the constellation of Leo, the lion, is by Johannes Hevelius.

Historically, stars have been important to civilizations throughout the world. They have been part of religious practices and used for celestial navigation and orientation. Many ancient astronomers believed that stars were permanently affixed to a heavenly sphere and that they were immutable. By convention, astronomers grouped stars into constellations and used them to track the motions of the planets and the inferred position of the Sun. The motion of the Sun against the background stars (and the horizon) was used to create calendars, which could be used to regulate agricultural practices. The Gregorian calendar, currently used nearly everywhere in the world, is a solar calendar based on the angle of the Earth's rotational axis relative to its local star, the Sun.

The constellation of Leo as it can be seen by the naked eye. Lines have been added.

The oldest accurately dated star chart was the result of ancient Egyptian astronomy in 1534 BC. The earliest known star catalogues were compiled by the ancient Babylonian astronomers of Mesopotamia in the late 2nd millennium BC, during the Kassite Period (*ca.* 1531–1155 BC).

The first star catalogue in Greek astronomy was created by Aristillus in approximately 300 BC, with the help of Timocharis. The star catalog of Hipparchus (2nd century BC) included 1020 stars, and was used to assemble Ptolemy's star catalogue. Hipparchus is known for the discovery of the first recorded *nova* (new star). Many of the constellations and star names in use today derive from Greek astronomy.

In spite of the apparent immutability of the heavens, Chinese astronomers were aware that new stars could appear. In 185 AD, they were the first to observe and write about a supernova, now known as the SN 185. The brightest stellar event in recorded history was the SN 1006 supernova, which was observed in 1006 and written about by the Egyptian astronomer Ali ibn Ridwan and several Chinese astronomers. The SN 1054 supernova, which gave birth to the Crab Nebula, was also observed by Chinese and Islamic astronomers.

Medieval Islamic astronomers gave Arabic names to many stars that are still used today and they invented numerous astronomical instruments that could compute the positions of the stars. They built the first large observatory research institutes, mainly for the purpose of producing *Zij* star catalogues. Among these, the *Book of Fixed Stars* (964) was written by the Persian astronomer Abd al-Rahman al-Sufi, who observed a number of stars, star clusters (including the Omicron Velorum and Brocchi's Clusters) and galaxies (including the Andromeda Galaxy). According to A. Zahoor, in the 11th century, the Persian polymath scholar Abu Rayhan Biruni described the Milky

Way galaxy as a multitude of fragments having the properties of nebulous stars, and also gave the latitudes of various stars during a lunar eclipse in 1019.

According to Josep Puig, the Andalusian astronomer Ibn Bajjah proposed that the Milky Way was made up of many stars that almost touched one another and appeared to be a continuous image due to the effect of refraction from sublunary material, citing his observation of the conjunction of Jupiter and Mars on 500 AH (1106/1107 AD) as evidence. Early European astronomers such as Tycho Brahe identified new stars in the night sky (later termed *novae*), suggesting that the heavens were not immutable. In 1584 Giordano Bruno suggested that the stars were like the Sun, and may have other planets, possibly even Earth-like, in orbit around them, an idea that had been suggested earlier by the ancient Greek philosophers, Democritus and Epicurus, and by medieval Islamic cosmologists such as Fakhr al-Din al-Razi. By the following century, the idea of the stars being the same as the Sun was reaching a consensus among astronomers. To explain why these stars exerted no net gravitational pull on the Solar System, Isaac Newton suggested that the stars were equally distributed in every direction, an idea prompted by the theologian Richard Bentley.

The Italian astronomer Geminiano Montanari recorded observing variations in luminosity of the star Algol in 1667. Edmond Halley published the first measurements of the proper motion of a pair of nearby "fixed" stars, demonstrating that they had changed positions since the time of the ancient Greek astronomers Ptolemy and Hipparchus.

William Herschel was the first astronomer to attempt to determine the distribution of stars in the sky. During the 1780s he established a series of gauges in 600 directions and counted the stars observed along each line of sight. From this he deduced that the number of stars steadily increased toward one side of the sky, in the direction of the Milky Way core. His son John Herschel repeated this study in the southern hemisphere and found a corresponding increase in the same direction. In addition to his other accomplishments, William Herschel is also noted for his discovery that some stars do not merely lie along the same line of sight, but are also physical companions that form binary star systems.

The science of stellar spectroscopy was pioneered by Joseph von Fraunhofer and Angelo Secchi. By comparing the spectra of stars such as Sirius to the Sun, they found differences in the strength and number of their absorption lines—the dark lines in a stellar spectra caused by the atmosphere's absorption of specific frequencies. In 1865 Secchi began classifying stars into spectral types. However, the modern version of the stellar classification scheme was developed by Annie J. Cannon during the 1900s.

Alpha Centauri A and B over limb of Saturn

The first direct measurement of the distance to a star (61 Cygni at 11.4 light-years) was made in 1838 by Friedrich Bessel using the parallax technique. Parallax measurements demonstrated the vast separation of the stars in the heavens. Observation of double stars gained increasing importance during the 19th century. In 1834, Friedrich Bessel observed changes in the proper motion of the star Sirius and inferred a hidden companion. Edward Pickering discovered the first spectroscopic binary in 1899 when he observed the periodic splitting of the spectral lines of the star Mizar in a 104-day period. Detailed observations of many binary star systems were collected by astronomers such as William Struve and S. W. Burnham, allowing the masses of stars to be determined from computation of orbital elements. The first solution to the problem of deriving an orbit of binary stars from telescope observations was made by Felix Savary in 1827. The twentieth century saw increasingly rapid advances in the scientific study of stars. The photograph became a valuable astronomical tool. Karl Schwarzschild discovered that the color of a star and, hence, its temperature, could be determined by comparing the visual magnitude against the photographic magnitude. The development of the photoelectric photometer allowed precise measurements of magnitude at multiple wavelength intervals. In 1921 Albert A. Michelson made the first measurements of a stellar diameter using an interferometer on the Hooker telescope at Mount Wilson Observatory.

Important theoretical work on the physical structure of stars occurred during the first decades of the twentieth century. In 1913, the Hertzsprung-Russell diagram was developed, propelling the astrophysical study of stars. Successful models were developed to explain the interiors of stars and stellar evolution. Cecilia Payne-Gaposchkin first proposed that stars were made primarily of hydrogen and helium in her 1925 PhD thesis. The spectra of stars were further understood through advances in quantum physics. This allowed the chemical composition of the stellar atmosphere to be determined.

With the exception of supernovae, individual stars have primarily been observed in the Local Group, and especially in the visible part of the Milky Way (as demonstrated by the detailed star catalogues available for our galaxy). But some stars have been observed in the M100 galaxy of the Virgo Cluster, about 100 million light years from the Earth. In the Local Supercluster it is possible to see star clusters, and current telescopes could in principle observe faint individual stars in the Local Group. However, outside the Local Supercluster of galaxies, neither individual stars nor clusters of stars have been observed. The only exception is a faint image of a large star cluster containing hundreds of thousands of stars located at a distance of one billion light years—ten times further than the most distant star cluster previously observed.

Designations

This view contains blue stars known as "Blue stragglers", for their apparent location on the Hertzsprung–Russell diagram

The concept of a constellation was known to exist during the Babylonian period. Ancient sky watchers imagined that prominent arrangements of stars formed patterns, and they associated these with particular aspects of nature or their myths. Twelve of these formations lay along the band of the ecliptic and these became the basis of astrology. Many of the more prominent individual stars were also given names, particularly with Arabic or Latin designations.

As well as certain constellations and the Sun itself, individual stars have their own myths. To the Ancient Greeks, some "stars", known as planets (Greek (planētēs), meaning "wanderer"), represented various important deities, from which the names of the planets Mercury, Venus, Mars, Jupiter and Saturn were taken. (Uranus and Neptune were also Greek and Roman gods, but neither planet was known in Antiquity because of their low brightness. Their names were assigned by later astronomers.)

Circa 1600, the names of the constellations were used to name the stars in the corresponding regions of the sky. The German astronomer Johann Bayer created a series of star maps and applied Greek letters as designations to the stars in each constellation. Later a numbering system based on the star's right ascension was invented and added to John Flamsteed's star catalogue in his book *"Historia coelestis Britannica"* (the 1712 edition), whereby this numbering system came to be called *Flamsteed designation* or *Flamsteed numbering*.

The only internationally recognized authority for naming celestial bodies is the International Astronomical Union (IAU). A number of private companies sell names of stars, which the British Library calls an unregulated commercial enterprise. The IAU has disassociated itself from this commercial practice, and these names are neither recognized by the IAU nor used by them. One such star-naming company is the International Star Registry, which, during the 1980s, was accused of deceptive practice for making it appear that the assigned name was official. This now-discontinued ISR practice was informally labeled a scam and a fraud, and the New York City Department of Consumer Affairs issued a violation against ISR for engaging in a deceptive trade practice.

Units of measurement

Although stellar parameters can be expressed in SI units or CGS units, it is often most convenient to express mass, luminosity, and radii in solar units, based on the characteristics of the Sun:

solar mass:	$M_\odot = 1.9891 \times 10^{30}$ kg
solar luminosity:	$L_\odot = 3.827 \times 10^{26}$ W
solar radius	$R_\odot = 6.960 \times 10^{8}$ m

Large lengths, such as the radius of a giant star or the semi-major axis of a binary star system, are often expressed in terms of the astronomical unit —approximately equal to the mean distance between the Earth and the Sun (150 million km or 93 million miles).

Formation and Evolution

Stars condense from regions of space of higher density, yet those regions are less dense than within a vacuum chamber. These regions - known as *molecular clouds* - consist mostly of hydrogen, with about 23 to 28 percent helium and a few percent heavier elements. One example of such a star-forming region is the Orion Nebula. Most stars form in groups of dozens to hundreds of thousands of stars. Massive stars in these groups may powerfully illuminate those clouds, ionizing the hydrogen, and creating H II regions. Such feedback effects, from star formation, may ultimately disrupt the cloud and prevent further star formation.

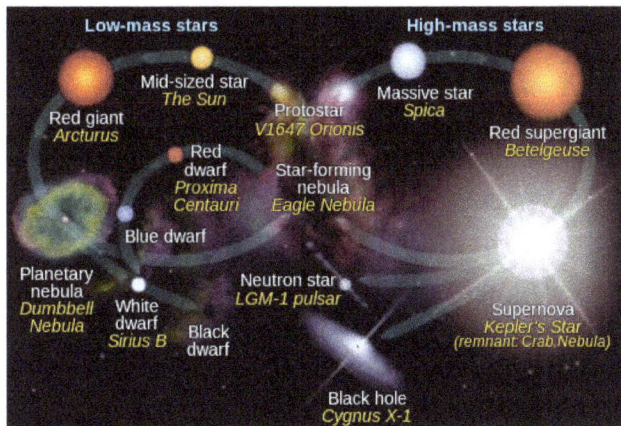

Stellar evolution of low-mass (left cycle) and high-mass (right cycle) stars, with examples in italics

All stars spend the majority of their existence as *main sequence stars,* fueled primarily by the nuclear fusion of hydrogen into helium within their cores. However, stars of different masses have markedly different properties at various stages of their development. The ultimate fate of more massive stars differs from that of less massive stars, as do their luminosities and the impact they have on their environment. Accordingly, astronomers often group stars by their mass:

- *Very low mass stars,* with masses below $0.5\ M_\odot$, are fully convective and distribute helium evenly throughout the whole star while on the main sequence. Therefore, they never undergo shell burning, never become red giants, which cease fusing and become helium white dwarfs and slowly cool after exhausting their hydrogen. However, as the lifetime of $0.5\ M_\odot$ stars is longer than the age of the universe, no such star has yet reached the white dwarf stage.

- *Low mass stars* (including the Sun), with a mass between $0.5\ M_\odot$ and 1.8–$2.5\ M_\odot$ depending on composition, do become red giants as their core hydrogen is depleted and they begin to burn helium in core in a helium flash; they develop a degenerate carbon-oxygen core later on the asymptotic giant branch; they finally blow off their outer shell as a planetary nebula and leave behind their core in the form of a white dwarf.

- *Intermediate-mass stars,* between 1.8–$2.5\ M_\odot$ and 5–$10\ M_\odot$, pass through evolutionary stages similar to low mass stars, but after a relatively short period on the RGB they ignite helium without a flash and spend an extended period in the red clump before forming a degenerate carbon-oxygen core.

- *Massive stars* generally have a minimum mass of 7–$10\ M_\odot$ (possibly as low as 5–$6\ M_\odot$). After exhausting the hydrogen at the core these stars become supergiants and go on to fuse elements heavier than helium. They end their lives when their cores collapse and they explode as supernovae.

Star Formation

The formation of a star begins with gravitational instability within a molecular cloud, caused by regions of higher density - often triggered by compression of clouds by radiation from massive stars, expanding bubbles in the interstellar medium, the collision of different molecular clouds, or the collision of galaxies (as in a starburst galaxy). When a region reaches a sufficient density of matter to satisfy the criteria for Jeans instability, it begins to collapse under its own gravitational force.

Artist's conception of the birth of a star within a dense molecular cloud.

As the cloud collapses, individual conglomerations of dense dust and gas form "Bok globules". As a globule collapses and the density increases, the gravitational energy converts into heat and the temperature rises. When the protostellar cloud has approximately reached the stable condition of hydrostatic equilibrium, a protostar forms at the core. These pre–main sequence stars are often surrounded by a protoplanetary disk and powered mainly by the conversion of gravitational energy. The period of gravitational contraction lasts about 10 to 15 million years.

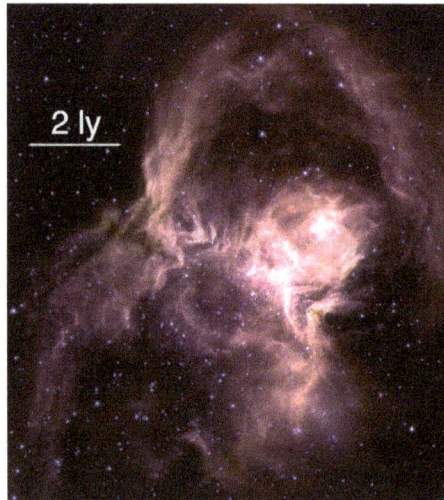

A cluster of approximately 500 young stars lies within the nearby W40 stellar nursery.

Early stars of less than $2\ M_\odot$ are called T Tauri stars, while those with greater mass are Herbig Ae/ Be stars. These newly formed stars emit jets of gas along their axis of rotation, which may reduce the angular momentum of the collapsing star and result in small patches of nebulosity known as Herbig–Haro objects. These jets, in combination with radiation from nearby massive stars, may help to drive away the surrounding cloud from which the star was formed.

Early in their development, T Tauri stars follow the Hayashi track—they contract and decrease in luminosity while remaining at roughly the same temperature. Less massive T Tauri stars follow this track to the main sequence, while more massive stars turn onto the Henyey track.

Most stars are observed to be members of binary star systems, and the properties of those binaries are the result of the conditions in which they formed. A gas cloud must lose its angular momentum in order to collapse and form a star. The fragmentation of the cloud into multiple stars distributes some of that angular momentum. The primordial binaries transfer some angular momentum by gravitational interactions during close encounters with other stars in young stellar clusters. These interactions tend to split apart more widely separated (soft) binaries while causing hard binaries to become more tightly bound. This produces the separation of binaries into their two observed populations distributions.

Main Sequence

Stars spend about 90% of their existence fusing hydrogen into helium in high-temperature and high-pressure reactions near the core. Such stars are said to be on the main sequence, and are called dwarf stars. Starting at zero-age main sequence, the proportion of helium in a star's core will steadily increase, the rate of nuclear fusion at the core will slowly increase, as will the star's temperature and luminosity. The Sun, for example, is estimated to have increased in luminosity by about 40% since it reached the main sequence 4.6 billion (4.6×10^9) years ago.

Every star generates a stellar wind of particles that causes a continual outflow of gas into space. For most stars, the mass lost is negligible. The Sun loses $10^{-14} \, M_\odot$ every year, or about 0.01% of its total mass over its entire lifespan. However, very massive stars can lose 10^{-7} to $10^{-5} \, M_\odot$ each year, significantly affecting their evolution. Stars that begin with more than 50 $M_\odot$ can lose over half their total mass while on the main sequence.

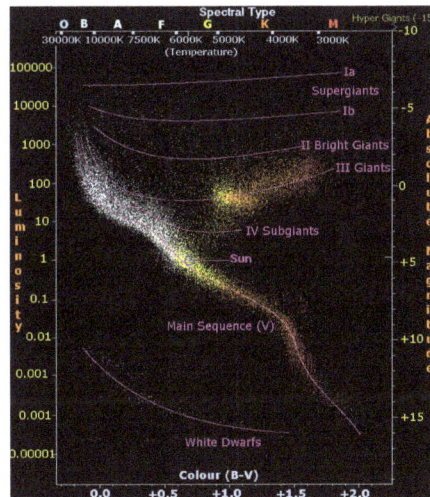

An example of a Hertzsprung–Russell diagram for a set of stars that includes the Sun (center).

The time a star spends on the main sequence depends primarily on the amount of fuel it has and the rate at which it fuses it. The Sun's is expected to live 10 billion (10^{10}) years. Massive stars consume their fuel very rapidly and are short-lived. Low mass stars consume their fuel very slowly.

Stars less massive than 0.25 $M_\odot$, called red dwarfs, are able to fuse nearly all of their mass while stars of about 1 $M_\odot$ can only fuse about 10% of their mass. The combination of their slow fuel-consumption and relatively large usable fuel supply allows low mass stars to last about one trillion (10^{12}) years; the most extreme of 0.08 $M_\odot$) will last for about 12 trillion years. Red dwarfs become hotter and more luminous as they accumulate helium. When they eventually run out of hydrogen, they contract into a white dwarf and decline in temperature. However, since the lifespan of such stars is greater than the current age of the universe (13.8 billion years), no stars under about 0.85 $M_\odot$ are expected to have moved off the main sequence.

Besides mass, the elements heavier than helium can play a significant role in the evolution of stars. Astronomers label all elements heavier than helium "metals", and call the chemical concentration of these elements in a star, its metallicity. A star's metallicity can influence the time the star takes to burn its fuel, and controls the formation of its magnetic fields, which affects the strength of its stellar wind. Older, population II stars have substantially less metallicity than the younger, population I stars due to the composition of the molecular clouds from which they formed. Over time, such clouds become increasingly enriched in heavier elements as older stars die and shed portions of their atmospheres.

Post–main Sequence

As stars of at least 0.4 $M_\odot$ exhaust their supply of hydrogen at their core, they start to fuse hydrogen in a shell outside the helium core. Their outer layers expand and cool greatly as they form a red giant. In about 5 billion years, when the Sun enters the helium burning phase, it will expand to a maximum radius of roughly 1 astronomical unit (150 million kilometres), 250 times its present size, and lose 30% of its current mass.

As the hydrogen shell burning produces more helium, the core increases in mass and temperature. In a red giant of up to 2.25 $M_\odot$, the mass of the helium core becomes degenerate prior to helium fusion. Finally, when the temperature increases sufficiently, helium fusion begins explosively in what is called a helium flash, and the star rapidly shrinks in radius, increases its surface temperature, and moves to the horizontal branch of the HR diagram. For more massive stars, helium core fusion starts before the core becomes degenerate, and the star spends some time in the red clump, slowly burning helium, before the outer convective envelope collapses and the star then moves to the horizontal branch.

After the star has fused the helium of its core, the carbon product fuses producing a hot core with an outer shell of fusing helium. The star then follows an evolutionary path called the asymptotic giant branch (AGB) that parallels the other described red giant phase, but with a higher luminosity. The more massive AGB stars may undergo a brief period of carbon fusion before the core becomes degenerate.

Massive Stars

During their helium-burning phase, stars of more than nine solar masses expand to form red supergiants. When this fuel is exhausted at the core, they continue to fuse elements heavier than helium.

The core contracts and the temperature and pressure rises enough to fuse carbon. This process continues, with the successive stages being fueled by neon oxygen and silicon.

Near the end of the star's life, fusion continues along a series of onion-layer shells within a massive star. Each shell fuses a different element, with the outermost shell fusing hydrogen; the next shell fusing helium, and so forth.

The final stage occurs when a massive star begins producing iron. Since iron nuclei are more tightly bound than any heavier nuclei, any fusion beyond iron does not produce a net release of energy. To a very limited degree such a process proceeds, but it consumes energy. Likewise, since they are more tightly bound than all lighter nuclei, such energy cannot be released by fission. In relatively old, very massive stars, a large core of inert iron will accumulate in the center of the star. The heavier elements in these stars can work their way to the surface, forming evolved objects known as Wolf-Rayet stars that have a dense stellar wind which sheds the outer atmosphere.

Collapse

As a star's core shrinks, the intensity of radiation from that surface increases, creating such radiation pressure on the outer shell of gas that it will push those layers away, forming a planetary nebula. If what remains after the outer atmosphere has been shed is less than $1.4\,M_\odot$, it shrinks to a relatively tiny object about the size of Earth, known as a white dwarf. White dwarfs lack the mass for further gravitational compression to take place. The electron-degenerate matter inside a white dwarf is no longer a plasma, even though stars are generally referred to as being spheres of plasma. Eventually, white dwarfs fade into black dwarfs over a very long period of time.

In larger stars, fusion continues until the iron core has grown so large (more than $1.4\,M_\odot$) that it can no longer support its own mass. This core will suddenly collapse as its electrons are driven into its protons, forming neutrons, neutrinos, and gamma rays in a burst of electron capture and inverse beta decay. The shockwave formed by this sudden collapse causes the rest of the star to explode in a supernova. Supernovae become so bright that they may briefly outshine the star's entire home galaxy. When they occur within the Milky Way, supernovae have historically been observed by naked-eye observers as "new stars" where none seemingly existed before.

A supernova explosion blows away the star's outer layers, leaving a remnant such as the Crab Nebula. The core is compressed into a neutron star, which sometimes manifests itself as a pulsar or X-ray burster. In the case of the largest stars, the remnant is a black hole greater than $4\,M_\odot$)s. In a neutron star the matter is in a state known as neutron-degenerate matter, with a more exotic form of degenerate matter, QCD matter, possibly present in the core. Within a black hole, the matter is in a state that is not currently understood.

The blown-off outer layers of dying stars include heavy elements, which may be recycled during the formation of new stars. These heavy elements allow the formation of rocky planets. The outflow from supernovae and the stellar wind of large stars play an important part in shaping the interstellar medium.

Binary Stars

The post–main-sequence evolution of binary stars may be significantly different from the evolution of single stars of the same mass. If stars in a binary system are sufficiently close, when one of the stars expands to become a red giant it may overflow its Roche lobe, the region around a star where material is gravitationally bound to that star, leading to transfer of material to the other. When the Roche lobe is violated, a variety of phenomena can result, including contact binaries, common-envelope binaries, cataclysmic variables, and type Ia supernovae.

Distribution

A white dwarf star in orbit around Sirius (artist's impression).

In addition to isolated stars, a multi-star system can consist of two or more gravitationally bound stars that orbit each other. The simplest and most common multi-star system is a binary star, but systems of three or more stars are also found. For reasons of orbital stability, such multi-star systems are often organized into hierarchical sets of binary stars. Larger groups called star clusters also exist. These range from loose stellar associations with only a few stars, up to enormous globular clusters with hundreds of thousands of stars. Such systems orbit our Milky Way galaxy.

It has been a long-held assumption that the majority of stars occur in gravitationally bound, multiple-star systems. This is particularly true for very massive O and B class stars, where 80% of the stars are believed to be part of multiple-star systems. The proportion of single star systems increases with decreasing star mass, so that only 25% of red dwarfs are known to have stellar companions. As 85% of all stars are red dwarfs, most stars in the Milky Way are likely single from birth.

Stars are not spread uniformly across the universe, but are normally grouped into galaxies along with interstellar gas and dust. A typical galaxy contains hundreds of billions of stars, and there are more than 100 billion (10^{11}) galaxies in the observable universe. In 2010, one estimate of the number of stars in the observable universe was 300 sextillion (3×10^{23}). While it is often believed that stars only exist within galaxies, intergalactic stars have been discovered.

The nearest star to the Earth, apart from the Sun, is Proxima Centauri, which is 39.9 trillion kilometres, or 4.2 light-years. Travelling at the orbital speed of the Space Shuttle (8 kilometres per second—almost 30,000 kilometres per hour), it would take about 150,000 years to arrive. This it typical of stellar separations in galactic discs. Stars can be much closer to each other in the centres of galaxies and in globular clusters, or much farther apart in galactic halos.

Due to the relatively vast distances between stars outside the galactic nucleus, collisions between stars are thought to be rare. In denser regions such as the core of globular clusters or the galactic center, collisions can be more common. Such collisions can produce what are known as blue stragglers. These abnormal stars have a higher surface temperature than the other main sequence stars with the same luminosity of the cluster to which it belongs.

Characteristics

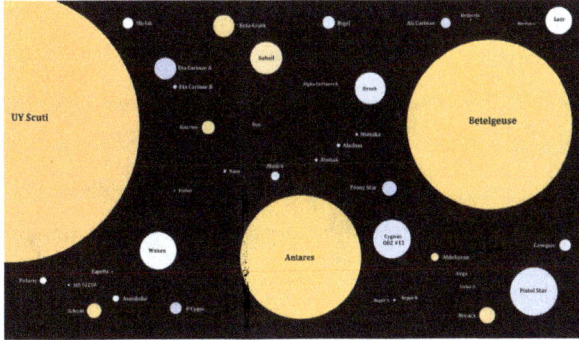

Some of the well-known stars with their apparent colors and relative sizes.

Almost everything about a star is determined by its initial mass, including such characteristics as luminosity, size, evolution, lifespan, and its eventual fate.

Age

Most stars are between 1 billion and 10 billion years old. Some stars may even be close to 13.8 billion years old—the observed age of the universe. The oldest star yet discovered, HD 140283, nicknamed Methuselah star, is an estimated 14.46 ± 0.8 billion years old. (Due to the uncertainty in the value, this age for the star does not conflict with the age of the Universe, determined by the Planck satellite as 13.799 ± 0.021).

The more massive the star, the shorter its lifespan, primarily because massive stars have greater pressure on their cores, causing them to burn hydrogen more rapidly. The most massive stars last an average of a few million years, while stars of minimum mass (red dwarfs) burn their fuel very slowly and can last tens to hundreds of billions of years.

Chemical Composition

When stars form in the present Milky Way galaxy they are composed of about 71% hydrogen and 27% helium, as measured by mass, with a small fraction of heavier elements. Typically the portion of heavy elements is measured in terms of the iron content of the stellar atmosphere, as iron is a common element and its absorption lines are relatively easy to measure. The portion of heavier elements may be an indicator of the likelihood that the star has a planetary system.

The star with the lowest iron content ever measured is the dwarf HE1327-2326, with only 1/200,000th the iron content of the Sun. By contrast, the super-metal-rich star μ Leonis has nearly double the abundance of iron as the Sun, while the planet-bearing star 14 Herculis has nearly triple the iron. There also exist chemically peculiar stars that show unusual abundances of certain elements in their spectrum; especially chromium and rare earth elements. Stars with cooler outer atmospheres, including the Sun, can form various diatomic and polyatomic molecules.

Diameter

Due to their great distance from the Earth, all stars except the Sun appear to the unaided eye as shining points in the night sky that twinkle because of the effect of the Earth's atmosphere. The

Sun is also a star, but it is close enough to the Earth to appear as a disk instead, and to provide daylight. Other than the Sun, the star with the largest apparent size is R Doradus, with an angular diameter of only 0.057 arcseconds.

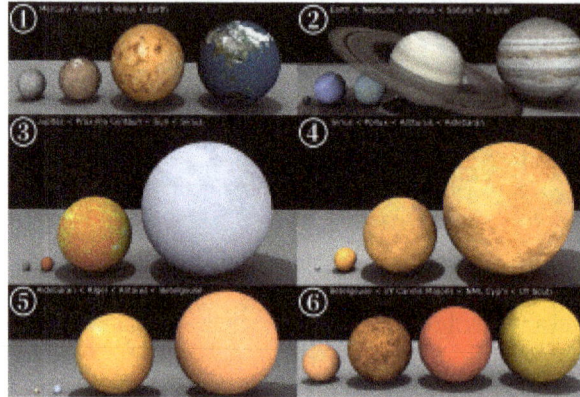

Stars vary widely in size. In each image in the sequence, the right-most object appears as the left-most object in the next panel. The Earth appears at right in panel 1 and the Sun is second from the right in panel 3. The rightmost star at panel 6 is UY Scuti, the largest known star.

The disks of most stars are much too small in angular size to be observed with current ground-based optical telescopes, and so interferometer telescopes are required to produce images of these objects. Another technique for measuring the angular size of stars is through occultation. By precisely measuring the drop in brightness of a star as it is occulted by the Moon (or the rise in brightness when it reappears), the star's angular diameter can be computed.

Stars range in size from neutron stars, which vary anywhere from 20 to 40 km (25 mi) in diameter, to supergiants like Betelgeuse in the Orion constellation, which has a diameter approximately 1,070 times that of the Sun—about 1,490,171,880 km (925,949,878 mi). Betelgeuse, however, has a much lower density than the Sun.

Kinematics

The motion of a star relative to the Sun can provide useful information about the origin and age of a star, as well as the structure and evolution of the surrounding galaxy. The components of motion of a star consist of the radial velocity toward or away from the Sun, and the traverse angular movement, which is called its proper motion.

The Pleiades, an open cluster of stars in the constellation of Taurus.
These stars share a common motion through space.

Radial velocity is measured by the doppler shift of the star's spectral lines, and is given in units of km/s. The proper motion of a star, its parallax, is determined by precise astrometric measurements in units of milli-arc seconds (mas) per year. With knowledge of the star's parallax and its distance, the proper motion velocity can be calculated. Together with the radial velocity, the total velocity can be calculated. Stars with high rates of proper motion are likely to be relatively close to the Sun, making them good candidates for parallax measurements.

When both rates of movement are known, the space velocity of the star relative to the Sun or the galaxy can be computed. Among nearby stars, it has been found that younger population I stars have generally lower velocities than older, population II stars. The latter have elliptical orbits that are inclined to the plane of the galaxy. A comparison of the kinematics of nearby stars has allowed astronomers to trace their origin to common points in giant molecular clouds, and are referred to as stellar associations.

Magnetic Field

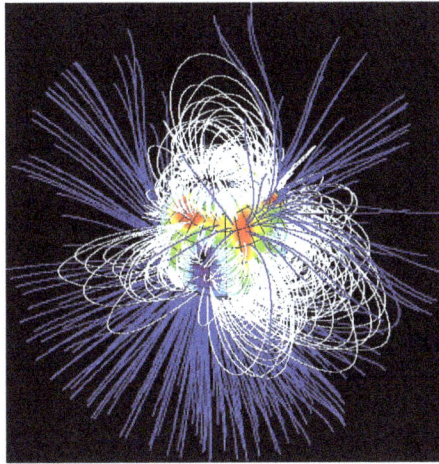

Surface magnetic field of SU Aur (a young star of T Tauri type), reconstructed by means of Zeeman-Doppler imaging

The magnetic field of a star is generated within regions of the interior where convective circulation occurs. This movement of conductive plasma functions like a dynamo, wherein the movement of electrical charges induce magnetic fields, as does a mechanical dynamo. Those magnetic fields have a great range that extend throughout and beyond the star. The strength of the magnetic field varies with the mass and composition of the star, and the amount of magnetic surface activity depends upon the star's rate of rotation. This surface activity produces starspots, which are regions of strong magnetic fields and lower than normal surface temperatures. Coronal loops are arching magnetic field flux lines that rise from a star's surface into the star's outer atmosphere, its corona. The coronal loops can be seen due to the plasma they conduct along their length. Stellar flares are bursts of high-energy particles that are emitted due to the same magnetic activity.

Young, rapidly rotating stars tend to have high levels of surface activity because of their magnetic field. The magnetic field can act upon a star's stellar wind, functioning as a brake to gradually slow the rate of rotation with time. Thus, older stars such as the Sun have a much slower rate of rotation and a lower level of surface activity. The activity levels of slowly rotating stars tend to vary in a cyclical manner and can shut down altogether for periods of time. During the Maunder minimum, for example, the Sun underwent a 70-year period with almost no sunspot activity.

Mass

One of the most massive stars known is Eta Carinae, which, with 100–150 times as much mass as the Sun, will have a lifespan of only several million years. Studies of the most massive open clusters suggests 150 $M_\odot$ as an upper limit for stars in the current era of the universe. This represents an empirical value for the theoretical limit on the mass of forming stars due to increasing radiation pressure on the accreting gas cloud. Several stars in the R136 cluster in the Large Magellanic Cloud have been measured with larger masses, but it has been determined that they could have been created through the collision and merger of massive stars in close binary systems, sidestepping the 150 $M_\odot$ limit on massive star formation.

The reflection nebula NGC 1999 is brilliantly illuminated by V380 Orionis (center), a variable star with about 3.5 times the mass of the Sun. The black patch of sky is a vast hole of empty space and not a dark nebula as previously thought.

The first stars to form after the Big Bang may have been larger, up to 300 $M_\odot$, due to the complete absence of elements heavier than lithium in their composition. This generation of supermassive population III stars is likely to have existed in the very early universe (i.e., they are observed to have a high redshift), and may have started the production of chemical elements heavier than hydrogen that are needed for the later formation of planets and life. In June 2015, astronomers reported evidence for Population III stars in the Cosmos Redshift 7 galaxy at $z = 6.60$.

With a mass only 80 times that of Jupiter (M_J), 2MASS J0523-1403 is the smallest known star undergoing nuclear fusion in its core. For stars with metallicity similar to the Sun, the theoretical minimum mass the star can have and still undergo fusion at the core, is estimated to be about 75 M_J. When the metallicity is very low, however, the minimum star size seems to be about 8.3% of the solar mass, or about 87 M_J. Smaller bodies called brown dwarfs, occupy a poorly defined grey area between stars and gas giants.

The combination of the radius and the mass of a star determines its surface gravity. Giant stars have a much lower surface gravity than do main sequence stars, while the opposite is the case for degenerate, compact stars such as white dwarfs. The surface gravity can influence the appearance of a star's spectrum, with higher gravity causing a broadening of the absorption lines.

Rotation

The rotation rate of stars can be determined through spectroscopic measurement, or more exactly determined by tracking their starspots. Young stars can have a rotation greater than 100 km/s at the equator. The B-class star Achernar, for example, has an equatorial velocity of about 225 km/s or greater, causing its equator to be slung outward and giving it an equatorial diameter that is more than 50% greater than between the poles. This rate of rotation is just below the critical velocity of 300 km/s at which speed the star would break apart. By contrast, the Sun rotates once every 25 – 35 days, with an equatorial velocity of 1.994 km/s. A main sequence star's magnetic field and the stellar wind serve to slow its rotation by a significant amount as it evolves on the main sequence.

Degenerate stars have contracted into a compact mass, resulting in a rapid rate of rotation. However they have relatively low rates of rotation compared to what would be expected by conservation of angular momentum—the tendency of a rotating body to compensate for a contraction in size by increasing its rate of spin. A large portion of the star's angular momentum is dissipated as a result of mass loss through the stellar wind. In spite of this, the rate of rotation for a pulsar can be very rapid. The pulsar at the heart of the Crab nebula, for example, rotates 30 times per second. The rotation rate of the pulsar will gradually slow due to the emission of radiation.

Temperature

The surface temperature of a main sequence star is determined by the rate of energy production of its core and by its radius, and is often estimated from the star's color index. The temperature is normally given in terms of an effective temperature, which is the temperature of an idealized black body that radiates its energy at the same luminosity per surface area as the star. Note that the effective temperature is only a representative of the surface, as the temperature increases toward the core. The temperature in the core region of a star is several million kelvins.

The stellar temperature will determine the rate of ionization of various elements, resulting in characteristic absorption lines in the spectrum. The surface temperature of a star, along with its visual absolute magnitude and absorption features, is used to classify a star.

Massive main sequence stars can have surface temperatures of 50,000 K. Smaller stars such as the Sun have surface temperatures of a few thousand K. Red giants have relatively low surface temperatures of about 3,600 K; but they also have a high luminosity due to their large exterior surface area.

Radiation

The energy produced by stars, a product of nuclear fusion, radiates to space as both electromagnetic radiation and particle radiation. The particle radiation emitted by a star is manifested as the stellar wind, which streams from the outer layers as electrically charged protons and alpha and beta particles. Although almost massless, there also exists a steady stream of neutrinos emanating from the star's core.

The production of energy at the core is the reason stars shine so brightly: every time two or more atomic nuclei fuse together to form a single atomic nucleus of a new heavier element, gamma ray photons are released from the nuclear fusion product. This energy is converted to other forms of electromagnetic energy of lower frequency, such as visible light, by the time it reaches the star's outer layers.

The color of a star, as determined by the most intense frequency of the visible light, depends on the temperature of the star's outer layers, including its photosphere. Besides visible light, stars also emit forms of electromagnetic radiation that are invisible to the human eye. In fact, stellar electromagnetic radiation spans the entire electromagnetic spectrum, from the longest wavelengths of radio waves through infrared, visible light, ultraviolet, to the shortest of X-rays, and gamma rays. From the standpoint of total energy emitted by a star, not all components of stellar electromagnetic radiation are significant, but all frequencies provide insight into the star's physics.

Using the stellar spectrum, astronomers can also determine the surface temperature, surface gravity, metallicity and rotational velocity of a star. If the distance of the star is found, such as by measuring the parallax, then the luminosity of the star can be derived. The mass, radius, surface gravity, and rotation period can then be estimated based on stellar models. (Mass can be calculated for stars in binary systems by measuring their orbital velocities and distances. Gravitational microlensing has been used to measure the mass of a single star.) With these parameters, astronomers can also estimate the age of the star.

Luminosity

The luminosity of a star is the amount of light and other forms of radiant energy it radiates per unit of time. It has units of power. The luminosity of a star is determined by its radius and surface temperature. Many stars do not radiate uniformly across their entire surface. The rapidly rotating star Vega, for example, has a higher energy flux (power per unit area) at its poles than along its equator.

Patches of the star's surface with a lower temperature and luminosity than average are known as starspots. Small, *dwarf* stars such as our Sun generally have essentially featureless disks with only small starspots. *Giant* stars have much larger, more obvious starspots, and they also exhibit strong stellar limb darkening. That is, the brightness decreases towards the edge of the stellar disk. Red dwarf flare stars such as UV Ceti may also possess prominent starspot features.

Magnitude

The apparent brightness of a star is expressed in terms of its apparent magnitude. It is a function of the star's luminosity, its distance from Earth, and the altering of the star's light as it passes through Earth's atmosphere. Intrinsic or absolute magnitude is directly related to a star's luminosity, and is what the apparent magnitude a star would be if the distance between the Earth and the star were 10 parsecs (32.6 light-years).

Number of stars brighter than magnitude	
Apparent magnitude	**Number of stars**
0	4
1	15
2	48
3	171
4	513
5	1,602
6	4,800
7	14,000

Both the apparent and absolute magnitude scales are logarithmic units: one whole number difference in magnitude is equal to a brightness variation of about 2.5 times (the 5th root of 100 or approximately 2.512). This means that a first magnitude star (+1.00) is about 2.5 times brighter than a second magnitude (+2.00) star, and about 100 times brighter than a sixth magnitude star (+6.00). The faintest stars visible to the naked eye under good seeing conditions are about magnitude +6.

On both apparent and absolute magnitude scales, the smaller the magnitude number, the brighter the star; the larger the magnitude number, the fainter the star. The brightest stars, on either scale, have negative magnitude numbers. The variation in brightness (ΔL) between two stars is calculated by subtracting the magnitude number of the brighter star (m_b) from the magnitude number of the fainter star (m_f), then using the difference as an exponent for the base number 2.512; that is to say:

$$\Delta m = m_f - m_b$$

$$2.512^{\Delta m} = \Delta L$$

Relative to both luminosity and distance from Earth, a star's absolute magnitude (M) and apparent magnitude (m) are not equivalent; for example, the bright star Sirius has an apparent magnitude of −1.44, but it has an absolute magnitude of +1.41.

The Sun has an apparent magnitude of −26.7, but its absolute magnitude is only +4.83. Sirius, the brightest star in the night sky as seen from Earth, is approximately 23 times more luminous than the Sun, while Canopus, the second brightest star in the night sky with an absolute magnitude of −5.53, is approximately 14,000 times more luminous than the Sun. Despite Canopus being vastly more luminous than Sirius, however, Sirius appears brighter than Canopus. This is because Sirius is merely 8.6 light-years from the Earth, while Canopus is much farther away at a distance of 310 light-years.

As of 2006, the star with the highest known absolute magnitude is LBV 1806-20, with a magnitude of −14.2. This star is at least 5,000,000 times more luminous than the Sun. The least luminous stars that are currently known are located in the NGC 6397 cluster. The faintest red dwarfs in the cluster were magnitude 26, while a 28th magnitude white dwarf was also discovered. These faint stars are so dim that their light is as bright as a birthday candle on the Moon when viewed from the Earth.

Classification

Surface temperature ranges for different stellar classes		
Class	**Temperature**	**Sample star**
O	33,000 K or more	Zeta Ophiuchi
B	10,500–30,000 K	Rigel
A	7,500–10,000 K	Altair
F	6,000–7,200 K	Procyon A
G	5,500–6,000 K	Sun
K	4,000–5,250 K	Epsilon Indi
M	2,600–3,850 K	Proxima Centauri

The current stellar classification system originated in the early 20th century, when stars were classified from *A* to *Q* based on the strength of the hydrogen line. It thought that the hydrogen line strength was a simple linear function of temperature. Rather, it was more complicated; it strengthened with increasing temperature, it peaked near 9000 K, and then declined at greater temperatures. When the classifications were reordered by temperature, it more closely resembled the modern scheme.

Stars are given a single-letter classification according to their spectra, ranging from type *O*, which are very hot, to *M*, which are so cool that molecules may form in their atmospheres. The main classifications in order of decreasing surface temperature are: *O, B, A, F, G, K,* and *M*. A variety of rare spectral types are given special classifications. The most common of these are types *L* and *T*, which classify the coldest low-mass stars and brown dwarfs. Each letter has 10 sub-divisions, numbered from 0 to 9, in order of decreasing temperature. However, this system breaks down at extreme high temperatures as classes *O0* and *O1* may not exist.

In addition, stars may be classified by the luminosity effects found in their spectral lines, which correspond to their spatial size and is determined by their surface gravity. These range from *0* (hypergiants) through *III* (giants) to *V* (main sequence dwarfs); some authors add *VII* (white dwarfs). Most stars belong to the main sequence, which consists of ordinary hydrogen-burning stars. These fall along a narrow, diagonal band when graphed according to their absolute magnitude and spectral type. The Sun is a main sequence *G2V* yellow dwarf of intermediate temperature and ordinary size.

Additional nomenclature, in the form of lower-case letters added to the end of the spectral type to indicate peculiar features of the spectrum. For example, an "*e*" can indicate the presence of emission lines; "*m*" represents unusually strong levels of metals, and "*var*" can mean variations in the spectral type.

White dwarf stars have their own class that begins with the letter *D*. This is further sub-divided into the classes *DA, DB, DC, DO, DZ,* and *DQ*, depending on the types of prominent lines found in the spectrum. This is followed by a numerical value that indicates the temperature.

Variable Stars

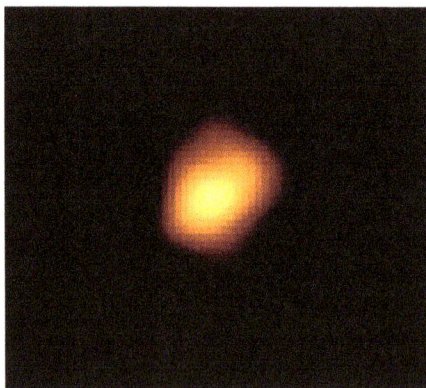

The asymmetrical appearance of Mira, an oscillating variable star.

Variable stars have periodic or random changes in luminosity because of intrinsic or extrinsic properties. Of the intrinsically variable stars, the primary types can be subdivided into three principal groups.

During their stellar evolution, some stars pass through phases where they can become pulsating variables. Pulsating variable stars vary in radius and luminosity over time, expanding and contracting with periods ranging from minutes to years, depending on the size of the star. This category includes Cepheid and Cepheid-like stars, and long-period variables such as Mira.

Eruptive variables are stars that experience sudden increases in luminosity because of flares or mass ejection events. This group includes protostars, Wolf-Rayet stars, and flare stars, as well as giant and supergiant stars.

Cataclysmic or explosive variable stars are those that undergo a dramatic change in their properties. This group includes novae and supernovae. A binary star system that includes a nearby white dwarf can produce certain types of these spectacular stellar explosions, including the nova and a Type 1a supernova. The explosion is created when the white dwarf accretes hydrogen from the companion star, building up mass until the hydrogen undergoes fusion. Some novae are also recurrent, having periodic outbursts of moderate amplitude.

Stars can also vary in luminosity because of extrinsic factors, such as eclipsing binaries, as well as rotating stars that produce extreme starspots. A notable example of an eclipsing binary is Algol, which regularly varies in magnitude from 2.3 to 3.5 over a period of 2.87 days.

Structure

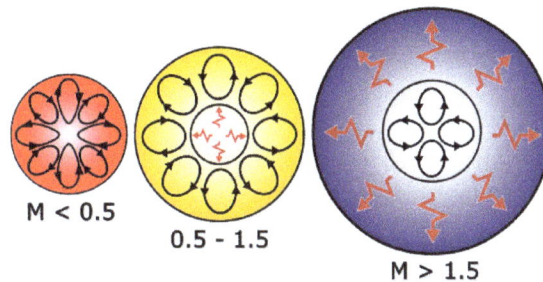

M < 0.5   0.5 - 1.5   M > 1.5

Internal structures of main sequence stars, convection zones with arrowed cycles and radiative zones with red flashes. To the left a **low-mass** red dwarf, in the center a **mid-sized** yellow dwarf, and, at the right, a **massive** blue-white main sequence star.

The interior of a stable star is in a state of hydrostatic equilibrium: the forces on any small volume almost exactly counterbalance each other. The balanced forces are inward gravitational force and an outward force due to the pressure gradient within the star. The pressure gradient is established by the temperature gradient of the plasma; the outer part of the star is cooler than the core. The temperature at the core of a main sequence or giant star is at least on the order of 10^7 K. The resulting temperature and pressure at the hydrogen-burning core of a main sequence star are sufficient for nuclear fusion to occur and for sufficient energy to be produced to prevent further collapse of the star.

As atomic nuclei are fused in the core, they emit energy in the form of gamma rays. These photons interact with the surrounding plasma, adding to the thermal energy at the core. Stars on the main sequence convert hydrogen into helium, creating a slowly but steadily increasing proportion of helium in the core. Eventually the helium content becomes predominant, and energy production ceases at the core. Instead, for stars of more than $0.4\,M_\odot$, fusion occurs in a slowly expanding shell around the degenerate helium core.

In addition to hydrostatic equilibrium, the interior of a stable star will also maintain an energy balance of thermal equilibrium. There is a radial temperature gradient throughout the interior that results in a flux of energy flowing toward the exterior. The outgoing flux of energy leaving any layer within the star will exactly match the incoming flux from below.

The radiation zone is the region of the stellar interior where the flux of energy outward is dependent on radiative heat transfer, since convective heat transfer is inefficient in that zone. In this region the plasma will not be perturbed, and any mass motions will die out. If this is not the case, however, then the plasma becomes unstable and convection will occur, forming a convection zone. This can occur, for example, in regions where very high energy fluxes occur, such as near the core or in areas with high opacity (making radiatative heat transfer inefficient) as in the outer envelope.

The occurrence of convection in the outer envelope of a main sequence star depends on the star's mass. Stars with several times the mass of the Sun have a convection zone deep within the interior and a radiative zone in the outer layers. Smaller stars such as the Sun are just the opposite, with the convective zone located in the outer layers. Red dwarf stars with less than $0.4\, M_\odot$ are convective throughout, which prevents the accumulation of a helium core. For most stars the convective zones will also vary over time as the star ages and the constitution of the interior is modified.

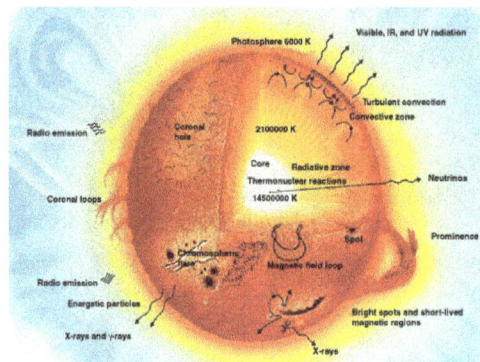

This diagram shows a cross-section of the Sun.

The photosphere is that portion of a star that is visible to an observer. This is the layer at which the plasma of the star becomes transparent to photons of light. From here, the energy generated at the core becomes free to propagate into space. It is within the photosphere that sun spots, regions of lower than average temperature, appear.

Above the level of the photosphere is the stellar atmosphere. In a main sequence star such as the Sun, the lowest level of the atmosphere, just above the photosphere, is the thin chromosphere region, where spicules appear and stellar flares begin. Above this is the transition region, where the temperature rapidly increases within a distance of only 100 km (62 mi). Beyond this is the corona, a volume of super-heated plasma that can extend outward to several million kilometres. The existence of a corona appears to be dependent on a convective zone in the outer layers of the star. Despite its high temperature, and the corona emits very little light, due to its low gas density. The corona region of the Sun is normally only visible during a solar eclipse.

From the corona, a stellar wind of plasma particles expands outward from the star, until it interacts with the interstellar medium. For the Sun, the influence of its solar wind extends throughout a bubble-shaped region called the heliosphere.

Nuclear fusion Reaction Pathways

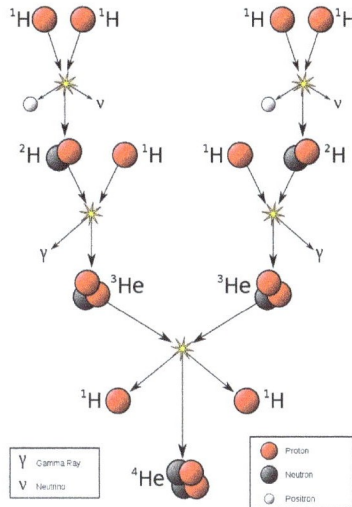

Overview of the proton-proton chain

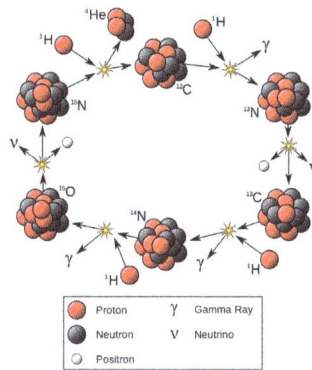

The carbon-nitrogen-oxygen cycle

A variety of nuclear fusion reactions take place in the cores of stars, that depend upon their mass and composition. When nuclei fuse, the mass of the fused product is less than the mass of the original parts. This lost mass is converted to electromagnetic energy, according to the mass-energy equivalence relationship $E = mc^2$.

The hydrogen fusion process is temperature-sensitive, so a moderate increase in the core temperature will result in a significant increase in the fusion rate. As a result, the core temperature of main sequence stars only varies from 4 million kelvin for a small M-class star to 40 million kelvin for a massive O-class star.

In the Sun, with a 10-million-kelvin core, hydrogen fuses to form helium in the proton-proton chain reaction:

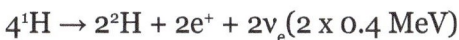

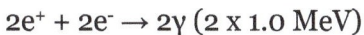

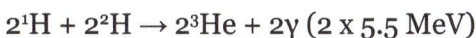

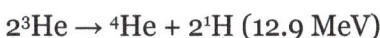

$4^1H \rightarrow 2\,^2H + 2e^+ + 2\nu_e (2 \times 0.4 \text{ MeV})$

$2e^+ + 2e^- \rightarrow 2\gamma (2 \times 1.0 \text{ MeV})$

$2^1H + 2\,^2H \rightarrow 2\,^3He + 2\gamma (2 \times 5.5 \text{ MeV})$

$2^3He \rightarrow {}^4He + 2\,^1H (12.9 \text{ MeV})$

These reactions result in the overall reaction:

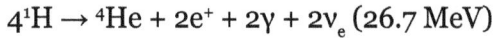

$$4^1H \rightarrow {}^4He + 2e^+ + 2\gamma + 2\nu_e \ (26.7 \text{ MeV})$$

where e^+ is a positron, γ is a gamma ray photon, ν_e is a neutrino, and H and He are isotopes of hydrogen and helium, respectively. The energy released by this reaction is in millions of electron volts, which is actually only a tiny amount of energy. However enormous numbers of these reactions occur constantly, producing all the energy necessary to sustain the star's radiation output. In comparison, the combustion of two hydrogen gas molecules with one oxygen gas molecule releases only 5.7 eV.

Minimum stellar mass required for fusion	
Element	Solar masses
Hydrogen	0.01
Helium	0.4
Carbon	5
Neon	8

In more massive stars, helium is produced in a cycle of reactions catalyzed by carbon called the carbon-nitrogen-oxygen cycle.

In evolved stars with cores at 100 million kelvin and masses between 0.5 and 10 $M_\odot$, helium can be transformed into carbon in the triple-alpha process that uses the intermediate element beryllium:

$$^4He + {}^4He + 92 \text{ keV} \rightarrow {}^{8*}Be$$

$$^4He + {}^{8*}Be + 67 \text{ keV} \rightarrow {}^{12*}C$$

$$^{12*}C \rightarrow {}^{12}C + \gamma + 7.4 \text{ MeV}$$

For an overall reaction of:

$$3^4He \rightarrow {}^{12}C + \gamma + 7.2 \text{ MeV}$$

In massive stars, heavier elements can also be burned in a contracting core through the neon burning process and oxygen burning process. The final stage in the stellar nucleosynthesis process is the silicon burning process that results in the production of the stable isotope iron-56, an endothermic process that consumes energy, and so further energy can only be produced through gravitational collapse.

The example below shows the amount of time required for a star of 20 $M_\odot$ to consume all of its nuclear fuel. As an O-class main sequence star, it would be 8 times the solar radius and 62,000 times the Sun's luminosity.

Fuel material	Temperature (million kelvins)	Density (kg/cm³)	Burn duration (τ in years)
H	37	0.0045	8.1 million
He	188	0.97	1.2 million
C	870	170	976
Ne	1,570	3,100	0.6
O	1,980	5,550	1.25
S/Si	3,340	33,400	0.0315

Nebula

A nebula (Latin for "cloud"; pl. nebulae, nebulæ, or nebulas) is an interstellar cloud of dust, hydrogen, helium and other ionized gases. Originally, *nebula* was a name for any diffuse astronomical object, including galaxies beyond the Milky Way. The Andromeda Galaxy, for instance, was once referred to as the *Andromeda Nebula* (and spiral galaxies in general as "spiral nebulae") before the true nature of galaxies was confirmed in the early 20th century by Vesto Slipher, Edwin Hubble and others.

The "Pillars of Creation" from the Eagle Nebula. Evidence from the Spitzer Telescope suggests that the pillars may already have been destroyed by a supernova explosion, but the light showing us the destruction will not reach the Earth for another millennium.

Most nebulae are of vast size, even millions of light years in diameter. Contrary to fictional depictions where starships hide in nebulae as thick as cloud banks, in reality a nebula that is barely visible to the human eye from Earth would appear larger, but no brighter, from close by. The Orion Nebula, the brightest nebula in the sky that occupies a region twice the diameter of the full Moon, can be viewed with the naked eye but was missed by early astronomers. Although denser than the space surrounding them, most nebulae are far less dense than any vacuum created on Earth – a nebular cloud the size of the Earth would have a total mass of only a few kilograms. Many nebulae are visible due to their fluorescence caused by the embedded hot stars, while others are so diffuse they can only be detected with long exposures and special filters. Some nebulae are variably illuminated by T Tauri variable stars. Nebulae are often star-forming regions, such as in the "Pillars of Creation" in the Eagle Nebula. In these regions the formations of gas, dust, and other materials "clump" together to form denser regions, which attract further matter, and eventually will become dense enough to form stars. The remaining material is then believed to form planets and other planetary system objects.

Observational History

Around 150 AD, Claudius Ptolemaeus (Ptolemy) recorded, in books VII-VIII of his *Almagest*, five stars that appeared nebulous. He also noted a region of nebulosity between the constellations Ursa Major and Leo that was not associated with any star. The first true nebula, as distinct from a star cluster, was mentioned by the Persian astronomer, Abd al-Rahman al-Sufi, in his *Book of Fixed Stars* (964). He noted "a little cloud" where the Andromeda Galaxy is located. He also cataloged the Omicron Velorum star cluster as a "nebulous star" and other nebulous objects, such as Brocchi's Cluster. The supernova that created the Crab Nebula, the SN 1054, was observed by Arabic and Chinese astronomers in 1054.

Portion of the Carina Nebula

In 1610, Nicolas-Claude Fabri de Peiresc discovered the Orion Nebula using a telescope. This nebula was also observed by Johann Baptist Cysat in 1618. However, the first detailed study of the Orion Nebula was not performed until 1659 by Christiaan Huygens, who also believed himself to be the first person to discover this nebulosity.

In 1715, Edmund Halley published a list of six nebulae. This number steadily increased during the century, with Jean-Philippe de Cheseaux compiling a list of 20 (including eight not previously known) in 1746. From 1751–53, Nicolas Louis de Lacaille cataloged 42 nebulae from the Cape of Good Hope, most of which were previously unknown. Charles Messier then compiled a catalog of 103 "nebulae" (now called Messier objects, which included what are now known to be galaxies) by 1781; his interest was detecting comets, and these were objects that might be mistaken for them.

The number of nebulae was then greatly expanded by the efforts of William Herschel and his sister Caroline Herschel. Their *Catalogue of One Thousand New Nebulae and Clusters of Stars* was published in 1786. A second catalog of a thousand was published in 1789 and the third and final catalog of 510 appeared in 1802. During much of their work, William Herschel believed that these nebulae were merely unresolved clusters of stars. In 1790, however, he discovered a star surrounded by nebulosity and concluded that this was a true nebulosity, rather than a more distant cluster.

Beginning in 1864, William Huggins examined the spectra of about 70 nebulae. He found that roughly a third of them had the emission spectrum of a gas. The rest showed a continuous spectrum and thus were thought to consist of a mass of stars. A third category was added in 1912 when Vesto Slipher showed that the spectrum of the nebula that surrounded the star Merope matched the spectra of the Pleiades open cluster. Thus the nebula radiates by reflected star light.

About 1922, following the Great Debate, it had become clear that many "nebulae" were in fact galaxies far from our own.

Slipher and Edwin Hubble continued to collect the spectra from many diffuse nebulae, finding 29 that showed emission spectra and 33 that had the continuous spectra of star light. In 1922, Hubble announced that nearly all nebulae are associated with stars, and their illumination comes from star light. He also discovered that the emission spectrum nebulae are nearly always associated with stars having spectral classifications of B1 or hotter (including all O-type main sequence stars), while nebulae with continuous spectra appear with cooler stars. Both Hubble and Henry Norris Russell concluded that the nebulae surrounding the hotter stars are transformed in some manner.

Formation

The Triangulum Emission Garren Nebula NGC 604

Many nebulae or stars form from the gravitational collapse of gas in the interstellar medium. As the material collapses under its own weight, massive stars may form in the center, and their ultraviolet radiation ionizes the surrounding gas, making it visible at optical wavelengths. Examples of these types of nebulae are the Rosette Nebula and the Pelican Nebula. The size of these nebulae, known as H II regions, varies depending on the size of the original cloud of gas. New stars are formed in the nebulae. The formed stars are sometimes known as a young, loose cluster.

Other nebulae form as the result of supernova explosions; the death throes of massive, short-lived stars. The materials thrown off from the supernova explosion are then ionized by the energy and the compact object that its core produces. One of the best examples of this is the Crab Nebula, in Taurus. The supernova event was recorded in the year 1054 and is labelled SN 1054. The compact object that was created after the explosion lies in the center of the Crab Nebula and its core is now a neutron star.

Still other nebulae form as planetary nebulae. This is the final stage of a low-mass star's life, like Earth's Sun. Stars with a mass up to 8–10 solar masses evolve into red giants and slowly lose their outer layers during pulsations in their atmospheres. When a star has lost enough material, its temperature increases and the ultraviolet radiation it emits can ionize the surrounding nebula that it has thrown off. Our Sun will produce a planetary nebula and its core will remain behind in the form of white dwarf.

Classical Types

Objects named nebulae belong to four major groups. Before their nature was understood, galaxies ("spiral nebulae") and star clusters too distant to be resolved as stars were also classified as nebulae, but no longer are.

- H II regions, large diffuse nebulae containing ionized hydrogen
- Planetary nebulae
- Supernova remnant (e.g., Crab Nebula)
- Dark nebula

Not all cloud-like structures are named nebulae; Herbig–Haro objects are an example.

Diffuse Nebulae

Most nebulae can be described as diffuse nebulae, which means that they are extended and contain no well-defined boundaries. Diffuse nebulae can be divided into emission nebula, reflection nebulae and "dark nebulae." In visible light nebulae may be divided into emission nebulae that emit spectral line radiation from excited or ionized gas (mostly ionized hydrogen); they are often called HII regions (the term "HII" refers to ionized hydrogen). Reflection nebulae are visible primarily due to the light they reflect. Reflection nebulae themselves do not emit significant amounts of visible light, but are near stars and reflect light from them. Similar nebulae not illuminated by stars do not exhibit visible radiation, but may be detected as opaque clouds blocking light from luminous objects behind them; they are called "dark nebulae".

The Carina Nebula is a diffuse nebula

Although these nebulae have different visibility at optical wavelengths, they are all bright sources of infrared emission, chiefly from dust within the nebulae.

Planetary Nebulae

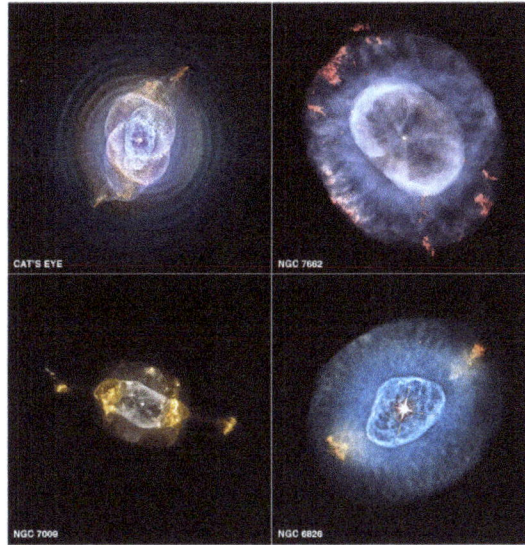

Four different planetary nebulae

Planetary nebulae form when low-mass asymptotic giant branch stars nova. A star that novas pushes the outer layers of the star's mass outward forming gaseous shells, while leaving behind the star's core in the form of a white dwarf. The hot white dwarf illuminates the expelled gases producing emission nebulae with spectra similar to those of emission nebulae found in star formation regions. Technically they are HII regions, because most hydrogen will be ionized, but they are denser and more compact than the nebulae found in star formation regions. Planetary nebulae were given their name by the first astronomical observers who were initially unable to distinguish them from planets, and who tended to confuse them with planets, which were of more interest to them. Our Sun is expected to spawn a planetary nebula about 12 billion years after its formation.

Protoplanetary Nebula

A protoplanetary nebula (PPN) is an astronomical object which is at the short-lived episode during a star's rapid stellar evolution between the late asymptotic giant branch (LAGB) phase and the following planetary nebula (PN) phase. During the AGB phase, the star undergoes mass loss, emitting a circumstellar shell of hydrogen gas. When this phase comes to an end, the star enters the PPN phase.

The PPN is energized by the central star, causing it to emit strong infrared radiation and become a reflection nebula. Collimated stellar winds from the central star shape and shock the shell into an axially symmetric form, while producing a fast moving molecular wind. The exact point when a PPN becomes a planetary nebula (PN) is defined by the temperature of the central star. The PPN phase continues until the central star reaches a temperature of 30,000 K, after which it is hot enough to ionize the surrounding gas.

Supernova Remnants

A supernova occurs when a high-mass star reaches the end of its life. When nuclear fusion in the core of the star stops, the star collapses. The gas falling inward either rebounds or gets so strongly heated that it expands outwards from the core, thus causing the star to explode. The expanding shell of gas forms a supernova remnant, a special diffuse nebula. Although much of the optical and X-ray emission from supernova remnants originates from ionized gas, a great amount of the radio emission is a form of non-thermal emission called synchrotron emission. This emission originates from high-velocity electrons oscillating within magnetic fields.

References

- Buta, Ronald James; Corwin, Harold G.; Odewahn, Stephen C. (2007). The de Vaucouleurs atlas of galaxies. Cambridge University Press. p. 301. ISBN 0-521-82048-0.

- Evans, James (1998). The History and Practice of Ancient Astronomy. Oxford University Press. pp. 296–7. ISBN 978-0-19-509539-5. Retrieved 2008-02-04.

- Hermann Hunger, ed. (1992). Astrological reports to Assyrian kings. State Archives of Assyria. 8. Helsinki University Press. ISBN 951-570-130-9.

- Sarma, K. V. (1997) "Astronomy in India" in Selin, Helaine (editor) Encyclopaedia of the History of Science, Technology, and Medicine in Non-Western Cultures, Kluwer Academic Publishers, ISBN 0-7923-4066-3.

- Lindberg, David C. (2007). The Beginnings of Western Science (2nd ed.). Chicago: The University of Chicago Press. p. 257. ISBN 978-0-226-48205-7.

- Cochrane, Ev (1997). Martian Metamorphoses: The Planet Mars in Ancient Myth and Tradition. Aeon Press. ISBN 0-9656229-0-8. Retrieved 2008-02-07.

- Zerubavel, Eviatar (1989). The Seven Day Circle: The History and Meaning of the Week. University of Chicago Press. p. 14. ISBN 0-226-98165-7. Retrieved 2008-02-07.

- Weintraub, David A. (2014), Is Pluto a Planet?: A Historical Journey through the Solar System, Princeton University Press, p. 226, ISBN 1400852978

- Zeilik, Michael A.; Gregory, Stephan A. (1998). Introductory Astronomy & Astrophysics (4th ed.). Saunders College Publishing. p. 67. ISBN 0-03-006228-4.

- Kivelson, Margaret Galland; Bagenal, Fran (2007). "Planetary Magnetospheres". In Lucyann Mcfadden; Paul Weissman; Torrence Johnson. Encyclopedia of the Solar System. Academic Press. p. 519. ISBN 978-0-12-088589-3.

- Canup, Robin M.; Ward, William R. (2008-12-30). Origin of Europa and the Galilean Satellites. University of Arizona Press. p. 59. arXiv:0812.4995. Bibcode:2009euro.book...59C. ISBN 978-0-8165-2844-8.

- Plutarch (2006). The Complete Works Volume 3: Essays and Miscellanies. Chapter 3: Echo Library. p. 66. ISBN 978-1-4068-3224-2.

- See text quoted from Wright's An original theory or new hypothesis of the Universe in Dyson, F. (1979). Disturbing the Universe. Pan Books. p. 245. ISBN 0-330-26324-2.

- Knapp, G. R. (1999). Star Formation in Early Type Galaxies. Astronomical Society of the Pacific. Bibcode:1998astro.ph..8266K. ISBN 1-886733-84-8. OCLC 41302839.

- North, John (1995). The Norton History of Astronomy and Cosmology. New York and London: W.W. Norton & Company. pp. 30–31. ISBN 0-393-03656-1.

- Murdin, P. (November 2000). "Aristillus (c. 200 BC)". Encyclopedia of Astronomy and Astrophysics. Bibcode:2000eaa..bookE3440.. doi:10.1888/0333750888/3440. ISBN 0-333-75088-8.

- Lyall, Francis; Larsen, Paul B. (2009). "Chapter 7: The Moon and Other Celestial Bodies". Space Law: A Treatise. Ashgate Publishing, Ltd. p. 176. ISBN 0-7546-4390-5.

- Plait, Philip C. (2002). Bad astronomy: misconceptions and misuses revealed, from astrology to the moon landing "hoax". John Wiley and Sons. pp. 237–240. ISBN 0-471-40976-6.

3

Branches of Astronomy

Astronomy is an interdisciplinary subject. It spreads to other fields as well. Some of the branches of astronomy covered in this chapter are astrometry, astrophysics, galactic astronomy, infrared astronomy and neutrino astronomy. Astronomy is best understood in confluence with the major topics listed in the following chapter.

Astrometry

Astrometry is the branch of astronomy that involves precise measurements of the positions and movements of stars and other celestial bodies. The information obtained by astrometric measurements provides information on the kinematics and physical origin of the Solar System and our galaxy, the Milky Way.

History

The history of astrometry is linked to the history of star catalogues, which gave astronomers reference points for objects in the sky so they could track their movements. This can be dated back to Hipparchus, who around 190 BC used the catalogue of his predecessors Timocharis and Aristillus to discover Earth's precession. In doing so, he also developed the brightness scale still in use today. Hipparchus compiled a catalogue with at least 850 stars and their positions. Hipparchus's successor, Ptolemy, included a catalogue of 1,022 stars in his work the *Almagest*, giving their location, coordinates, and brightness.

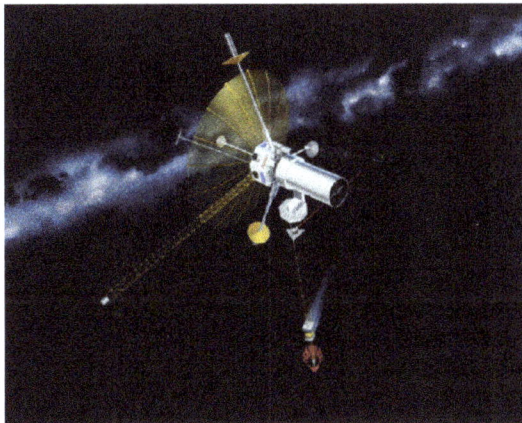

Concept art for the TAU spacecraft, a 1980s era study which would have used an interstellar precursor probe to expand the baseline for calculating stellar parallax in support of Astrometry

In the 10th century, Abd al-Rahman al-Sufi carried out observations on the stars and described their positions, magnitudes and star color, and gave drawings for each constellation, in his *Book of*

Fixed Stars. Ibn Yunus observed more than 10,000 entries for the Sun's position for many years using a large astrolabe with a diameter of nearly 1.4 metres. His observations on eclipses were still used centuries later in Simon Newcomb's investigations on the motion of the Moon, while his other observations inspired Laplace's *Obliquity of the Ecliptic* and *Inequalities of Jupiter and Saturn*. In the 15th century, the Timurid astronomer Ulugh Beg compiled the *Zij-i-Sultani*, in which he catalogued 1,019 stars. Like the earlier catalogs of Hipparchus and Ptolemy, Ulugh Beg's catalogue is estimated to have been precise to within approximately 20 minutes of arc.

In the 16th century, Tycho Brahe used improved instruments, including large mural instruments, to measure star positions more accurately than previously, with a precision of 15–35 arcsec. Taqi al-Din measured the right ascension of the stars at the Istanbul observatory of Taqi al-Din using the "observational clock" he invented. When telescopes became commonplace, setting circles sped measurements

James Bradley first tried to measure stellar parallaxes in 1729. The stellar movement proved too insignificant for his telescope, but he instead discovered the aberration of light and the nutation of the Earth's axis. His cataloguing of 3222 stars was refined in 1807 by Friedrich Bessel, the father of modern astrometry. He made the first measurement of stellar parallax: 0.3 arcsec for the binary star 61 Cygni.

Being very difficult to measure, only about 60 stellar parallaxes had been obtained by the end of the 19th century, mostly by use of the filar micrometer. Astrographs using astronomical photographic plates sped the process in the early 20th century. Automated plate-measuring machines and more sophisticated computer technology of the 1960s allowed more efficient compilation of star catalogues. In the 1980s, charge-coupled devices (CCDs) replaced photographic plates and reduced optical uncertainties to one milliarcsecond. This technology made astrometry less expensive, opening the field to an amateur audience.

In 1989, the European Space Agency's Hipparcos satellite took astrometry into orbit, where it could be less affected by mechanical forces of the Earth and optical distortions from its atmosphere. Operated from 1989 to 1993, Hipparcos measured large and small angles on the sky with much greater precision than any previous optical telescopes. During its 4-year run, the positions, parallaxes, and proper motions of 118,218 stars were determined with an unprecedented degree of accuracy. A new "Tycho catalog" drew together a database of 1,058,332 to within 20-30 mas (milliarcseconds). Additional catalogues were compiled for the 23,882 double/multiple stars and 11,597 variable stars also analyzed during the Hipparcos mission.

Today, the catalogue most often used is USNO-B1.0, an all-sky catalogue that tracks proper motions, positions, magnitudes and other characteristics for over one billion stellar objects. During the past 50 years, 7,435 Schmidt camera plates were used to complete several sky surveys that make the data in USNO-B1.0 accurate to within 0.2 arcsec.

Applications

Apart from the fundamental function of providing astronomers with a reference frame to report their observations in, astrometry is also fundamental for fields like celestial mechanics, stellar dynamics and galactic astronomy. In observational astronomy, astrometric techniques help identify stellar objects by their unique motions. It is instrumental for keeping time, in that UTC is basically the atomic time synchronized to Earth's rotation by means of exact observations. Astrometry is an important step in the cosmic distance ladder because it establishes parallax distance estimates for stars in the Milky Way.

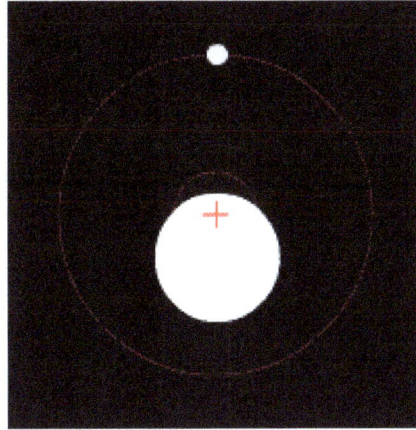

Diagram showing how a smaller object (such as an extrasolar planet) orbiting a larger object (such as a star) could produce changes in position and velocity of the latter as they orbit their common center of mass (red cross).

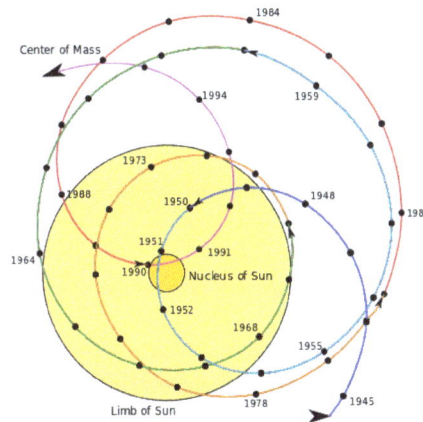

Motion of barycenter of solar system relative to the Sun.

Astrometry has also been used to support claims of extrasolar planet detection by measuring the displacement the proposed planets cause in their parent star's apparent position on the sky, due to their mutual orbit around the center of mass of the system. Although, as of 2009, none of the extrasolar planets detected by ground-based astrometry has been verified in subsequent studies, astrometry is expected to be more accurate in space missions that are not affected by the distorting effects of the Earth's atmosphere. NASA's planned Space Interferometry Mission (SIM PlanetQuest) (now cancelled) was to utilize astrometric techniques to detect terrestrial planets orbiting 200 or so of the nearest solar-type stars, and the European Space Agency's Gaia Mission, launched in 2013, which will be applying astrometric techniques in its stellar census.

Astrometric measurements are used by astrophysicists to constrain certain models in celestial mechanics. By measuring the velocities of pulsars, it is possible to put a limit on the asymmetry of supernova explosions. Also, astrometric results are used to determine the distribution of dark matter in the galaxy.

Astronomers use astrometric techniques for the tracking of near-Earth objects. Astrometry is responsible for the detection of many record-breaking Solar System objects. To find such objects astrometrically, astronomers use telescopes to survey the sky and large-area cameras to take pictures at various determined intervals. By studying these images, they can detect Solar System objects by their movements relative to the background stars, which remain fixed. Once a movement per unit time is observed, astronomers compensate for the parallax caused by Earth's motion during this time and the heliocentric distance to this object is calculated. Using this distance and other photographs, more information about the object, including its orbital elements, can be obtained.

50000 Quaoar and 90377 Sedna are two Solar System objects discovered in this way by Michael E. Brown and others at Caltech using the Palomar Observatory's Samuel Oschin telescope of 48 inches (1.2 m) and the Palomar-Quest large-area CCD camera. The ability of astronomers to track the positions and movements of such celestial bodies is crucial to the understanding of the Solar System and its interrelated past, present, and future with others in the Universe.

Statistics

A fundamental aspect of astrometry is error correction. Various factors introduce errors into the measurement of stellar positions, including atmospheric conditions, imperfections in the instruments and errors by the observer or the measuring instruments. Many of these errors can be reduced by various techniques, such as through instrument improvements and compensations to the data. The results are then analyzed using statistical methods to compute data estimates and error ranges.

Astrophysics

Astrophysics is the branch of astronomy that employs the principles of physics and chemistry "to ascertain the nature of the heavenly bodies, rather than their positions or motions in space." Among the objects studied are the Sun, other stars, galaxies, extrasolar planets, the interstellar medium and the cosmic microwave background. Their emissions are examined across all parts of the electromagnetic spectrum, and the properties examined include luminosity, density, temperature, and chemical composition. Because astrophysics is a very broad subject, *astrophysicists* typically apply many disciplines of physics, including mechanics, electromagnetism, statistical mechanics, thermodynamics, quantum mechanics, relativity, nuclear and particle physics, and atomic and molecular physics.

In practice, modern astronomical research often involves a substantial amount of work in the realms of theoretical and observational physics. Some areas of study for astrophysicists include their attempts to determine: the properties of dark matter, dark energy, and black holes; whether or not time travel is possible, wormholes can form, or the multiverse exists; and the origin and ultimate fate of the universe. Topics also studied by theoretical astrophysicists include: Solar System formation and evolution; stellar dynamics and evolution; galaxy formation and evolution; magnetohydrodynamics; large-scale structure of matter in the universe; origin of cosmic rays; general relativity and physical cosmology, including string cosmology and astroparticle physics.

History

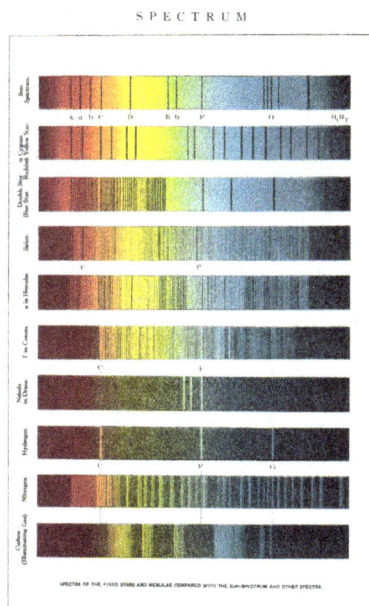

Early 20th-century comparison of elemental, solar, and stellar spectra

Although astronomy is as ancient as recorded history itself, it was long separated from the study of terrestrial physics. In the Aristotelian worldview, bodies in the sky appeared to be unchanging spheres whose only motion was uniform motion in a circle, while the earthly world was the realm which underwent growth and decay and in which natural motion was in a straight line and ended when the moving object reached its goal. Consequently, it was held that the celestial region was made of a fundamentally different kind of matter from that found in the terrestrial sphere; either Fire as maintained by Plato, or Aether as maintained by Aristotle. During the 17th century, natural philosophers such as Galileo, Descartes, and Newton began to maintain that the celestial and terrestrial regions were made of similar kinds of material and were subject to the same natural laws. Their challenge was that the tools had not yet been invented with which to prove these assertions.

For much of the nineteenth century, astronomical research was focused on the routine work of measuring the positions and computing the motions of astronomical objects. A new astronomy, soon to be called astrophysics, began to emerge when William Hyde Wollaston and Joseph von Fraunhofer independently discovered that, when decomposing the light from the Sun, a multitude of dark lines (regions where there was less or no light) were observed in the spectrum. By 1860 the physicist, Gustav Kirchhoff, and the chemist, Robert Bunsen, had demonstrated that the dark lines in the solar spectrum corresponded to bright lines in the spectra of known gases, specific lines corresponding to unique chemical elements. Kirchhoff deduced that the dark lines in the solar spectrum are caused by absorption by chemical elements in the Solar atmosphere. In this way it was proved that the chemical elements found in the Sun and stars were also found on Earth.

Among those who extended the study of solar and stellar spectra was Norman Lockyer, who in 1868 detected bright, as well as dark, lines in solar spectra. Working with the chemist, Edward Frankland, to investigate the spectra of elements at various temperatures and pressures, he could not associate a yellow line in the solar spectrum with any known elements. He thus claimed the line represented a new element, which was called helium, after the Greek Helios, the Sun personified.

In 1885, Edward C. Pickering undertook an ambitious program of stellar spectral classification at Harvard College Observatory, in which a team of woman computers, notably Williamina Fleming, Antonia Maury, and Annie Jump Cannon, classified the spectra recorded on photographic plates. By 1890, a catalog of over 10,000 stars had been prepared that grouped them into thirteen spectral types. Following Pickering's vision, by 1924 Cannon expanded the catalog to nine volumes and over a quarter of a million stars, developing the Harvard Classification Scheme which was accepted for worldwide use in 1922.

In 1895, George Ellery Hale and James E. Keeler, along with a group of ten associate editors from Europe and the United States, established *The Astrophysical Journal: An International Review of Spectroscopy and Astronomical Physics*. It was intended that the journal would fill the gap between journals in astronomy and physics, providing a venue for publication of articles on astronomical applications of the spectroscope; on laboratory research closely allied to astronomical physics, including wavelength determinations of metallic and gaseous spectra and experiments on radiation and absorption; on theories of the Sun, Moon, planets, comets, meteors, and nebulae; and on instrumentation for telescopes and laboratories.

In 1925 Cecilia Helena Payne (later Cecilia Payne-Gaposchkin) wrote an influential doctoral dissertation at Radcliffe College, in which she applied ionization theory to stellar atmospheres to relate the spectral classes to the temperature of stars. Most significantly, she discovered that hydrogen and helium were the principal components of stars. This discovery was so unexpected that her dissertation readers convinced her to modify the conclusion before publication. However, later research confirmed her discovery.

By the end of the 20th century, further study of stellar and experimental spectra advanced, particularly as a result of the advent of quantum physics.

Observational Astrophysics

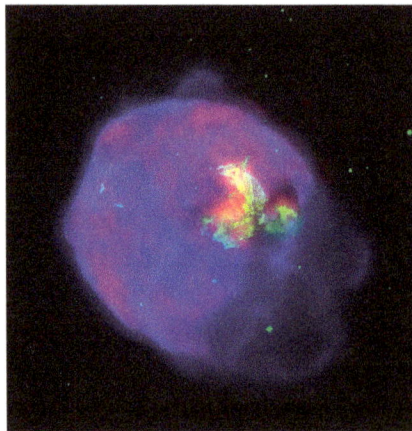

Supernova remnant LMC N 63A imaged in x-ray (blue), optical (green) and radio (red) wavelengths. The X-ray glow is from material heated to about ten million degrees Celsius by a shock wave generated by the supernova explosion.

Observational astronomy is a division of the astronomical science that is concerned with recording data, in contrast with theoretical astrophysics, which is mainly concerned with finding out the measurable implications of physical models. It is the practice of observing celestial objects by using telescopes and other astronomical apparatus.

The majority of astrophysical observations are made using the electromagnetic spectrum.

- Radio astronomy studies radiation with a wavelength greater than a few millimeters. Example areas of study are radio waves, usually emitted by cold objects such as interstellar gas and dust clouds; the cosmic microwave background radiation which is the redshifted light from the Big Bang; pulsars, which were first detected at microwave frequencies. The study of these waves requires very large radio telescopes.

- Infrared astronomy studies radiation with a wavelength that is too long to be visible to the naked eye but is shorter than radio waves. Infrared observations are usually made with telescopes similar to the familiar optical telescopes. Objects colder than stars (such as planets) are normally studied at infrared frequencies.

- Optical astronomy is the oldest kind of astronomy. Telescopes paired with a charge-coupled device or spectroscopes are the most common instruments used. The Earth's atmosphere interferes somewhat with optical observations, so adaptive optics and space telescopes are used to obtain the highest possible image quality. In this wavelength range, stars are highly visible, and many chemical spectra can be observed to study the chemical composition of stars, galaxies and nebulae.

- Ultraviolet, X-ray and gamma ray astronomy study very energetic processes such as binary pulsars, black holes, magnetars, and many others. These kinds of radiation do not penetrate the Earth's atmosphere well. There are two methods in use to observe this part of the electromagnetic spectrum—space-based telescopes and ground-based imaging air Cherenkov telescopes (IACT). Examples of Observatories of the first type are RXTE, the Chandra X-ray Observatory and the Compton Gamma Ray Observatory. Examples of IACTs are the High Energy Stereoscopic System (H.E.S.S.) and the MAGIC telescope.

Other than electromagnetic radiation, few things may be observed from the Earth that originate from great distances. A few gravitational wave observatories have been constructed, but gravitational waves are extremely difficult to detect. Neutrino observatories have also been built, primarily to study our Sun. Cosmic rays consisting of very high energy particles can be observed hitting the Earth's atmosphere.

Observations can also vary in their time scale. Most optical observations take minutes to hours, so phenomena that change faster than this cannot readily be observed. However, historical data on some objects is available, spanning centuries or millennia. On the other hand, radio observations may look at events on a millisecond timescale (millisecond pulsars) or combine years of data (pulsar deceleration studies). The information obtained from these different timescales is very different.

The study of our very own Sun has a special place in observational astrophysics. Due to the tremendous distance of all other stars, the Sun can be observed in a kind of detail unparalleled by any other star. Our understanding of our own Sun serves as a guide to our understanding of other stars.

The topic of how stars change, or stellar evolution, is often modeled by placing the varieties of star types in their respective positions on the Hertzsprung–Russell diagram, which can be viewed as representing the state of a stellar object, from birth to destruction.

Theoretical Astrophysics

Theoretical astrophysicists use a wide variety of tools which include analytical models (for example, polytropes to approximate the behaviors of a star) and computational numerical simulations. Each has some advantages. Analytical models of a process are generally better for giving insight into the heart of what is going on. Numerical models can reveal the existence of phenomena and effects that would otherwise not be seen.

Stream lines on this simulation of a supernova show the flow of matter behind the shock wave giving clues as to the origin of pulsars

Theorists in astrophysics endeavor to create theoretical models and figure out the observational consequences of those models. This helps allow observers to look for data that can refute a model or help in choosing between several alternate or conflicting models.

Theorists also try to generate or modify models to take into account new data. In the case of an inconsistency, the general tendency is to try to make minimal modifications to the model to fit the data. In some cases, a large amount of inconsistent data over time may lead to total abandonment of a model.

Topics studied by theoretical astrophysicists include: stellar dynamics and evolution; galaxy formation and evolution; magnetohydrodynamics; large-scale structure of matter in the universe; origin of cosmic rays; general relativity and physical cosmology, including string cosmology and astroparticle physics. Astrophysical relativity serves as a tool to gauge the properties of large scale structures for which gravitation plays a significant role in physical phenomena investigated and as the basis for black hole (*astro*)physics and the study of gravitational waves.

Some widely accepted and studied theories and models in astrophysics, now included in the Lambda-CDM model, are the Big Bang, cosmic inflation, dark matter, dark energy and fundamental theories of physics. Wormholes are examples of hypotheses which are yet to be proven (or disproven).

Popularization

The roots of astrophysics can be found in the seventeenth century emergence of a unified physics, in which the same laws applied to the celestial and terrestrial realms. There were scientists who were qualified in both physics and astronomy who laid the firm foundation for the current science of astrophysics. In modern times, students continue to be drawn to astrophysics due to its popularization by the Royal Astronomical Society and notable educators such as prominent professors

Subrahmanyan Chandrasekhar, Stephen Hawking, Hubert Reeves, Carl Sagan and Neil deGrasse Tyson. The efforts of the early, late, and present scientists continue to attract young people to study the history and science of astrophysics.

Extragalactic Astronomy

Galaxies in the Hubble Deep Field.

Extragalactic astronomy is the branch of astronomy concerned with objects outside the Milky Way galaxy. In other words, it is the study of all astronomical objects which are not covered by galactic astronomy, and is considered the next level of galactic astronomy.

As instrumentation has improved, more distant objects can now be examined in detail. It is therefore useful to sub-divide this branch into Near-Extragalactic Astronomy and Far-Extragalactic Astronomy. The former deals with objects such as the galaxies of the Local Group, which are close enough to allow very detailed analyses of their contents (e.g. supernova remnants, stellar associations).

Some topics include:

- Galaxy groups
- Galaxy clusters
- Superclusters
- Galaxy filaments
- Quasars
- Radio galaxies
- Supernovae
- Intergalactic stars

- Intergalactic dust
- Intergalactic dust clouds
- the observable universe

Galactic Astronomy

Galactic astronomy is the study of the Milky Way galaxy and all its contents. This is in contrast to extragalactic astronomy, which is the study of everything outside our galaxy, including all other galaxies.

Galactic astronomy should not be confused with galaxy formation and evolution, which is the general study of galaxies, their formation, structure, components, dynamics, interactions, and the range of forms they take.

The Milky Way galaxy, where the Solar System belongs, is in many ways the best studied galaxy, although important parts of it are obscured from view in visible wavelengths by regions of cosmic dust. The development of radio astronomy, infrared astronomy and submillimetre astronomy in the 20th Century allowed the gas and dust of the Milky Way to be mapped for the first time.

Subcategories

A standard set of subcategories is used by astronomical journals to split up the subject of Galactic Astronomy:

1. abundances – the study of the location of elements heavier than helium

2. bulge – the study of the bulge around the center of the Milky Way

3. center – the study of the central region of the Milky Way

4. disk – the study of the Milky Way disk (the plane upon which most galactic objects are aligned)

5. evolution – the evolution of the Milky Way

6. formation – the formation of the Milky Way

7. fundamental parameters – the fundamental parameters of the Milky Way (mass, size etc.)

8. globular clusters – globular clusters within the Milky Way

9. halo – the large halo around the Milky Way

10. kinematics and dynamics – the motions of stars and clusters

11. nucleus – the region around the black hole at the center of the Milky Way (Sagittarius A*)

12. open clusters and associations – open clusters and associations of stars

13. solar neighbourhood – nearby stars

14. stellar content – numbers and types of stars in the Milky Way

15. structure – the structure (spiral arms etc.)

Infrared Astronomy

Carina Nebula in infrared light captured by the Hubble's Wide Field Camera 3.

Infrared astronomy is the branch of astronomy and astrophysics that studies astronomical objects visible in infrared (IR) radiation. The wavelength of infrared light ranges from 0.75 to 300 micrometers. Infrared falls in between visible radiation, which ranges from 380 to 750 nanometers, and submillimeter waves.

Infrared astronomy began in the 1830s, a few decades after the discovery of infrared light by William Herschel in 1800. Early progress was limited, and it was not until the early 20th century that conclusive detections of astronomical objects other than the Sun and Moon were detected in infrared light. After a number of discoveries were made in the 1950s and 1960s in radio astronomy, astronomers realized the information available outside of the visible wavelength range, and modern infrared astronomy was established.

Infrared and optical astronomy are often practiced using the same telescopes, as the same mirrors or lenses are usually effective over a wavelength range that includes both visible and infrared light. Both fields also use solid state detectors, though the specific type of solid state detectors used are different. Infrared light is absorbed at many wavelengths by water vapor in the Earth's atmosphere, so most infrared telescopes are at high elevations in dry places, above as much of the atmosphere as possible. There are also infrared observatories in space, including the Spitzer Space Telescope and the Herschel Space Observatory.

History

The discovery of infrared radiation is attributed to William Herschel, who performed an experiment where he placed a thermometer in sunlight of different colors after it passed through a prism. He noticed that the temperature increase induced by sunlight was highest *outside* the visible spectrum, just beyond the red color. That the temperature increase was highest at infrared wavelengths was due to the spectral index of the prism rather than properties of the Sun, but the fact that there was any temperature increase at all prompted Herschel to deduce that there was invisible radiation from the Sun. He dubbed this radiation "calorific rays", and went on to show that it could be reflected, transmitted, and absorbed just like visible light.

SOFIA is an infrared telescope in an aircraft, shown here in a 2009 test

High on the Chajnantor Plateau, the Atacama Large Millimeter Array provides an extraordinary place for infrared astronomy.

Efforts were made starting in the 1830s and continuing through the 19th century to detect infrared radiation from other astronomical sources. Radiation from the Moon was first detected in 1873 by William Parsons, 3rd Earl of Rosse. Ernest Fox Nichols used a modified Crookes radiometer in an attempt to detect infrared radiation from Arcturus and Vega, but Nichols deemed the results inconclusive. Even so, the ratio of flux he reported for the two stars is consistent with the modern value, so George Rieke gives Nichols credit for the first detection of a star other than our own in the infrared.

The field of infrared astronomy continued to develop slowly in the early 20th century, as Seth Barnes Nicholson and Edison Pettit developed thermopile detectors capable of accurate infrared photometry and sensitive to a few hundreds of stars. The field was mostly neglected by traditional astronomers though until the 1960s, with most scientists who practiced infrared astronomy having actually been trained physicists. The success of radio astronomy during the 1950s and 1960s, combined with the improvement of infrared detector technology, prompted more astronomers to take notice, and infrared astronomy became well established as a subfield of astronomy.

Modern Infrared Astronomy

Infrared radiation with wavelengths just longer than visible light, known as near-infrared, behaves in a very similar way to visible light, and can be detected using similar solid state devices. For this reason, the near infrared region of the spectrum is commonly incorporated as part of the "optical" spectrum, along with the near ultraviolet. Many optical telescopes, such as those at Keck Observatory, operate effectively in the near infrared as well as at visible wavelengths. The far-infrared extends to submillimeter wavelengths, which are observed by telescopes such as the James Clerk Maxwell Telescope at Mauna Kea Observatory.

Hubble infrared view of the Tarantula Nebula.

Artist impression of galaxy W2246-0526, a single galaxy glowing in infrared light as intensely as 350 trillion suns.

Like all other forms of electromagnetic radiation, infrared is utilized by astronomers to study the universe. Indeed, infrared measurements taken by the 2MASS and WISE astronomical surveys have been particularly effective at unveiling previously undiscovered star clusters. Examples of such embedded star clusters are FSR 1424, FSR 1432, Camargo 394, Camargo 399, Majaess 30, and Majaess 99. Infrared telescopes, which includes most major optical telescopes as well as a few dedicated infrared telescopes, need to be chilled with liquid nitrogen and shielded from warm objects. The reason for this is that objects with temperatures of a few hundred Kelvin emit most of their thermal energy at infrared wavelengths. If infrared detectors were not kept cooled, the radiation from the detector itself would contribute noise that would dwarf the radiation from any celestial source. This is particularly important in the mid-infrared and far-infrared regions of the spectrum.

ALMA Observatory's antennas appear to take in the sight of the Milky Way.

To achieve higher angular resolution, some infrared telescopes are combined to form astronomical interferometers. The effective resolution of an interferometer is set by the distance between the telescopes, rather than the size of the individual telescopes. When used together with adaptive optics, infrared interferometers, such as two 10 meter telescopes at Keck Observatory or the four 8.2 meter telescopes that make up the Very Large Telescope Interferometer, can achieve high angular resolution.

The principal limitation on infrared sensitivity from ground-based telescopes is the Earth's atmosphere. Water vapor absorbs a significant amount of infrared radiation, and the atmosphere itself emits at infrared wavelengths. For this reason, most infrared telescopes are built in very dry places at high altitude, so that they are above most of the water vapor in the atmosphere. Suitable locations on Earth include Mauna Kea Observatory at 4205 meters above sea level, the Paranal Observatory at 2635 meters in Chile and regions of high altitude ice-desert such as Dome C in Antarctic. Even at high altitudes, the transparency of the Earth's atmosphere is limited except in infrared windows, or wavelengths where the Earth's atmosphere is transparent. The main infrared windows are listed below:

Spectrum	Wavelength (micrometres)	Astronomical bands	Telescopes
Near Infrared	0.65 to 1.0	R and I bands	All major optical telescopes
Near Infrared	1.1 to 1.4	J band	Most major optical telescopes and most dedicated infrared telescopes
Near Infrared	1.5 to 1.8	H band	Most major optical telescopes and most dedicated infrared telescopes
Near Infrared	2.0 to 2.4	K band	Most major optical telescopes and most dedicated infrared telescopes
Near Infrared	3.0 to 4.0	L band	Most dedicated infrared telescopes and some optical telescopes
Near Infrared	4.6 to 5.0	M band	Most dedicated infrared telescopes and some optical telescopes
Mid Infrared	7.5 to 14.5	N band	Most dedicated infrared telescopes and some optical telescopes
Mid Infrared	17 to 25	Q band	Some dedicated infrared telescopes and some optical telescopes
Far Infrared	28 to 40	Z band	Some dedicated infrared telescopes and some optical telescopes
Far Infrared	330 to 370		Some dedicated infrared telescopes and some optical telescopes
Far Infrared	450	submillimeter	Submillimeter telescopes

As is the case for visible light telescopes, space is the ideal place for infrared telescopes. In space, images from infrared telescopes can achieve higher resolution, as they do not suffer from blurring caused by the Earth's atmosphere, and are also free from absorption caused by the Earth's atmosphere. Current infrared telescopes in space include the Herschel Space Observatory, the Spitzer Space Telescope, and the Wide-field Infrared Survey Explorer. Since putting telescopes in orbit is expensive, there are also airborne observatories, such as the Stratospheric Observatory for Infrared Astronomy and the Kuiper Airborne Observatory. These observatories place telescopes above most, but not all, of the atmosphere, which means there is absorption of infrared light from space by water vapor in the atmosphere.

Infrared Technology

One of the most common infrared detector arrays used at research telescopes is HgCdTe arrays. These operate well between 0.6 and 5 micrometre wavelengths. For longer wavelength observations or higher sensitivity other detectors may be used, including other narrow gap semiconductor detectors, low temperature bolometer arrays or photon-counting Superconducting Tunnel Junction arrays.

Special requirements for infrared astronomy include: very low dark currents to allow long integration times, associated low noise readout circuits and sometimes very high pixel counts.

Low temperature is often achieved by a coolant, which can run out. Space missions have either ended or shifted to "warm" observations when the coolant supply used up. For example, WISE ran out of coolant in October 2010, about ten months after being launched.

Neutrino Astronomy

Neutrino telescope

Neutrino astronomy is the branch of astronomy that observes astronomical objects with neutrino detectors in special observatories. Neutrinos are created as a result of certain types of radioactive decay, or nuclear reactions such as those that take place in the Sun, in nuclear reactors, or when cosmic rays hit atoms. Due to their weak interactions with matter, neutrinos offer a unique opportunity to observe processes that are inaccessible to optical telescopes.

The field of neutrino astronomy is still very much in its infancy – the only confirmed extraterrestrial sources so far are the Sun and supernova SN1987A.

History

Neutrinos were first recorded in 1956 by Clyde Cowan and Frederick Reines from a nuclear reactor. Their discovery was acknowledged with a Nobel Prize for physics in 1995.

In 1968, Raymond Davis, Jr. and John N. Bahcall successfully detected the first solar neutrinos in the Homestake experiment. Davis, along with Japanese physicist Masatoshi Koshiba were jointly awarded half of the 2002 Nobel Prize in Physics "for pioneering contributions to astrophysics, in particular for the detection of cosmic neutrinos (the other half went to Riccardo Giacconi for corresponding pioneering contributions which have led to the discovery of cosmic X-ray sources)."

This was followed by the first atmospheric neutrino detection in 1965 by two groups almost simultaneously. One was led by Frederick Reines who operated a liquid scintillator in the East Rand gold mine in South Africa at an 8.8 km water depth equivalent. The other was a Bombay-Osaka-Durham collaboration that operated in the Indian Kolar Gold Field mine at an equivalent water depth of 7.5 km. Although the KGF group detected neutrino candidates two months later than Reines, they were given formal priority due to publishing their findings two weeks earlier.

The first generation of undersea neutrino telescope projects began with the proposal by Moisey Markov in 1960 "...to install detectors deep in a lake or a sea and to determine the location of charged particles with the help of Cherenkov radiation."

The first underwater neutrino telescope began as the DUMAND project. DUMAND stands for Deep Underwater Muon and Neutrino Detector. The project began in 1976 and although it was eventually cancelled in 1995, it acted as a precursor to many of the following telescopes in the following decades.

The Baikal Neutrino Telescope is installed in the southern part of Lake Baikal in Russia. The detector is located at a depth of 1.1 km and began surveys in 1980. In 1993, it was the first to deploy three strings to reconstruct the muon trajectories as well as the first to record atmospheric neutrinos underwater.

AMANDA (Antarctic Muon And Neutrino Detector Array) used the 3 km thick ice layer at the South Pole and was located several hundred meters from the Amundsen-Scott station. Holes 60 cm in diameter were drilled with pressurized hot water in which strings with optical modules were deployed before the water refroze. The depth proved to be insufficient to be able to reconstruct the trajectory due to the scattering of light on air bubbles. A second group of 4 strings were added in 1995/96 to a depth of about 2000 m that was sufficient for track reconstruction. The AMANDA array was subsequently upgraded until January 2000 when it consisted of 19 strings with a total of 667 optical modules at a depth range between 1500 m and 2000 m. AMANDA would eventually be the predecessor to IceCube in 2005.

21st Century

After the decline of DUMAND the participating groups split into three branches to explore deep sea options in the Mediterranean Sea. ANATES was anchored to the sea floor in the region off Toulon at the French Mediterranean coast. It consists of 12 strings, each carrying 25 "storeys" equipped with three optical modules, an electronic container, and calibration devices down to a maximum depth of 2475 m.

NEMO (NEutrino Mediterranean Observatory) was pursued by Italian groups to investigate the feasibility of a cubic-kilometer scale deep-sea detector. A suitable site at a depth of 3.5 km about 100 km off Capo Passero at the South-Eastern coast of Sicily has been identified. From 2007-2011 the first prototyping phase tested a "mini-tower" with 4 bars deployed for several weeks near Catania at a depth of 2 km. The second phase as well as plans to deploy the full-size prototype tower will be pursued in the KM3NeT framework.

The NESTOR Project was installed in 2004 to a depth of 4 km and operated for one month until a failure of the cable to shore forced it to be terminated. The data taken still successfully demonstrated the detector's functionality and provided a measurement of the atmospheric muon flux. The proof of concept will be implemented in the KM3Net framework.

The second generation of deep-sea neutrino telescope projects reach or even exceed the size originally conceived by the DUMAND pioneers. IceCube, located at the South Pole and incorporating its predecessor AMANDA, was completed in December 2010. It currently consists of 5160 digital optical modules installed on 86 strings at depths of 1450 to 2550 m in the Antarctic ice. The KM3NeT in the Mediterranean Sea and the GVD are in their preparatory/prototyping phase. IceCube instruments 1 km^3 of ice. GVD is also planned to cover 1 km^3 but at a much higher energy threshold. KM3NeT is planned to cover several km^3. Both KM3NeT and GVD could be completed by 2017 and it is expected that all three will form a global neutrino observatory.

Detection Methods

Since neutrinos interact only very rarely with matter, the enormous flux of solar neutrinos racing through the Earth is sufficient to produce only 1 interaction for 10^{36} target atoms, and each interaction produces only a few photons or one transmuted atom. The observation of neutrino interactions requires a large detector mass, along with a sensitive amplification system.

Given the very weak signal, sources of background noise must be reduced as much as possible. The detectors must be shielded by a large shield mass, and so are constructed deep underground, or underwater. They record upward going muons in charged current muon neutrino interactions. Upward because no other known particle can traverse the entire Earth. The detector must be at least 1 km deep to suppress downward traveling muons, and are subject to an irreducible background of extraterrestric neutrinos interacting in the Earth's atmosphere. This background also provides a standard calibration source. Sources of radioactive isotopes must also be controlled as they produce energetic particles when they decay. The detectors consist of an array of photomultiplier tubes (PMTs) housed in transparent pressure spheres which are suspended in a large volume of water or ice. The PMTs record the arrival time and amplitude of the Cherenkov light emitted by muons or particle cascades. The trajectory can then usually be reconstructed by triangulation if at least three "strings" are used to detect the events.

Applications

When astronomical bodies, such as the Sun, are studied using light, only the surface of the object can be directly observed. Any light produced in the core of a star will interact with gas particles in the outer layers of the star, taking hundreds of thousands of years to make it to the surface, making it impossible to observe the core directly. Since neutrinos are also created in the cores of stars (as a result of stellar fusion), the core can be observed using neutrino astronomy. Other sources of neutrinos- such as neutrinos released by supernovae- have been detected. There are currently goals to detect neutrinos from other sources, such as Active Galactic Nuclei (AGN), as well as Gamma-ray bursts and Starburst galaxies. Neutrino astronomy may also indirectly detect dark matter.

References

- Lankford, John (1997). "Astrometry". History of astronomy: an encyclopedia. Taylor & Francis. p. 49. ISBN 0-8153-0322-X.

- Kovalevsky, Jean; Seidelmann, P. Kenneth (2004). Fundamentals of Astrometry. Cambridge University Press. pp. 2–3. ISBN 0-521-64216-7.

- Tekeli, Sevim (1997). "Taqi al-Din". Encyclopaedia of the History of Science, Technology, and Medicine in Non-Western Cultures. Kluwer Academic Publishers. ISBN 0-7923-4066-3.

- Lloyd, G.E.R. (1968). Aristotle: The Growth and Structure of His Thought. Cambridge: Cambridge University Press. pp. 134–5. ISBN 0-521-09456-9.

- Galilei, Galileo (1989), Van Helden, Albert, ed., Sidereus Nuncius or The Sidereal Messenger, Chicago: University of Chicago Press, pp. 21, 47, ISBN 0-226-27903-0

- Westfall, Richard S. (1980), Never at Rest: A Biography of Isaac Newton, Cambridge: Cambridge University Press, pp. 731–732, ISBN 0-521-27435-4

- Ladislav Kvasz (2013). "Galileo, Descartes, and Newton – Founders of the Language of Physics" (PDF). Institute of Philosophy, Academy of Sciences of the Czech Republic. Retrieved 2015-07-18.

- Hetherington, Norriss S.; McCray, W. Patrick, Weart, Spencer R., ed., Spectroscopy and the Birth of Astrophysics, American Institute of Physics, Center for the History of Physics, retrieved July 19, 2015

- The science.ca team (2015). "Hubert Reeves – Astronomy, Astrophysics and Space Science". GCS Research Society. Retrieved 2015-07-17.

4

Observational Astronomy and its Branches

Observational astronomy is concerned with recording data; it is the practice of observing objects by using telescopes and other astronomical tools. This chapter is a compilation of the various branches of astronomy such as observational astronomy, gravitational wave astronomy, visible light astronomy and radio astronomy.

Observational Astronomy

Observational astronomy is a division of the astronomical science that is concerned with recording data, in contrast with theoretical astrophysics, which is mainly concerned with finding out the measurable implications of physical models. It is the practice of observing celestial objects by using telescopes and other astronomical apparatus.

Mayall telescope at Kitt Peak National Observatory west

As a science, the study of astronomy is somewhat hindered in that direct experiments with the properties of the distant universe are not possible. However, this is partly compensated by the fact that astronomers have a vast number of visible examples of stellar phenomena that can be examined. This allows for observational data to be plotted on graphs, and general trends recorded. Nearby examples of specific phenomena, such as variable stars, can then be used to infer the behavior of more distant representatives. Those distant yardsticks can then be employed to measure other phenomena in that neighborhood, including the distance to a galaxy.

A gathering to observe the Perseids.

Galileo Galilei turned a telescope to the heavens and recorded what he saw. Since that time, observational astronomy has made steady advances with each improvement in telescope technology.

Subdivisions of Observational Astronomy

A traditional division of observational astronomy is given by the region of the electromagnetic spectrum observed:

- Optical astronomy is the part of astronomy that uses optical components (mirrors, lenses and solid-state detectors) to observe light from near infrared to near ultraviolet wavelengths. Visible-light astronomy (using wavelengths that can be detected with the eyes, about 400 - 700 nm) falls in the middle of this range.

- Infrared astronomy deals with the detection and analysis of infrared radiation (this typically refers to wavelengths longer than the detection limit of silicon solid-state detectors, about 1 μm wavelength). The most common tool is the reflecting telescope but with a detector sensitive to infrared wavelengths. Space telescopes are used at certain wavelengths where the atmosphere is opaque, or to eliminate noise (thermal radiation from the atmosphere).

- Radio astronomy detects radiation of millimetre to decametre wavelength. The receivers are similar to those used in radio broadcast transmission but much more sensitive.

- High-energy astronomy includes X-ray astronomy, gamma-ray astronomy, and extreme UV astronomy.

Methods

In addition to using electromagnetic radiation, modern astrophysicists can also make observations using neutrinos, cosmic rays or gravitational waves. Observing a source using multiple methods is known as multi-messenger astronomy.

Optical and radio astronomy can be performed with ground-based observatories, because the atmosphere is relatively transparent at the wavelengths being detected. Observatories are usually located at high altitudes so as to minimise the absorption and distortion caused by the Earth's atmosphere. Some wavelengths of infrared light are heavily absorbed by water vapor, so many infrared observatories are located in dry places at high altitude, or in space.

Ultra HD photography taken at La Silla Observatory.

The atmosphere is opaque at the wavelengths used by X-ray astronomy, gamma-ray astronomy, UV astronomy and (except for a few wavelength "windows") far infrared astronomy, so observations must be carried out mostly from balloons or space observatories. Powerful gamma rays can, however be detected by the large air showers they produce, and the study of cosmic rays is a rapidly expanding branch of astronomy.

Important Factors

Sunset over Mauna Kea Observatories.

For much of the history of observational astronomy, almost all observation was performed in the visual spectrum with optical telescopes. While the Earth's atmosphere is relatively transparent in this portion of the electromagnetic spectrum, most telescope work is still dependent on seeing conditions and air transparency, and is generally restricted to the night time. The seeing conditions depend on the turbulence and thermal variations in the air. Locations that are frequently cloudy or suffer from atmospheric turbulence limit the resolution of observations. Likewise the presence of the full Moon can brighten up the sky with scattered light, hindering observation of faint objects.

For observation purposes, the optimal location for an optical telescope is undoubtedly in outer space. There the telescope can make observations without being affected by the atmosphere. However, at present it remains costly to lift telescopes into orbit. Thus the next best locations are certain mountain peaks that have a high number of cloudless days and generally possess good atmospheric conditions (with good seeing conditions). The peaks of the islands of Mauna Kea, Hawaii and La Palma possess these properties, as to a lesser extent do inland sites such as Llano de Chajnantor, Paranal, Cerro Tololo and La Silla in Chile. These observatory locations have attracted an assemblage of powerful telescopes, totalling many billion US dollars of investment.

The darkness of the night sky is an important factor in optical astronomy. With the size of cities and human populated areas ever expanding, the amount of artificial light at night has also increased. These artificial lights produce a diffuse background illumination that makes observation of faint astronomical features very difficult without special filters. In a few locations such as the state of Arizona and in the United Kingdom, this has led to campaigns for the reduction of light pollution. The use of hoods around street lights not only improves the amount of light directed toward the ground, but also helps reduce the light directed toward the sky.

Atmospheric effects (astronomical seeing) can severely hinder the resolution of a telescope. Without some means of correcting for the blurring effect of the shifting atmosphere, telescopes larger than about 15–20 cm in aperture can not achieve their theoretical resolution at visible wavelengths. As a result, the primary benefit of using very large telescopes has been the improved light-gathering capability, allowing very faint magnitudes to be observed. However the resolution handicap has begun to be overcome by adaptive optics, speckle imaging and interferometric imaging, as well as the use of space telescopes.

Measuring Results

Astronomers have a number of observational tools that they can use to make measurements of the heavens. For objects that are relatively close to the Sun and Earth, direct and very precise position measurements can be made against a more distant (and thereby nearly stationary) background. Early observations of this nature were used to develop very precise orbital models of the various planets, and to determine their respective masses and gravitational perturbations. Such measurements led to the discovery of the planets Uranus, Neptune, and (indirectly) Pluto. They also resulted in an erroneous assumption of a fictional planet Vulcan within the orbit of Mercury (but the explanation of the precession of Mercury's orbit by Einstein is considered one of the triumphs of his general relativity theory).

Developments and Diversity

In addition to examination of the universe in the optical spectrum, astronomers have increasingly been able to acquire information in other portions of the electromagnetic spectrum. The earliest such non-optical measurements were made of the thermal properties of the Sun. Instruments employed during a solar eclipse could be used to measure the radiation from the corona.

Fully-steerable radio telescope in Green Bank, West Virginia.

Radio Astronomy

With the discovery of radio waves, radio astronomy began to emerge as a new discipline in astronomy. The long wavelengths of radio waves required much larger collecting dishes in order to make images with good resolution, and later led to the development of the multi-dish interferometer for making high-resolution aperture synthesis radio images (or "radio maps"). The development of the microwave horn receiver led to the discovery of the microwave background radiation associated with the big bang.

Radio astronomy has continued to expand its capabilities, even using radio astronomy satellites to produce interferometers with baselines much larger than the size of the Earth. However, the ever-expanding use of the radio spectrum for other uses is gradually drowning out the faint radio signals from the stars. For this reason, in the future radio astronomy might be performed from shielded locations, such as the far side of the Moon.

Late 20th Century Developments

The last part of the twentieth century saw rapid technological advances in astronomical instrumentation. Optical telescopes were growing ever larger, and employing adaptive optics to partly negate atmospheric blurring. New telescopes were launched into space, and began observing the universe in the infrared, ultraviolet, x-ray, and gamma ray parts of the electromagnetic spectrum, as well as observing cosmic rays. Interferometer arrays produced the first extremely high-resolution images using aperture synthesis at radio, infrared and optical wavelengths. Orbiting instruments such as the Hubble Space Telescope produced rapid advances in astronomical knowledge, acting as the workhorse for visible-light observations of faint objects. New space instruments under development are expected to directly observe planets around other stars, perhaps even some Earth-like worlds.

In addition to telescopes, astronomers have begun using other instruments to make observations.

Other Instruments

Neutrino astronomy is the branch of astronomy that observes astronomical objects with neutrino detectors in special observatories, usually huge underground tanks. Nuclear reactions in stars and supernova explosions produce very large numbers of neutrinos, a very few of which may be detected by a neutrino telescope. Neutrino astronomy is motivated by the possibility of observing processes that are inaccessible to optical telescopes, such as the Sun's core.

Gravitational wave detectors are being designed that may capture events such as collisions of massive objects such as neutron stars or black holes.

Robotic spacecraft are also being increasingly used to make highly detailed observations of planets within the solar system, so that the field of planetary science now has significant cross-over with the disciplines of geology and meteorology.

Observation Tools

Skalnaté pleso observatory, Slovakia.

One of the Oldest Observatories in South America is the Quito Astronomical Observatory, founded in 1873 and located 12 minutes south of the Equator in Quito, Ecuador. The Quito Astronomical Observatory is the National Observatory of Ecuador and is located in the Historic Center of Quito and is managed by the National Polytechnic School.

Telescopes

The key instrument of nearly all modern observational astronomy is the telescope. This serves the dual purposes of gathering more light so that very faint objects can be observed, and magnifying the image so that small and distant objects can be observed. Optical astronomy requires telescopes that use optical components of great precision. Typical requirements for grinding and polishing a curved mirror, for example, require the surface to be within a fraction of a wavelength of light of a particular conic shape. Many modern "telescopes" actually consist of arrays of telescopes working together to provide higher resolution through aperture synthesis.

An amateur astrophotography setup with an automated guide system connected to a laptop.

Large telescopes are housed in domes, both to protect them from the weather and to stabilize the environmental conditions. For example, if the temperature is different from one side of the telescope to the other, the shape of the structure changes, due to thermal expansion pushing optical elements out of position. This can affect the image. For this reason, the domes are usually bright white (titanium dioxide) or unpainted metal. Domes are often opened around sunset, long before observing can begin, so that air can circulate and bring the entire telescope to the same temperature as the surroundings. To prevent wind-buffet or other vibrations affecting observations, it is standard practice to mount the telescope on a concrete pier whose foundations are entirely separate from those of the surrounding dome and building.

To do almost any scientific work requires that telescopes track objects as they wheel across the visible sky. In other words, they must smoothly compensate for the rotation of the Earth. Until the advent of computer controlled drive mechanisms, the standard solution was some form of equatorial mount, and for small telescopes this is still the norm. However, this is a structurally poor design and becomes more and more cumbersome as the diameter and weight of the telescope increases. The world's largest equatorial mounted telescope is the 200 inch (5.1 m) Hale Telescope, whereas recent 8–10 m telescopes use the structurally better altazimuth mount, and are actually physically *smaller* than the Hale, despite the larger mirrors. As of 2006, there are design projects underway for gigantic alt-az telescopes: the Thirty Metre Telescope , and the 100 m diameter Overwhelmingly Large Telescope.

Amateur astronomers use such instruments as the Newtonian reflector, the Refractor and the increasingly popular Maksutov telescope.

Photography

The photograph has served a critical role in observational astronomy for over a century, but in the last 30 years it has been largely replaced for imaging applications by digital sensors such as CCDs and CMOS chips. Specialist areas of astronomy such as photometry and interferometry have utilised electronic detectors for a much longer period of time. Astrophotography uses specialised photographic film (or usually a glass plate coated with photographic emulsion), but there are a number of drawbacks, particularly a low quantum efficiency, of the order of 3%, whereas CCDs can be tuned for a QE >90% in a narrow band. Almost all modern telescope instruments are electronic arrays, and older telescopes have been either been retrofitted with these instruments or closed down. Glass plates are still used in some applications, such as surveying, because the resolution possible with a chemical film is much higher than any electronic detector yet constructed.

Advantages

Prior to the invention of photography, all astronomy was done with the naked eye. However, even before films became sensitive enough, scientific astronomy moved entirely to film, because of the overwhelming advantages:

- The human eye discards what it sees from split-second to split-second, but photographic film gathers more and more light for as long as the shutter is open.

- The resulting image is permanent, so many astronomers can use the same data.

- It is possible to see objects as they change over time (SN 1987A is a spectacular example).

Blink Comparator

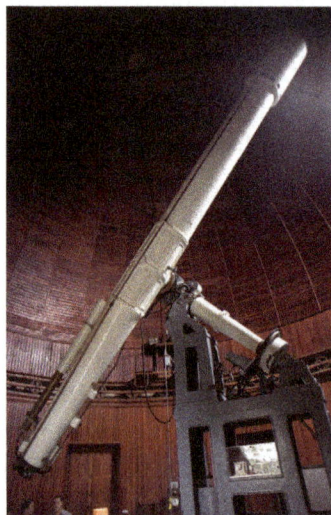

50 cm refracting telescope at Nice Observatory.

The blink comparator is an instrument that is used to compare two nearly identical photographs made of the same section of sky at different points in time. The comparator alternates illumination of the two plates, and any changes are revealed by blinking points or streaks. This instrument has been used to find asteroids, comets, and variable stars.

Micrometer

The position or cross-wire micrometer is an implement that has been used to measure double stars. This consists of a pair of fine, movable lines that can be moved together or apart. The telescope lens is lined up on the pair and oriented using position wires that lie at right angles to the star separation. The movable wires are then adjusted to match the two star positions. The separation of the stars is then read off the instrument, and their true separation determined based on the magnification of the instrument.

Spectrograph

A vital instrument of observational astronomy is the spectrograph. The absorption of specific wavelengths of light by elements allows specific properties of distant bodies to be observed. This capability has resulted in the discovery of the element of helium in the Sun's emission spectrum, and has allowed astronomers to determine a great deal of information concerning distant stars, galaxies, and other celestial bodies. Doppler shift (particularly "redshift") of spectra can also be used to determine the radial motion or distance with respect to the Earth.

Early spectrographs employed banks of prisms that split light into a broad spectrum. Later the grating spectrograph was developed, which reduced the amount of light loss compared to prisms and provided higher spectral resolution. The spectrum can be photographed in a long exposure, allowing the spectrum of faint objects (such as distant galaxies) to be measured.

Stellar photometry came into use in 1861 as a means of measuring stellar colors. This technique measured the magnitude of a star at specific frequency ranges, allowing a determination of the overall color, and therefore temperature of a star. By 1951 an internationally standardized system of UBV-magnitudes (*U*ltraviolet-*B*lue-*V*isual) was adopted.

Photoelectric Photometry

Photoelectric photometry using the CCD is now frequently used to make observations through a telescope. These sensitive instruments can record the image nearly down to the level of individual photons, and can be designed to view in parts of the spectrum that are invisible to the eye. The ability to record the arrival of small numbers of photons over a period of time can allow a degree of computer correction for atmospheric effects, sharpening up the image. Multiple digital images can also be combined to further enhance the image. When combined with the adaptive optics technology, image quality can approach the theoretical resolution capability of the telescope.

Filters are used to view an object at particular frequencies or frequency ranges. Multilayer film filters can provide very precise control of the frequencies transmitted and blocked, so that, for example, objects can be viewed at a particular frequency emitted only by excited hydrogen atoms. Filters can also be used to partially compensate for the effects of light pollution by blocking out unwanted light. Polarization filters can also be used to determine if a source is emitting polarized light, and the orientation of the polarization.

Observing

Astronomers observe a wide range of astronomical sources, including high-redshift galaxies, AGNs, the afterglow from the Big Bang and many different types of stars and protostars.

The main platform at La Silla hosts a huge range of telescopes with which astronomers can explore the Universe.

A variety of data can be observed for each object. The position coordinates locate the object on the sky using the techniques of spherical astronomy, and the magnitude determines its brightness as seen from the Earth. The relative brightness in different parts of the spectrum yields information about the temperature and physics of the object. Photographs of the spectra allow the chemistry of the object to be examined.

Parallax shifts of a star against the background can be used to determine the distance, out to a limit imposed by the resolution of the instrument. The radial velocity of the star and changes in its position over time (proper motion) can be used to measure its velocity relative to the Sun. Variations in the brightness of the star give evidence of instabilities in the star's atmosphere, or else the presence of an occulting companion. The orbits of binary stars can be used to measure the relative masses of each companion, or the total mass of the system. Spectroscopic binaries can be found by observing doppler shifts in the spectrum of the star and its close companion.

Stars of identical masses that formed at the same time and under similar conditions typically have nearly identical observed properties. Observing a mass of closely associated stars, such as in a globular cluster, allows data to be assembled about the distribution of stellar types. These tables can then be used to infer the age of the association.

For distant galaxies and AGNs observations are made of the overall shape and properties of the galaxy, as well as the groupings where they are found. Observations of certain types of variable stars and supernovae of known luminosity, called standard candles, in other galaxies allows the inference of the distance to the host galaxy. The expansion of space causes the spectra of these galaxies to be shifted, depending on the distance, and modified by the Doppler effect of the galaxy's radial velocity. Both the size of the galaxy and its redshift can be used to infer something about the distance of the galaxy. Observations of large numbers of galaxies are referred to as redshift surveys, and are used to model the evolution of galaxy forms.

Gravitational-wave Astronomy

Gravitational-wave astronomy is an emerging branch of observational astronomy which aims to use gravitational waves (minute distortions of spacetime predicted by Einstein's theory of general relativity) to collect observational data about objects such as neutron stars and black holes, events such as supernovae, and processes including those of the early universe shortly after the Big Bang.

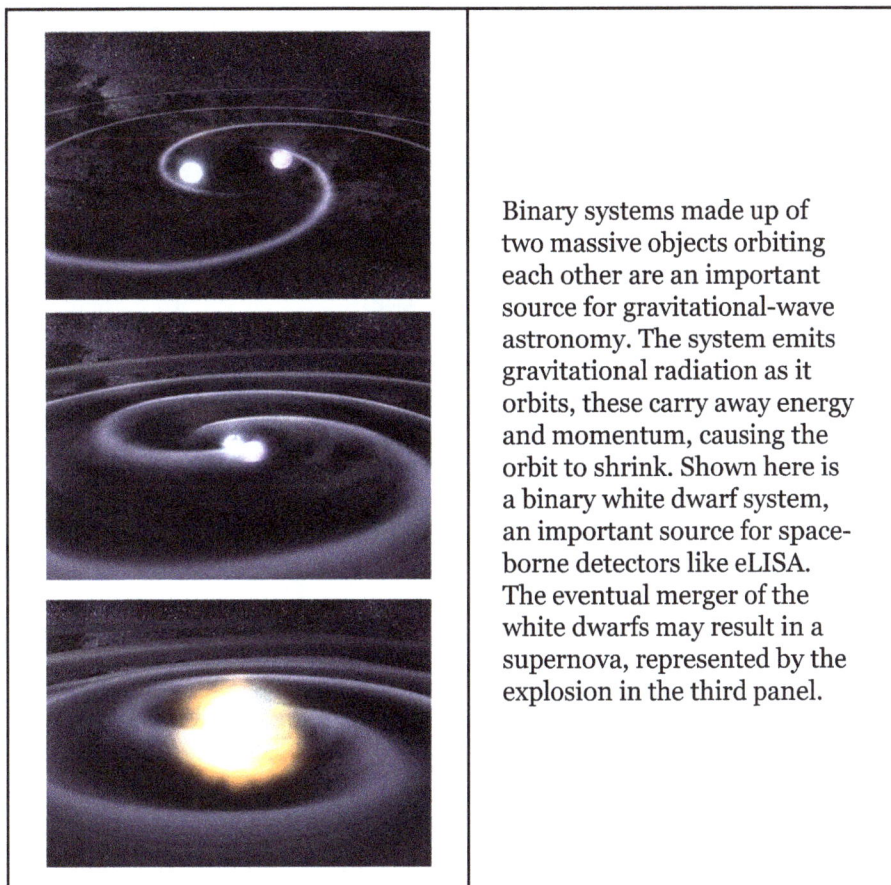

Binary systems made up of two massive objects orbiting each other are an important source for gravitational-wave astronomy. The system emits gravitational radiation as it orbits, these carry away energy and momentum, causing the orbit to shrink. Shown here is a binary white dwarf system, an important source for space-borne detectors like eLISA. The eventual merger of the white dwarfs may result in a supernova, represented by the explosion in the third panel.

Gravitational waves have a solid theoretical basis, founded upon the theory of relativity. They were first predicted by Einstein in 1916; although a specific consequence of general relativity, they are a common feature of all theories of gravity that obey special relativity. Indirect observational evidence for their existence first came in 1974 from measurements of the Hulse–Taylor binary pulsar, whose orbit evolves exactly as would be expected for gravitational wave emission. Richard Hulse and Joseph Taylor were awarded the 1993 Nobel Prize in Physics for this discovery. Subsequently, many other binary pulsars (including one double pulsar system) have been observed, all fitting gravitational-wave predictions.

On 11 February 2016 it was announced that LIGO had directly observed the first gravitational waves in September 2015. The second observation of gravitational waves was made on 26 December 2015 and announced on 15 June 2016.

Observations

Ordinary gravitational waves frequencies are very low and much harder to detect, while higher frequencies occur in more dramatic events and thus have become the first to be observed.

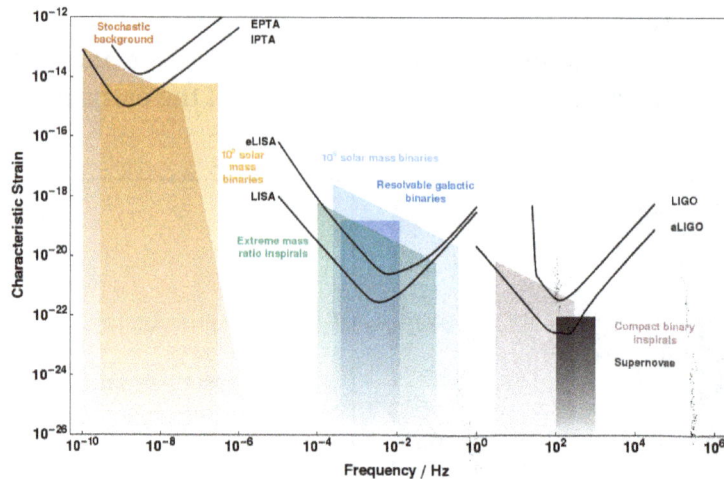

Noise curves for a selection of gravitational-wave detectors as a function of frequency. At very low frequencies are pulsar timing arrays, the European Pulsar Timing Array (EPTA) and the future International Pulsar Timing Array (IPTA); at low frequencies are space-borne detectors, the formerly proposed Laser Interferometer Space Antenna (LISA) and the currently proposed evolved Laser Interferometer Space Antenna (eLISA), and at high frequencies are ground-based detectors, the initial Laser Interferometer Gravitational-Wave Observatory (LIGO) and its advanced configuration (aLIGO). The characteristic strain of potential astrophysical sources are also shown. To be detectable the characteristic strain of a signal must be above the noise curve.

High Frequency

In 2015, the LIGO project was the first to directly observe gravitational waves using laser interferometers. The LIGO detectors observed gravitational waves from the merger of two stellar-mass black holes, matching predictions of general relativity. These observations demonstrated the existence of binary stellar-mass black hole systems, and were the first direct detection of gravitational waves and the first observation of a binary black hole merger. This finding has been characterized as revolutionary to science, because the verification of our ability to use gravitational-wave astronomy to progress in our search and exploration of dark matter and the big bang.

There are several current scientific collaborations for observing gravitational waves. There is a world-wide network of ground-based detectors, these are kilometre-scale laser interferometers including: the Laser Interferometer Gravitational-Wave Observatory (LIGO), a joint project between MIT, Caltech and the scientists of the LIGO Scientific Collaboration with detectors in Livingston, Louisiana and Hanford, Washington; Virgo, at the European Gravitational Observatory, Cascina, Italy; GEO600 in Sarstedt, Germany, and the Kamioka Gravitational Wave Detector (KAGRA), operated by the University of Tokyo in the Kamioka Observatory, Japan. LIGO and Virgo are currently being upgraded to their advanced configurations. Advanced LIGO began observations in 2015, detecting gravitational waves even though not having reached its design sensitivity yet; Advanced Virgo is expected to start observing in 2016. The more advanced KAGRA is scheduled for 2018. GEO600 is currently operational, but its sensitivity makes it unlikely to make an observation; its primary purpose is to trial technology.

Low Frequency

An alternative means of observation is using pulsar timing arrays (PTAs). There are three consortia, the European Pulsar Timing Array (EPTA), the North American Nanohertz Observatory for

Gravitational Waves (NANOGrav), and the Parkes Pulsar Timing Array (PPTA), which co-operate as the International Pulsar Timing Array. These use existing radio telescopes, but since they are sensitive to frequencies in the nanohertz range, many years of observation are needed to detect a signal and detector sensitivity improves gradually. Current bounds are approaching those expected for astrophysical sources.

Intermediate Frequencies

Further in the future, there is the possibility of space-borne detectors. The European Space Agency has selected a gravitational-wave mission for its L3 mission, due to launch 2034, the current concept is the evolved Laser Interferometer Space Antenna (eLISA). Also in development is the Japanese Deci-hertz Interferometer Gravitational wave Observatory (DECIGO).

Science Potential

Astronomy has traditionally relied on electromagnetic radiation. Astronomy originated with visible-light astronomy and what could be seen with the naked eye. As technology advanced, it became possible to observe other parts of the electromagnetic spectrum, from radio to gamma rays. Each new frequency band gave a new perspective on the Universe and heralded new discoveries. Late in the 20th century, the detection of solar neutrinos founded the field of neutrino astronomy, giving an insight into previously invisible phenomena, such as the inner workings of the Sun. The observation of gravitational waves provides a further means of making astrophysical observations.

Gravitational waves provide complementary information to that provided by other means. By combining observations of a single event made using different means, it is possible to gain a more complete understanding of the source's properties. This is known as multi-messenger astronomy. Gravitational waves can also be used to observe systems that are invisible (or almost impossible to detect) to measure by any other means, for example, they provide a unique method of measuring the properties of black holes.

Gravitational waves can be emitted by many systems, but, to produce detectable signals, the source must consist of extremely massive objects moving at a significant fraction of the speed of light. The main source is a binary of two compact objects. Example systems include:

- Compact binaries made up of two closely orbiting stellar-mass objects, such as white dwarfs, neutron stars or black holes. Wider binaries, which have lower orbital frequencies, are a source for detectors like LISA. Closer binaries produce a signal for ground-based detectors like LIGO. Ground-based detectors could potentially detect binaries containing an intermediate mass black hole of several hundred solar masses.

- Supermassive black hole binaries, consisting of two black holes with masses of 10^5–10^9 solar masses. Supermassive black holes are found at the centre of galaxies. When galaxies merge, it is expected that their central supermassive black holes merge too. These are potentially the loudest gravitational-wave signals. The most massive binaries are a source for PTAs. Less massive binaries (about a million solar masses) are a source for space-borne detectors like LISA.

- Extreme-mass-ratio systems of a stellar-mass compact object orbiting a supermassive black hole. These are sources for detectors like LISA. Systems with highly eccentric orbits produce a burst of gravitational radiation as they pass through the point of closest approach; systems with near-circular orbits, which are expected towards the end of the inspiral, emit continuously within LISA's frequency band. Extreme-mass-ratio inspirals can be observed over many orbits. This makes them excellent probes of the background spacetime geometry, allowing for precision tests of general relativity.

In addition to binaries, there are other potential sources:

- Supernovae generate high-frequency bursts of gravitational waves that could be detected with LIGO or Virgo.

- Rotating neutron stars are a source of continuous high-frequency waves if they possess axial asymmetry.

- Early universe processes, such as inflation or a phase transition.

- Cosmic strings could also emit gravitational radiation if they do exist. Discovery of these gravitational waves would confirm the existence of cosmic strings.

Gravitational waves interact only weakly with matter. This is what makes them difficult to detect. It also means that they can travel freely through the Universe, and are not absorbed or scattered like electromagnetic radiation. It is therefore possible to see to the center of dense systems, like the cores of supernovae or the Galactic Centre. It is also possible to see further back in time than with electromagnetic radiation, as the early universe was opaque to light prior to recombination, but transparent to gravitational waves.

The ability of gravitational waves to move freely through matter also means that gravitational-wave detectors, unlike telescopes, are not pointed to observe a single field of view but observe the entire sky. Detectors are more sensitive in some directions than others, which is one reason why it is beneficial to have a network of detectors.

In Cosmic Inflation

Cosmic inflation, a hypothesized period when the universe rapidly expanded 10^{-36} s after the Big Bang, would have given rise to gravitational waves; they would have left a characteristic imprint in the polarization of the CMB radiation. It is possible to calculate the properties of the primordial gravitational waves from measurements of the patterns in the microwave radiation, and use this to learn about the early universe. Again, the gravitational waves are not directly detected, but their presence must be inferred from other astronomical techniques.

Development

As a young area of research, gravitational-wave astronomy is still in development; however, there is consensus within the astrophysics community that this field will evolve to become an established component of 21st century multi-messenger astronomy.

Gravitational-wave observations complement observations in the electromagnetic spectrum. These waves also promise to yield information in ways not possible via detection and analysis of electromagnetic waves. Electromagnetic waves can be absorbed and re-radiated in ways that make extracting information about the source difficult. Gravitational waves, however, only interact weakly with matter, meaning that they are not scattered or absorbed. This should allow astronomers to view the center of a supernova, stellar nebulae, and even colliding galactic cores in new ways.

The LIGO Hanford Control Room

Ground-based detectors yield new information about the inspiral phase and mergers of binary stellar mass black holes, and binaries consisting of one such black hole and a neutron star (a candidate mechanism for some gamma ray bursts). They could also detect signals from core-collapse supernovae, and from periodic sources such as pulsars with small deformations. If there is truth to speculation about certain kinds of phase transitions or kink bursts from long cosmic strings in the very early universe (at cosmic times around 10^{-25} seconds), these could also be detectable. Space-based detectors like LISA should detect objects such as binaries consisting of two white dwarfs, and AM CVn stars (a white dwarf accreting matter from its binary partner, a low-mass helium star), and also observe the mergers of supermassive black holes and the inspiral of smaller objects (between one and a thousand solar masses) into such black holes. LISA should also be able to listen to the same kind of sources from the early universe as ground-based detectors, but at even lower frequencies and with greatly increased sensitivity.

Detecting emitted gravitational waves is a difficult endeavor. It involves ultra stable high quality lasers and detectors calibrated with a sensitivity of at least $2 \cdot 10^{-22}$ $Hz^{-1/2}$ as shown at the ground-based detector, GEO600. It has also been proposed that even from large astronomical events, such as supernova explosions, these waves are likely to degrade to vibrations as small as an atomic diameter.

Submillimetre Astronomy

Submillimetre astronomy or submillimeter astronomy is the branch of observational astronomy that is conducted at submillimetre wavelengths (i.e., terahertz radiation) of the electromagnetic spectrum. Astronomers place the submillimetre waveband between the far-infrared and microwave wavebands, typically taken to be between a few hundred micrometres and a millimetre. It is still common in submillimetre astronomy to quote wavelengths in 'microns', the old name for micrometre.

The Caltech Submillimeter Observatory at Mauna Kea Observatory was commissioned in 1988, and has a 10.4 m(34 ft) dish

Using submillimetre observations, astronomers examine molecular clouds and dark cloud cores with a goal of clarifying the process of star formation from earliest collapse to stellar birth. Submillimetre observations of these dark clouds can be used to determine chemical abundances and cooling mechanisms for the molecules which comprise them. In addition, submillimetre observations give information on the mechanisms for the formation and evolution of galaxies.

Submillimetre astronomy from the ground

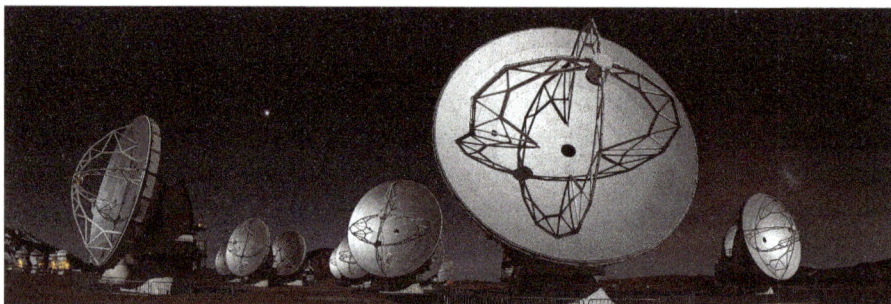

Panoramic view of the Chajnantor plateau, spanning about 180 degrees from north (on the left) to south (on the right) shows the antennas of the Atacama Large Millimeter Array.

The most significant limitation to the detection of astronomical emission at submillimetre wavelengths with ground based observatories is atmospheric emission, noise and attenuation. Like the infrared, the submillimetre atmosphere is dominated by numerous water vapour absorption bands and it is only through "windows" between these bands that observations are possible. The ideal submillimetre observing site is dry, cool, has stable weather conditions and is away from urban population centres. There are only a handful of such sites identified, they include Mauna Kea (Hawaii, United States), the Llano de Chajnantor Observatory on the Atacama Plateau (Chile), the South Pole, and Hanle in India (the Himalayan site of the Indian Astronomical Observatory). Comparisons show that all four sites are excellent for submillimetre astronomy, and of these sites Mauna Kea is the most established and arguably the most accessible. There has been some recent interest in high-altitude Arctic sites, particularly Summit Station in Greenland where the PWV (precipitable water vapor) measure is always better than at Mauna Kea (however Mauna Kea's equatorial latitude of 19 degrees means it can observe more of the southern skies than Greenland).

The Llano de Chajnantor Observatory site hosts the Atacama Pathfinder Experiment (APEX), the largest submillimetre telescope operating in the southern hemisphere, and the world's largest ground based astronomy project, the Atacama Large Millimeter Array (ALMA), an interferometer for submillimetre wavelength observations made of 54 12-metre and 12 7-metre radio telescopes. The Submillimeter Array (SMA) is another interferometer, located at Mauna Kea, consisting of eight 6-metre diameter radio telescopes. The largest existing submillimetre telescope, the James Clerk Maxwell Telescope, is also located on Mauna Kea.

Submillimetre Astronomy from Near-Space

With high-altitude balloons and aircraft, one can get above even more of the atmosphere. The BLAST experiment and SOFIA are two examples, respectively, although SOFIA can also handle near infrared observations.

Submillimetre Astronomy from Space

Comparison			
Name	Year	Wavelength	Aperture
Human Eye	-	0.39-0.75 μm	0.01 m
SWAS	1998	540 - 610 μm	0.55 - 0.7
Herschel	2009	55-672 μm	3.5 m

Space-based observations at the submillimetre wavelengths remove the ground-based limitations of atmospheric absorption. The Submillimeter Wave Astronomy Satellite (SWAS) was launched into low Earth orbit on December 5, 1998 as one of NASA's Small Explorer Program (SMEX) missions. The mission of the spacecraft is to make targeted observations of giant molecular clouds and dark cloud cores. The focus of SWAS is five spectral lines: water (H_2O), isotopic water ($H_2^{18}O$), isotopic carbon monoxide (^{13}CO), molecular oxygen (O_2), and neutral carbon (C I).

The SWAS satellite was repurposed in June, 2005 to provide support for the NASA *Deep Impact* mission. SWAS provided water production data on the comet until the end of August 2005.

The European Space Agency launched a space-based mission known as the Herschel Space Observatory (formerly called Far Infrared and Sub-millimetre Telescope or FIRST) in 2009. Herschel deploys the largest mirror ever launched into space and studies radiation in the far infrared and submillimetre wavebands. Rather than an Earth orbit, Herschel entered into a Lissajous orbit around L_2, the second Lagrangian point of the Earth-Sun system. L_2 is located approximately 1.5 million km from Earth and the placement of Herschel there lessens the interference by infrared and visible radiation from the Earth and Sun. Herschel's mission focuses primarily on the origins of galaxies and galactic formation.

Visible-light Astronomy

Visible-light astronomy encompasses a wide variety of observations via telescopes that are sensitive in the range of visible light (optical telescopes). It includes *imaging*, where a picture of some sort is made of the object; *photometry*, where the amount of light coming from an object is measured, *spectroscopy*, where the distribution of that light with respect to its wavelength is measured, and *polarimetry* where the polarisation state of that light is measured. An example of spectroscopy is the study of spectral lines to understand of what kind of matter light is going through. Visible astronomy also includes looking up at night (skygazing). Visible-light astronomy is part of optical astronomy, and differs from astronomies based on invisible types of light in the electromagnetic radiation spectrum, such as radio waves, infrared waves, ultraviolet waves, X-ray waves and gamma-ray waves.

Beginning

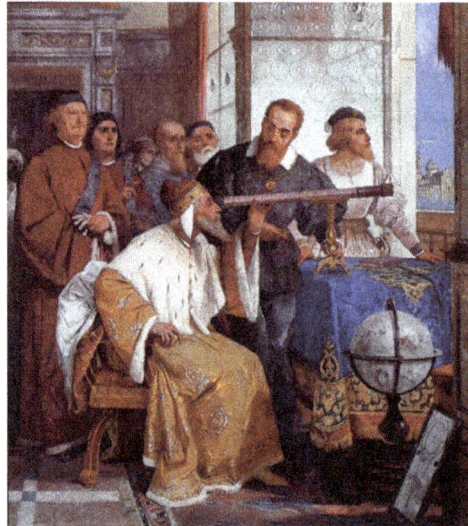

Fresco by Giuseppe Bertini depicting Galileo showing the Doge of Venice how to use the telescope

Based only on uncertain descriptions of the first practical telescope, invented by Hans Lippershey in the Netherlands in 1608, Galileo, in the following year, made a telescope with about 3x magnification. He later made improved versions with up to about 30x magnification. With a Galilean telescope the observer could see magnified, upright images on the earth—it was what is commonly known as a terrestrial telescope or a spyglass. He could also use it to observe the sky; for a time he was one of those who could construct telescopes good enough for that purpose. On 25 August 1609, he demonstrated one of his early telescopes, with a magnification of about 8 or 9, to Venetian lawmakers. His telescopes were also a profitable sideline for Galileo selling them to merchants who found them useful both at sea and as items of trade. He published his initial telescopic astronomical observations in March 1610 in a brief treatise entitled *Sidereus Nuncius* (*Starry Messenger*).

Effect of Ambient Brightness

The visibility of celestial objects in the night sky is affected by light pollution. The presence of the Moon in the night sky has historically hindered astronomical observation by increasing the amount of ambient lighting. With the advent of artificial light sources, however, light pollution has been a growing problem for viewing the night sky. Special filters and modifications to light fixtures can help to alleviate this problem, but for the best views, both professional and amateur optical astronomers seek viewing sites located far from major urban areas.

Ultraviolet Astronomy

A GALEX image of the spiral galaxy Messier 81 in ultraviolet light. Credit:GALEX/NASA/JPL-Caltech.

Ultraviolet astronomy is the observation of electromagnetic radiation at ultraviolet wavelengths between approximately 10 and 320 nanometres; shorter wavelengths—higher energy photons— are studied by X-ray astronomy and gamma ray astronomy. Light at these wavelengths is absorbed by the Earth's atmosphere, so observations at these wavelengths must be performed from the upper atmosphere or from space.

Ultraviolet line spectrum measurements are used to discern the chemical composition, densities, and temperatures of the interstellar medium, and the temperature and composition of hot young stars. UV observations can also provide essential information about the evolution of galaxies.

The ultraviolet Universe looks quite different from the familiar stars and galaxies seen in visible light. Most stars are actually relatively cool objects emitting much of their electromagnetic radiation in the visible or near-infrared part of the spectrum. Ultraviolet radiation is the signature of hotter objects, typically in the early and late stages of their evolution. If we could see the sky in ultraviolet light, most stars would fade in prominence. We would see some very young massive stars and some very old stars and galaxies, growing hotter and producing higher-energy radiation near their birth or death. Clouds of gas and dust would block our vision in many directions along the Milky Way.

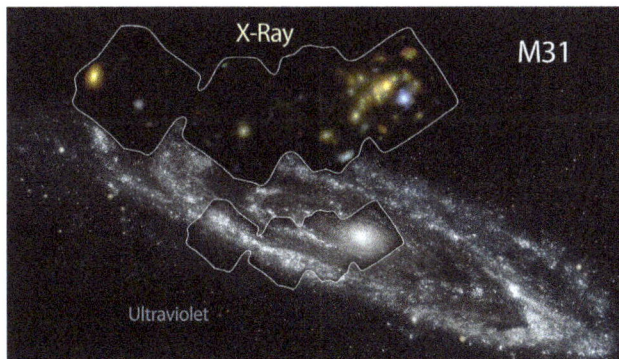

Andromeda Galaxy - in high-energy X-ray and ultraviolet light (released 5 January 2016).

The Hubble Space Telescope and FUSE have been the most recent major space telescopes to view the near and far UV spectrum of the sky, though other UV instruments have flown on sounding rockets and the Space Shuttle.

Charles Stuart Bowyer is generally given credit for starting this field.

Ultraviolet Space Telescopes

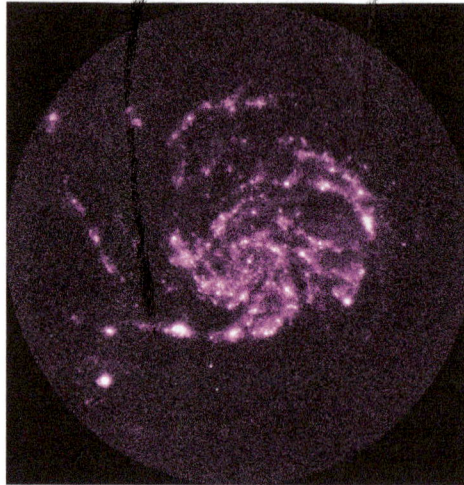

Astro 2 UIT captures M101 with ultraviolet shown in purple

- 🇺🇸- Far Ultraviolet Camera/Spectrograph on Apollo 16 (April 1972)

- 🇺🇸+ ESRO - TD-1A (135-286 nm; 1972–74)

- 🇺🇸- Orbiting Astronomical Observatory (#2:1968-73. #3:1972-81)

- 🟧- Orion 1 and Orion 2 Space Observatories (#1:1971; 200-380 nm spectra; #2:1973; 200-300 nm spectra)

- 🇺🇸+ 🟧- Astronomical Netherlands Satellite (150-330 nm, 1974–76)

- 🇺🇸+ ESA - International Ultraviolet Explorer (115-320 nm spectra, 1978–96)

- 🟥- Astron-1 (1983–89; 150-350 nm)

- 🟥- Glazar 1 & 2 on Mir (in Kvant-1, 1987-2001)

- 🇺🇸- EUVE (7-76 nm, 1992-2001)

- 🇺🇸- FUSE (90.5-119.5 nm, 1999-2007)

- 🇺🇸+ ESA - Extreme ultraviolet Imaging Telescope (on SOHO imaging sun at 17.1, 19.5, 28.4, and 30.4 nm)

- 🇺🇸- GALEX (135-280 nm, 2003-2013)

- 🇺🇸+ ESA - Hubble Space Telescope (Hubble STIS 1997–115–1030 nm) (Hubble WFC3 2009–200-1700 nm)

- 🇺🇸- Swift Gamma-Ray Burst Mission (170–650 nm spectra, 2004--)

- 🇺🇸- Hopkins Ultraviolet Telescope (flew in 1990 and 1995)

- 🇩🇪- ROSAT XUV (17-210eV) (30-6 nm, 1990-1999)

- 🇩🇪- Public Telescope (PST) (100-180 nm, Launch planned 2019)

- 🇮🇳- Astrosat (130-530 nm, launched in September 2015)

Types of Ultraviolet Astronomy

X-ray Astronomy

X-ray astronomy is an observational branch of astronomy which deals with the study of X-ray observation and detection from astronomical objects. X-radiation is absorbed by the Earth's atmosphere, so instruments to detect X-rays must be taken to high altitude by balloons, sounding rockets, and satellites. X-ray astronomy is the space science related to a type of space telescope that can see farther than standard light-absorption telescopes, such as the Mauna Kea Observatories, via x-ray radiation.

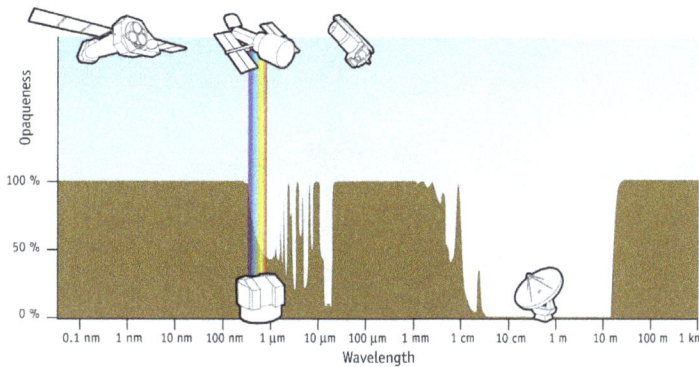

X-rays start at ~0.008 nm and extend across the electromagnetic spectrum to ~8 nm, over which the Earth's atmosphere is opaque.

X-ray emission is expected from astronomical objects that contain extremely hot gasses at temperatures from about a million kelvin (K) to hundreds of millions of kelvin (MK). Although X-rays have been observed emanating from the Sun since the 1940s, the discovery in 1962 of the first cosmic X-ray source was a surprise. This source is called Scorpius X-1 (Sco X-1), the first X-ray source found in the constellation Scorpius. The X-ray emission of Scorpius X-1 is 10,000 times greater than its visual emission, whereas that of the Sun is about a million times less. In addition, the energy output in X-rays is 100,000 times greater than the total emission of the Sun in all wavelengths. Based on discoveries in this new field of X-ray astronomy, starting with Scorpius X-1, Riccardo Giacconi received the Nobel Prize in Physics in 2002. It is now known that such X-ray sources as Sco X-1 are compact stars, such as neutron stars or black holes. Material falling into a black hole may emit X-rays, but the black hole itself does not. The energy source for the X-ray emission is gravity. Infalling gas and dust is heated by the strong gravitational fields of these and other celestial objects.

Many thousands of X-ray sources are known. In addition, the space between galaxies in galaxy clusters is filled with a very hot, but very dilute gas at a temperature between 10 and 100 megakelvins (MK). The total amount of hot gas is five to ten times the total mass in the visible galaxies.

Sounding Rocket Flights

The first sounding rocket flights for X-ray research were accomplished at the White Sands Missile Range in New Mexico with a V-2 rocket on January 28, 1949. A detector was placed in the nose cone section and the rocket was launched in a suborbital flight to an altitude just above the atmosphere.

X-rays from the Sun were detected by the U.S. Naval Research Laboratory Blossom experiment on board. An Aerobee 150 rocket was launched on June 12, 1962 and it detected the first X-rays from other celestial sources (Scorpius X-1).

The largest drawback to rocket flights is their very short duration (just a few minutes above the atmosphere before the rocket falls back to Earth) and their limited field of view. A rocket launched from the United States will not be able to see sources in the southern sky; a rocket launched from Australia will not be able to see sources in the northern sky.

X-ray Quantum Calorimeter (XQC) Project

A launch of the Black Brant 8 Microcalorimeter (XQC-2) at the turn of the century is a part of the joint undertaking by the University of Wisconsin-Madison and NASA's Goddard Space Flight Center known as the X-ray Quantum Calorimeter (XQC) project.

In astronomy, the interstellar medium (or **ISM**) is the gas and cosmic dust that pervade interstellar space: the matter that exists between the star systems within a galaxy. It fills interstellar space and blends smoothly into the surrounding intergalactic medium. The interstellar medium consists of an extremely dilute (by terrestrial standards) mixture of ions, atoms, molecules, larger dust grains, cosmic rays, and (galactic) magnetic fields. The energy that occupies the same volume, in the form of electromagnetic radiation, is the interstellar radiation field.

Of interest is the hot ionized medium (HIM) consisting of a coronal cloud ejection from star surfaces at 10^6-10^7 K which emits X-rays. The ISM is turbulent and full of structure on all spatial scales. Stars are born deep inside large complexes of molecular clouds, typically a few parsecs in size. During their lives and deaths, stars interact physically with the ISM. Stellar winds from young clusters of stars (often with giant or supergiant HII regions surrounding them) and shock waves created by supernovae inject enormous amounts of energy into their surroundings, which leads to hypersonic turbulence. The resultant structures are stellar wind bubbles and superbubbles of hot gas. The Sun is currently traveling through the Local Interstellar Cloud, a denser region in the low-density Local Bubble.

To measure the spectrum of the diffuse X-ray emission from the interstellar medium over the energy range 0.07 to 1 keV, NASA launched a Black Brant 9 from White Sands Missile Range, New Mexico on May 1, 2008. The Principal Investigator for the mission is Dr. Dan McCammon of the University of Wisconsin.

Balloons

Balloon flights can carry instruments to altitudes of up to 40 km above sea level, where they are above as much as 99.997% of the Earth's atmosphere. Unlike a rocket where data are collected during a brief few minutes, balloons are able to stay aloft for much longer. However, even at such altitudes, much of the X-ray spectrum is still absorbed. X-rays with energies less than 35 keV (5,600 aJ) cannot reach balloons. On July 21, 1964, the Crab Nebula supernova remnant was discovered to be a hard X-ray (15 – 60 keV) source by a scintillation counter flown on a balloon launched from Palestine, Texas, USA. This was likely the first balloon-based detection of X-rays from a discrete cosmic X-ray source.

High-energy Focusing Telescope

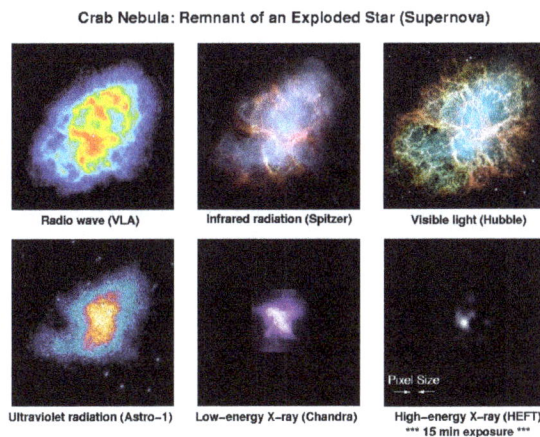

The Crab Nebula is a remnant of an exploded star. This image shows the Crab Nebula in various energy bands, including a hard X-ray image from the HEFT data taken during its 2005 observation run. Each image is 6′ wide.

The high-energy focusing telescope (HEFT) is a balloon-borne experiment to image astrophysical sources in the hard X-ray (20–100 keV) band. Its maiden flight took place in May 2005 from Fort Sumner, New Mexico, USA. The angular resolution of HEFT is ~1.5'. Rather than using a grazing-angle X-ray telescope, HEFT makes use of a novel tungsten-silicon multilayer coatings to extend the reflectivity of nested grazing-incidence mirrors beyond 10 keV. HEFT has an energy resolution of 1.0 keV full width at half maximum at 60 keV. HEFT was launched for a 25-hour balloon flight in May 2005. The instrument performed within specification and observed Tau X-1, the Crab Nebula.

High-resolution Gamma-ray and Hard X-ray Spectrometer (HIREGS)

A balloon-borne experiments called the High-resolution gamma-ray and hard X-ray spectrometer (HIREGS) made observed in X-ray and gamma-rays It was launched from McMurdo Station, Antarctica in December 1991. Steady winds carried the balloon on a circumpolar flight lasting about two weeks.

Rockoons

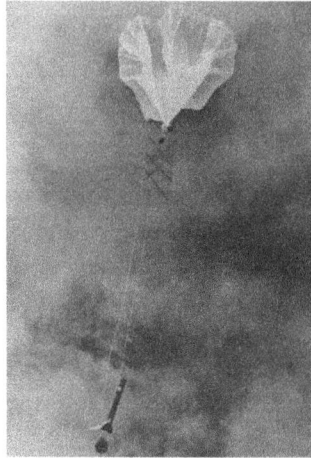

Navy Deacon rockoon photographed just after a shipboard launch in July 1956.

The rockoon (a portmanteau of rocket and balloon) was a solid fuel rocket that, rather than being immediately lit while on the ground, was first carried into the upper atmosphere by a gas-filled balloon. Then, once separated from the balloon at its maximum height, the rocket was automatically ignited. This achieved a higher altitude, since the rocket did not have to move through the lower thicker air layers that would have required much more chemical fuel.

The original concept of "rockoons" was developed by Cmdr. Lee Lewis, Cmdr. G. Halvorson, S. F. Singer, and James A. Van Allen during the Aerobee rocket firing cruise of the USS *Norton Sound* on March 1, 1949.

From July 17 to July 27, 1956, the Naval Research Laboratory (NRL) shipboard launched eight Deacon rockoons for solar ultraviolet and X-ray observations at ~30° N ~121.6° W, southwest of San Clemente Island, apogee: 120 km.

X-ray Astronomy Satellites

X-ray astronomy satellites study X-ray emissions from celestial objects. Satellites, which can detect and transmit data about the X-ray emissions are deployed as part of branch of space science known as X-ray astronomy. Satellites are needed because X-radiation is absorbed by the Earth's atmosphere, so instruments to detect X-rays must be taken to high altitude by balloons, sounding rockets, and satellites.

X-ray Telescopes and Mirrors

X-ray telescopes (XRTs) have varying directionality or imaging ability based on glancing angle reflection rather than refraction or large deviation reflection. This limits them to much narrower fields of view than visible or UV telescopes. The mirrors can be made of ceramic or metal foil.

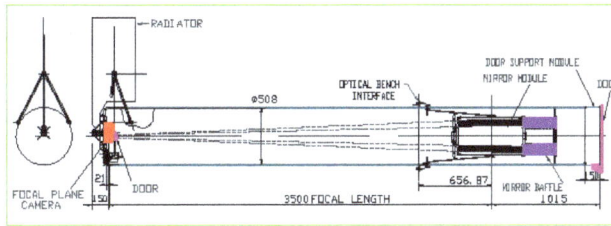

The Swift Gamma-Ray Burst Mission contains a grazing incidence Wolter I telescope (XRT) to focus X-rays onto a state-of-the-art CCD.

The first X-ray telescope in astronomy was used to observe the Sun. The first X-ray picture (taken with a grazing incidence telescope) of the Sun was taken in 1963, by a rocket-borne telescope. On April 19, 1960 the very first X-ray image of the sun was taken using a pinhole camera on an Aerobee-Hi rocket.

The utilization of X-ray mirrors for extrasolar X-ray astronomy simultaneously requires:

- the ability to determine the location at the arrival of an X-ray photon in two dimensions and

- a reasonable detection efficiency.

X-ray Astronomy Detectors

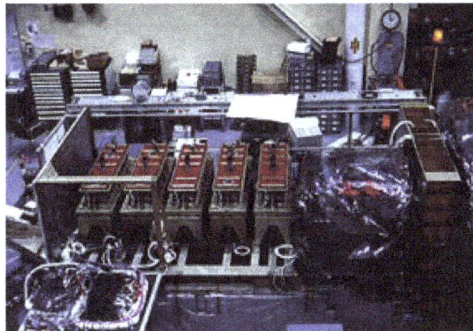

Proportional Counter Array on the Rossi X-ray Timing Explorer (RXTE) satellite.

X-ray astronomy detectors have been designed and configured primarily for energy and occasionally for wavelength detection using a variety of techniques usually limited to the technology of the time.

X-ray detectors collect individual X-rays (photons of X-ray electromagnetic radiation) and count the number of photons collected (intensity), the energy (0.12 to 120 keV) of the photons collected, wavelength (~0.008 to 8 nm), or how fast the photons are detected (counts per hour), to tell us about the object that is emitting them.

Astrophysical Sources of X-rays

Several types of astrophysical objects emit, fluoresce, or reflect X-rays, from galaxy clusters, through black holes in active galactic nuclei (AGN) to galactic objects such as supernova remnants, stars, and binary stars containing a white dwarf (cataclysmic variable stars and super soft X-ray sources), neutron star or black hole (X-ray binaries). Some solar system bodies emit X-rays, the most notable being the Moon, although most of the X-ray brightness of the Moon arises from reflected solar X-rays. A combination of many unresolved X-ray sources is thought to produce the observed X-ray background. The X-ray continuum can arise from bremsstrahlung, black-body radiation, synchrotron radiation, or what is called inverse Compton scattering of lower-energy photons by relativistic electrons, knock-on collisions of fast protons with atomic electrons, and atomic recombination, with or without additional electron transitions.

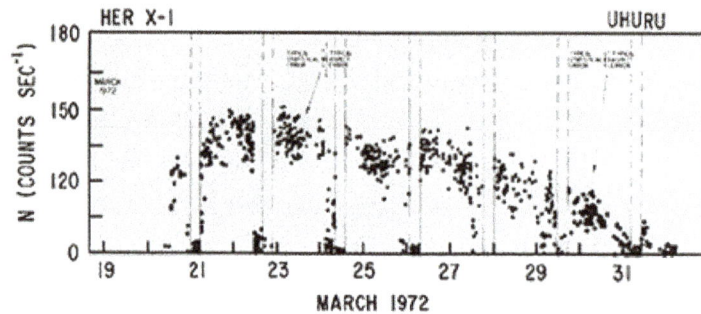

This light curve of Her X-1 shows long term and medium term variability. Each pair of vertical lines delineate the eclipse of the compact object behind its companion star. In this case, the companion is a two solar-mass star with a radius of nearly four times that of our Sun. This eclipse shows us the orbital period of the system, 1.7 days.

An intermediate-mass X-ray binary (IMXB) is a binary star system where one of the components is a neutron star or a black hole. The other component is an intermediate mass star.

Hercules X-1 is composed of a neutron star accreting matter from a normal star (HZ Herculis) probably due to Roche lobe overflow. X-1 is the prototype for the massive X-ray binaries although it falls on the borderline, ~2 $M_\odot$, between high- and low-mass X-ray binaries.

Celestial X-ray Sources

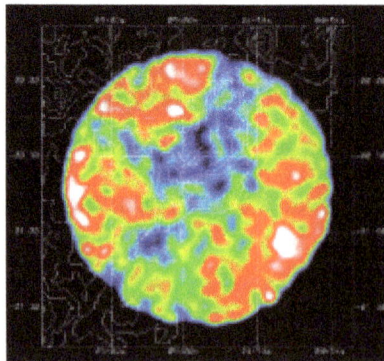

This ROSAT PSPC false-color image is of a portion of a nearby stellar wind superbubble (the Orion-Eridanus Superbubble) stretching across Eridanus and Orion.

The celestial sphere has been divided into 88 constellations. The International Astronomical Union (IAU) constellations are areas of the sky. Each of these contains remarkable X-ray sources. Some of them are have been identified from astrophysical modeling to be galaxies or black holes at the centers of galaxies. Some are pulsars. As with sources already successfully modeled by X-ray astrophysics, striving to understand the generation of X-rays by the apparent source helps to understand the Sun, the universe as a whole, and how these affect us on Earth. Constellations are an astronomical device for handling observation and precision independent of current physical theory or interpretation. Astronomy has been around for a long time. Physical theory changes with time. With respect to celestial X-ray sources, X-ray astrophysics tends to focus on the physical reason for X-ray brightness, whereas X-ray astronomy tends to focus on their classification, order of discovery, variability, resolvability, and their relationship with nearby sources in other constellations.

Within the constellations Orion and Eridanus and stretching across them is a soft X-ray "hot spot" known as the Orion-Eridanus Superbubble, the Eridanus Soft X-ray Enhancement, or simply the Eridanus Bubble, a 25° area of interlocking arcs of Hα emitting filaments. Soft X-rays are emitted by hot gas (T ~ 2–3 MK) in the interior of the superbubble. This bright object forms the background for the "shadow" of a filament of gas and dust. The filament is shown by the overlaid contours, which represent 100 micrometre emission from dust at a temperature of about 30 K as measured by IRAS. Here the filament absorbs soft X-rays between 100 and 300 eV, indicating that the hot gas is located behind the filament. This filament may be part of a shell of neutral gas that surrounds the hot bubble. Its interior is energized by ultraviolet (UV) light and stellar winds from hot stars in the Orion OB1 association. These stars energize a superbubble about 1200 lys across which is observed in the visual (Hα) and X-ray portions of the spectrum.

Proposed (Future) X-ray Observatory Satellites

There are several projects that are proposed for X-ray observatory satellites.

Explorational X-ray Astronomy

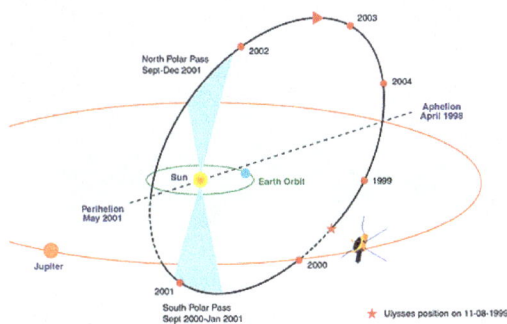

Ulysses' second orbit: it arrived at Jupiter on February 8, 1992, for a swing-by maneuver that increased its inclination to the ecliptic by 80.2 degrees.

Usually observational astronomy is considered to occur on Earth's surface (or beneath it in neutrino astronomy). The idea of limiting observation to Earth includes orbiting the Earth. As soon as the observer leaves the cozy confines of Earth, the observer becomes a deep space explorer. Except for Explorer 1 and Explorer 3 and the earlier satellites in the series, usually if a probe is going to be a deep space explorer it leaves the Earth or an orbit around the Earth.

For a satellite or space probe to qualify as a deep space X-ray astronomer/explorer or "astronobot"/explorer, all it needs to carry aboard is an XRT or X-ray detector and leave Earth orbit.

Ulysses is launched October 6, 1990, and reached Jupiter for its "gravitational slingshot" in February 1992. It passed the south solar pole in June 1994 and crossed the ecliptic equator in February 1995. The solar X-ray and cosmic gamma-ray burst experiment (GRB) had 3 main objectives: study and monitor solar flares, detect and localize cosmic gamma-ray bursts, and in-situ detection of Jovian aurorae. Ulysses was the first satellite carrying a gamma burst detector which went outside the orbit of Mars. The hard X-ray detectors operated in the range 15–150 keV. The detectors consisted of 23-mm thick × 51-mm diameter CsI(Tl) crystals mounted via plastic light tubes to photomultipliers. The hard detector changed its operating mode depending on (1) measured count rate, (2) ground command, or (3) change in spacecraft telemetry mode. The trigger level was generally set for 8-sigma above background and the sensitivity is 10^{-6} erg/cm^2 (1 nJ/m^2). When a burst trigger is recorded, the instrument switches to record high resolution data, recording it to a 32-kbit memory for a slow telemetry read out. Burst data consist of either 16 s of 8-ms resolution count rates or 64 s of 32-ms count rates from the sum of the 2 detectors. There were also 16 channel energy spectra from the sum of the 2 detectors (taken either in 1, 2, 4, 16, or 32 second integrations). During 'wait' mode, the data were taken either in 0.25 or 0.5 s integrations and 4 energy channels (with shortest integration time being 8 s). Again, the outputs of the 2 detectors were summed.

The Ulysses soft X-ray detectors consisted of 2.5-mm thick × 0.5 cm^2 area Si surface barrier detectors. A 100 mg/cm^2 beryllium foil front window rejected the low energy X-rays and defined a conical FOV of 75° (half-angle). These detectors were passively cooled and operate in the temperature range −35 to −55 °C. This detector had 6 energy channels, covering the range 5–20 keV.

X-Rays from Pluto

Theoretical X-ray Astronomy

Theoretical X-ray astronomy is a branch of theoretical astronomy that deals with the theoretical astrophysics and theoretical astrochemistry of X-ray generation, emission, and detection as applied to astronomical objects.

Like theoretical astrophysics, theoretical X-ray astronomy uses a wide variety of tools which include analytical models to approximate the behavior of a possible X-ray source and computational numerical simulations to approximate the observational data. Once potential observational consequences are available they can be compared with experimental observations. Observers can look for data that refutes a model or helps in choosing between several alternate or conflicting models.

Theorists also try to generate or modify models to take into account new data. In the case of an inconsistency, the general tendency is to try to make minimal modifications to the model to fit the data. In some cases, a large amount of inconsistent data over time may lead to total abandonment of a model.

Most of the topics in astrophysics, astrochemistry, astrometry, and other fields that are branches of astronomy studied by theoreticians involve X-rays and X-ray sources. Many of the beginnings for a theory can be found in an Earth-based laboratory where an X-ray source is built and studied.

Dynamos

Dynamo theory describes the process through which a rotating, convecting, and electrically conducting fluid acts to maintain a magnetic field. This theory is used to explain the presence of anomalously long-lived magnetic fields in astrophysical bodies. If some of the stellar magnetic fields are really induced by dynamos, then field strength might be associated with rotation rate.

Astronomical Models

Images released to celebrate the International Year of Light 2015 (Chandra X-Ray Observatory).

From the observed X-ray spectrum, combined with spectral emission results for other wavelength ranges, an astronomical model addressing the likely source of X-ray emission can be constructed. For example, with Scorpius X-1 the X-ray spectrum steeply drops off as X-ray energy increases up to 20 keV, which is likely for a thermal-plasma mechanism. In addition, there is no radio emission, and the visible continuum is roughly what would be expected from a hot plasma fitting the observed X-ray flux. The plasma could be a coronal cloud of a central object or a transient plasma, where the energy source is unknown, but could be related to the idea of a close binary.

In the Crab Nebula X-ray spectrum there are three features that differ greatly from Scorpius X-1: its spectrum is much harder, its source diameter is in light-years (ly)s, not astronomical units (AU), and its radio and optical synchrotron emission are strong. Its overall X-ray luminosity rivals the optical emission and could be that of a nonthermal plasma. However, the Crab Nebula appears as an X-ray source that is a central freely expanding ball of dilute plasma, where the energy content is 100 times the total energy content of the large visible and radio portion, obtained from the unknown source.

The "Dividing Line" as giant stars evolve to become red giants also coincides with the Wind and Coronal Dividing Lines. To explain the drop in X-ray emission across these dividing lines, a number of models have been proposed:

1. low transition region densities, leading to low emission in coronae,

2. high-density wind extinction of coronal emission,

3. only cool coronal loops become stable,

4. changes in a magnetic field structure to that an open topology, leading to a decrease of magnetically confined plasma, or

5. changes in the magnetic dynamo character, leading to the disappearance of stellar fields leaving only small-scale, turbulence-generated fields among red giants.

Analytical X-ray Astronomy

Analytical X-ray astronomy is applied to an astronomy puzzle in an attempt to provide an acceptable solution. Consider the following puzzle.

High-mass X-ray binaries (HMXBs) are composed of OB supergiant companion stars and compact objects, usually neutron stars (NS) or black holes (BH). Supergiant X-ray binaries (SGXBs) are HMXBs in which the compact objects orbit massive companions with orbital periods of a few days (3–15 d), and in circular (or slightly eccentric) orbits. SGXBs show typical the hard X-ray spectra of accreting pulsars and most show strong absorption as obscured HMXBs. X-ray luminosity (L_x) increases up to 10^{36} erg·s^{-1} (10^{29} watts).

The mechanism triggering the different temporal behavior observed between the classical SGXBs and the recently discovered supergiant fast X-ray transients (SFXT)s is still debated.

Aim: use the discovery of long orbits (>15 d) to help discriminate between emission models and perhaps bring constraints on the models.

Method: analyze archival data on various SGXBs such as has been obtained by INTEGRAL for candidates exhibiting long orbits. Build short- and long-term light curves. Perform a timing analysis in order to study the temporal behavior of each candidate on different time scales.

Compare various astronomical models:

- direct spherical accretion

- Roche-Lobe overflow via an accretion disk on the compact object.

Draw some conclusions: for example, the SGXB SAX J1818.6-1703 was discovered by BeppoSAX in 1998, identified as a SGXB of spectral type between O9I–B1I, which also displayed short and bright flares and an unusually very low quiescent level leading to its classification as a SFXT. The analysis indicated an unusually long orbital period: 30.0 ± 0.2 d and an elapsed accretion phase of ~6 d implying an elliptical orbit and possible supergiant spectral type between B0.5-1I with eccentricities e ~ 0.3–0.4. The large variations in the X-ray flux can be explained through accretion of macro-clumps formed within the stellar wind.

Choose which model seems to work best: for SAX J1818.6-1703 the analysis best fits the model that predicts SFXTs behave as SGXBs with different orbital parameters; hence, different temporal behavior.

Stellar X-ray Astronomy

Stellar X-ray astronomy is said to have started on April 5, 1974, with the detection of X-rays from Capella. A rocket flight on that date briefly calibrated its attitude control system when a star sensor pointed the payload axis at Capella (α Aur). During this period, X-rays in the range 0.2–1.6 keV were detected by an X-ray reflector system co-aligned with the star sensor. The X-ray luminosity of $L_x = 10^{31}$ erg·s^{-1} (10^{24} W) is four orders of magnitude above the Sun's X-ray luminosity.

Eta Carinae

New X-ray observations by the Chandra X-ray Observatory show three distinct structures: an outer, horseshoe-shaped ring about 2 light years in diameter, a hot inner core about 3 light-months in diameter, and a hot central source less than 1 light-month in diameter which may contain the superstar that drives the whole show. The outer ring provides evidence of another large explosion that occurred over 1,000 years ago. These three structures around Eta Carinae are thought to represent shock waves produced by matter rushing away from the superstar at supersonic speeds. The temperature of the shock-heated gas ranges from 60 MK in the central regions to 3 MK on the horseshoe-shaped outer structure. "The Chandra image contains some puzzles for existing ideas of how a star can produce such hot and intense X-rays," says Prof. Kris Davidson of the University of Minnesota. Davidson is principal investigator for the Eta Carina observations by the Hubble Space telescope. "In the most popular theory, X-rays are made by colliding gas streams from two stars so close together that they'd look like a point source to us. But what happens to gas streams that escape to farther distances? The extended hot stuff in the middle of the new image gives demanding new conditions for any theory to meet."

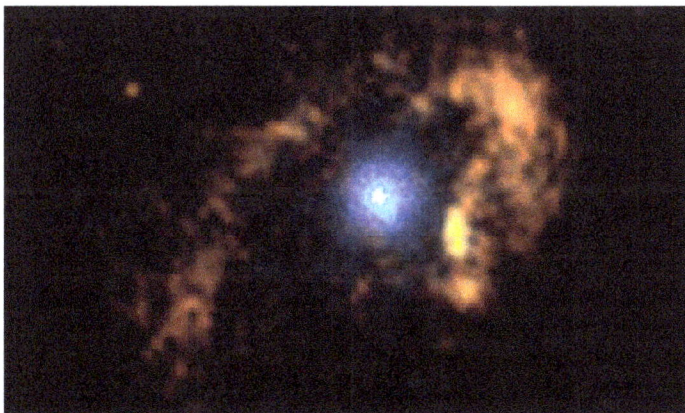

Classified as a Peculiar star, Eta Carinae exhibits a superstar at its center as seen in this image from Chandra X-ray Observatory. Credit: Chandra Science Center and NASA.

Stellar Coronae

Coronal stars, or stars within a coronal cloud, are ubiquitous among the stars in the cool half of the Hertzsprung-Russell diagram. Experiments with instruments aboard Skylab and Copernicus have been used to search for soft X-ray emission in the energy range ~0.14–0.284 keV from stellar coronae. The experiments aboard ANS succeeded in finding X-ray signals from Capella and Sirius (α CMa). X-ray emission from an enhanced solar-like corona was proposed for the first time. The high temperature of Capella's corona as obtained from the first coronal X-ray spectrum of Capella using HEAO 1 required magnetic confinement unless it was a free-flowing coronal wind.

In 1977 Proxima Centauri is discovered to be emitting high-energy radiation in the XUV. In 1978, α Cen was identified as a low-activity coronal source. With the operation of the Einstein observatory, X-ray emission was recognized as a characteristic feature common to a wide range of stars covering essentially the whole Hertzsprung-Russell diagram. The Einstein initial survey led to significant insights:

- X-ray sources abound among all types of stars, across the Hertzsprung-Russell diagram and across most stages of evolution,

- the X-ray luminosities and their distribution along the main sequence were not in agreement with the long-favored acoustic heating theories, but were now interpreted as the effect of magnetic coronal heating, and

- stars that are otherwise similar reveal large differences in their X-ray output if their rotation period is different.

To fit the medium-resolution spectrum of UX Ari, subsolar abundances were required.

Stellar X-ray astronomy is contributing toward a deeper understanding of

- magnetic fields in magnetohydrodynamic dynamos,

- the release of energy in tenuous astrophysical plasmas through various plasma-physical processes, and

- the interactions of high-energy radiation with the stellar environment.

Current wisdom has it that the massive coronal main sequence stars are late-A or early F stars, a conjecture that is supported both by observation and by theory.

Unstable Winds

Given the lack of a significant outer convection zone, theory predicts the absence of a magnetic dynamo in earlier A stars. In early stars of spectral type O and B, shocks developing in unstable winds are the likely source of X-rays.

Coolest M Dwarfs

Beyond spectral type M5, the classical $\alpha\omega$ dynamo can no longer operate as the internal structure of dwarf stars changes significantly: they become fully convective. As a distributed (or α^2) dynamo may become relevant, both the magnetic flux on the surface and the topology of the magnetic fields in the corona should systematically change across this transition, perhaps resulting in some discontinuities in the X-ray characteristics around spectral class dM5. However, observations do not seem to support this picture: long-time lowest-mass X-ray detection, VB 8 (M7e V), has shown steady emission at levels of X-ray luminosity $(L_x) \approx 10^{26}$ erg·s^{-1} (10^{19} W) and flares up to an order of magnitude higher. Comparison with other late M dwarfs shows a rather continuous trend.

Strong X-ray Emission from Herbig Ae/Be Stars

Herbig Ae/Be stars are pre-main sequence stars. As to their X-ray emission properties, some are

- reminiscent of hot stars,

- others point to coronal activity as in cool stars, in particular the presence of flares and very high temperatures.

The nature of these strong emissions has remained controversial with models including

- unstable stellar winds,

- colliding winds,

- magnetic coronae,

- disk coronae,

- wind-fed magnetospheres,

- accretion shocks,

- the operation of a shear dynamo,

- the presence of unknown late-type companions.

K Giants

The FK Com stars are giants of spectral type K with an unusually rapid rotation and signs of extreme activity. Their X-ray coronae are among the most luminous ($L_x \geq 10^{32}$ erg·s^{-1} or 10^{25} W) and the hottest known with dominant temperatures up to 40 MK. However, the current popular hypothesis involves a merger of a close binary system in which the orbital angular momentum of the companion is transferred to the primary.

Pollux is the brightest star in the constellation Gemini, despite its Beta designation, and the 17th brightest in the sky. Pollux is a giant orange K star that makes an interesting color contrast with its white "twin", Castor. Evidence has been found for a hot, outer, magnetically supported corona around Pollux, and the star is known to be an X-ray emitter.

Amateur X-ray Astronomy

Collectively, amateur astronomers observe a variety of celestial objects and phenomena sometimes with equipment that they build themselves. The United States Air Force Academy (USAFA) is the home of the US's only undergraduate satellite program, and has and continues to develop the

FalconLaunch sounding rockets. In addition to any direct amateur efforts to put X-ray astronomy payloads into space, there are opportunities that allow student-developed experimental payloads to be put on board commercial sounding rockets as a free-of-charge ride.

There are major limitations to amateurs observing and reporting experiments in X-ray astronomy: the cost of building an amateur rocket or balloon to place a detector high enough and the cost of appropriate parts to build a suitable X-ray detector.

History of X-ray Astronomy

NRL scientists J. D. Purcell, C. Y. Johnson, and Dr. F. S. Johnson are among those recovering instruments from a V-2 used for upper atmospheric research above the New Mexico desert. This is V-2 number 54, launched January 18, 1951, (photo by Dr. Richard Tousey, NRL).

In 1927, E.O. Hulburt of the US Naval Research Laboratory and associates Gregory Breit and Merle A. Tuve of the Carnegie Institution of Washington explored the possibility of equipping Robert H. Goddard's rockets to explore the upper atmosphere. "Two years later, he proposed an experimental program in which a rocket might be instrumented to explore the upper atmosphere, including detection of ultraviolet radiation and X-rays at high altitudes".

In the late 1930s, the presence of a very hot, tenuous gas surrounding the Sun was inferred indirectly from optical coronal lines of highly ionized species. The Sun has been known to be surrounded by a hot tenuous corona. In the mid-1940s radio observations revealed a radio corona around the Sun.

The beginning of the search for X-ray sources from above the Earth's atmosphere was on August 5, 1948 12:07 GMT. A US Army (formerly German) V-2 rocket as part of Project Hermes was launched from White Sands Proving Grounds. The first solar X-rays were recorded by T. Burnight.

Through the 1960s, 70s, 80s, and 90s, the sensitivity of detectors increased greatly during the 60 years of X-ray astronomy. In addition, the ability to focus X-rays has developed enormously—allowing the production of high-quality images of many fascinating celestial objects.

Major questions in X-ray Astronomy

As X-ray astronomy uses a major spectral probe to peer into source, it is a valuable tool in efforts to understand many puzzles.

Stellar Magnetic Fields

Magnetic fields are ubiquitous among stars, yet we do not understand precisely why, nor have we fully understood the bewildering variety of plasma physical mechanisms that act in stellar environments. Some stars, for example, seem to have magnetic fields, fossil stellar magnetic fields left over from their period of formation, while others seem to generate the field anew frequently.

Extrasolar X-ray Source Astrometry

With the initial detection of an extrasolar X-ray source, the first question usually asked is "What is the source?" An extensive search is often made in other wavelengths such as visible or radio for possible coincident objects. Many of the verified X-ray locations still do not have readily discernible sources. X-ray astrometry becomes a serious concern that results in ever greater demands for finer angular resolution and spectral radiance.

There are inherent difficulties in making X-ray/optical, X-ray/radio, and X-ray/X-ray identifications based solely on positional coincidents, especially with handicaps in making identifications, such as the large uncertainties in positional determinants made from balloons and rockets, poor source separation in the crowded region toward the galactic center, source variability, and the multiplicity of source nomenclature.

X-ray source counterparts to stars can be identified by calculating the angular separation between source centroids and position of the star. The maximum allowable separation is a compromise between a larger value to identify as many real matches as possible and a smaller value to minimize the probability of spurious matches. "An adopted matching criterion of 40" finds nearly all possible X-ray source matches while keeping the probability of any spurious matches in the sample to 3%."

Solar X-ray Astronomy

All of the detected X-ray sources at, around, or near the Sun are within or associated with the coronal cloud which is its outer atmosphere.

Coronal Heating Problem

In the area of solar X-ray astronomy, there is the coronal heating problem. The photosphere of the Sun has an effective temperature of 5,570 K yet its corona has an average temperature of $1–2 \times 10^6$ K. However, the hottest regions are $8–20 \times 10^6$ K. The high temperature of the corona shows that it is heated by something other than direct heat conduction from the photosphere.

It is thought that the energy necessary to heat the corona is provided by turbulent motion in the convection zone below the photosphere, and two main mechanisms have been proposed to explain coronal heating. The first is wave heating, in which sound, gravitational or magnetohydrodynamic waves are produced by turbulence in the convection zone. These waves travel upward and dissipate in the corona, depositing their energy in the ambient gas in the form of heat. The other is magnetic heating, in which magnetic energy is continuously built up by photospheric motion and released through magnetic reconnection in the form of large solar flares and myriad similar but smaller events—nanoflares.

Currently, it is unclear whether waves are an efficient heating mechanism. All waves except Alfvén waves have been found to dissipate or refract before reaching the corona. In addition, Alfvén waves do not easily dissipate in the corona. Current research focus has therefore shifted towards flare heating mechanisms.

Coronal Mass Ejection

A coronal mass ejection (CME) is an ejected plasma consisting primarily of electrons and protons (in addition to small quantities of heavier elements such as helium, oxygen, and iron), plus the entraining coronal closed magnetic field regions. Evolution of these closed magnetic structures in response to various photospheric motions over different time scales (convection, differential rotation, meridional circulation) somehow leads to the CME. Small-scale energetic signatures such as plasma heating (observed as compact soft X-ray brightening) may be indicative of impending CMEs.

The soft X-ray sigmoid (an S-shaped intensity of soft X-rays) is an observational manifestation of the connection between coronal structure and CME production. "Relating the sigmoids at X-ray (and other) wavelengths to magnetic structures and current systems in the solar atmosphere is the key to understanding their relationship to CMEs."

The first detection of a Coronal mass ejection (CME) as such was made on December 1, 1971 by R. Tousey of the US Naval Research Laboratory using OSO 7. Earlier observations of coronal transients or even phenomena observed visually during solar eclipses are now understood as essentially the same thing.

The largest geomagnetic perturbation, resulting presumably from a "prehistoric" CME, coincided with the first-observed solar flare, in 1859. The flare was observed visually by Richard Christopher Carrington and the geomagnetic storm was observed with the recording magnetograph at Kew Gardens. The same instrument recorded a crotchet, an instantaneous perturbation of the Earth's ionosphere by ionizing soft X-rays. This could not easily be understood at the time because it predated the discovery of X-rays (by Roentgen) and the recognition of the ionosphere (by Kennelly and Heaviside).

Exotic X-ray Sources

A microquasar is a smaller cousin of a quasar that is a radio emitting X-ray binary, with an often resolvable pair of radio jets. LSI+61°303 is a periodic, radio-emitting binary system that is also the gamma-ray source, CG135+01. Observations are revealing a growing number of recurrent X-ray transients, characterized by short outbursts with very fast rise times (tens of minutes) and typical durations of a few hours that are associated with OB supergiants and hence define a new class of massive X-ray binaries: Supergiant Fast X-ray Transients (SFXTs). Observations made by Chandra indicate the presence of loops and rings in the hot X-ray emitting gas that surrounds Messier 87. A magnetar is a type of neutron star with an extremely powerful magnetic field, the decay of which powers the emission of copious amounts of high-energy electromagnetic radiation, particularly X-rays and gamma rays.

X-ray Dark Stars

During the solar cycle, as shown in the sequence of images at right, at times the Sun is almost X-ray dark, almost an X-ray variable. Betelgeuse, on the other hand, appears to be always X-ray dark. Hardly any X-rays are emitted by red giants. There is a rather abrupt onset of X-ray emission around spectral type A7-F0, with a large range of luminosities developing across spectral class F. Altair is spectral type A7V and Vega is A0V. Altair's total X-ray luminosity is at least an order of magnitude larger than the X-ray luminosity for Vega. The outer convection zone of early F stars is expected to be very shallow and absent in A-type dwarfs, yet the acoustic flux from the interior reaches a maximum for late A and early F stars provoking investigations of magnetic activity in A-type stars along three principal lines. Chemically peculiar stars of spectral type Bp or Ap are appreciable magnetic radio sources, most Bp/Ap stars remain undetected, and of those reported early on as producing X-rays only few of them can be identified as probably single stars. X-ray observations offer the possibility to detect (X-ray dark) planets as they eclipse part of the corona of their parent star while in transit. "Such methods are particularly promising for low-mass stars as a Jupiter-like planet could eclipse a rather significant coronal area."

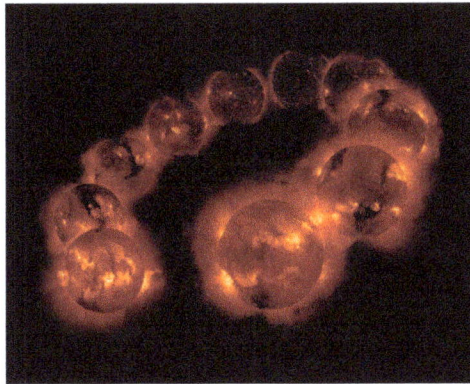

A solar cycle: a montage of ten years' worth of Yohkoh SXT images, demonstrating the variation in solar activity during a sunspot cycle, from after August 30, 1991, at the peak of cycle 22, to September 6, 2001, at the peak of cycle 23. Credit: the Yohkoh mission of Institute of Space and Astronautical Science (ISAS, Japan) and NASA (US).

X-ray Dark Planet/Comet

X-ray observations offer the possibility to detect (X-ray dark) planets as they eclipse part of the corona of their parent star while in transit. "Such methods are particularly promising for low-mass stars as a Jupiter-like planet could eclipse a rather significant coronal area."

As X-ray detectors have become more sensitive, they have observed that some planets and other normally X-ray non-luminescent celestial objects under certain conditions emit, fluoresce, or reflect X-rays.

Comet Lulin

NASA's Swift Gamma-Ray Burst Mission satellite was monitoring Comet Lulin as it closed to 63 Gm of Earth. For the first time, astronomers can see simultaneous UV and X-ray images of a comet. "The solar wind—a fast-moving stream of particles from the sun—interacts with the comet's broader cloud of atoms. This causes the solar wind to light up with X-rays, and that's what Swift's XRT sees", said Stefan Immler, of the Goddard Space Flight Center. This interaction, called charge exchange, results in X-rays from most comets when they pass within about three times Earth's distance from the Sun. Because Lulin is so active, its atomic cloud is especially dense. As a result, the X-ray-emitting region extends far sunward of the comet.

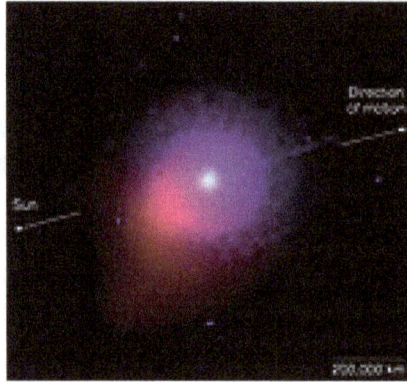

Comet Lulin was passing through the constellation Libra when Swift imaged it on January 28, 2009. This image merges data acquired by Swift's Ultraviolet/Optical Telescope (blue and green) and X-Ray Telescope (red). At the time of the observation, the comet was 99.5 million miles from Earth and 115.3 million miles from the Sun.

Single X-ray Stars

In addition to the Sun there are many unary stars or star systems throughout the galaxy that emit X-rays. β Hydri (G2 IV) is a normal single, post main-sequence subgiant star, T_{eff} = 5800 K. It exhibits coronal X-ray fluxes.

The benefit of studying single stars is that it allows measurements free of any effects of a companion or being a part of a multiple star system. Theories or models can be more readily tested. See, e.g., Betelgeuse, Red giants, and Vega and Altair.

Gamma-ray Astronomy

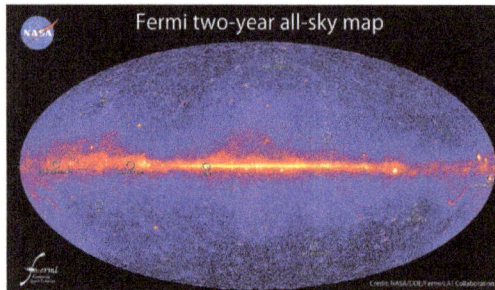

First survey of the sky at energies above 1 GeV, collected by the Fermi Gamma-ray Space Telescope in three years of observation (2009 to 2011)

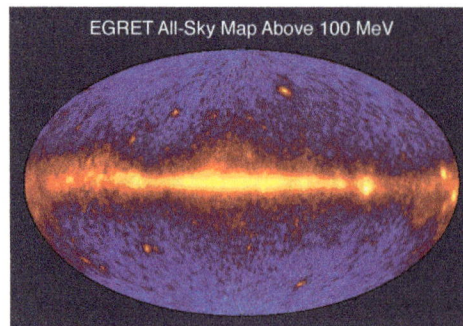

The sky at energies above 100 MeV observed by the Energetic Gamma Ray Experiment Telescope (EGRET) of the Compton Gamma Ray Observatory (CGRO) satellite (1991–2000)

Gamma-ray astronomy is the astronomical observation of gamma rays,[nb 1] the most energetic form of electromagnetic radiation, with photon energies above 100 keV. Radiation below 100 keV is classified as X-rays and is the subject of X-ray astronomy.

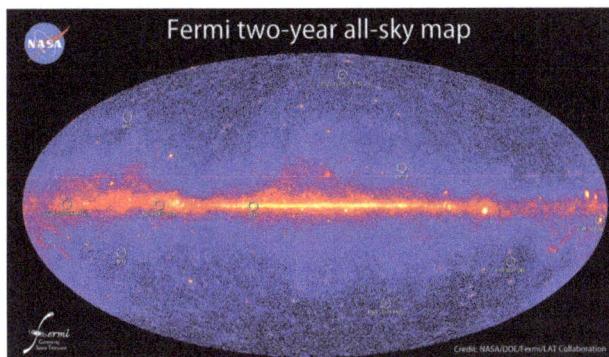

September 02 2011 Fermi Second catalog of Gamma Ray Sources constructed over 2 years. An all sky image showing energies greater than 1 biilion electron volts (1 GeV) ub. Brighter colors indicate gamma-ray sources.

Gamma rays in the MeV range are generated in solar flares (and even in the Earth's atmosphere), but gamma rays in the GeV range do not originate in the Solar System and are important in the study of extrasolar, and especially extra-galactic astronomy. The mechanisms emitting gamma rays are diverse, mostly identical with those emitting X-rays but at higher energies, including electron-positron annihilation, the Inverse Compton Effect, and in some cases also the decay of radioactive material (gamma decay) in space reflecting extreme events such as supernovae and hypernovae, and the behaviour of matter under extreme conditions, as in pulsars and blazars. The highest photon energies measured to date are in the TeV range, the record being held by the Crab Pulsar in 2004, yielding photons with as much as 80 TeV.

Detector Technology

Observation of gamma rays first became possible in the 1960s. Their observation is much more problematic than that of X-rays or of visible light, because gamma-rays are comparatively rare, even a "bright" source needing an observation time of several minutes before it is even detected, and because gamma rays are difficult to focus, resulting in a very low resolution. The most recent generation of gamma-ray telescopes (2000s) have a resolution of the order of 6 arc minutes in the GeV range (seeing the Crab Nebula as a single "pixel"), compared to 0.5 arc seconds seen in the low energy X-ray (1 keV) range by the Chandra X-ray Observatory (1999), and about 1.5 arc minutes in the high energy X-ray (100 keV) range seen by High-Energy Focusing Telescope (2005).

Very energetic gamma rays, with photon energies over ~30 GeV, can also be detected by ground based experiments. The extremely low photon fluxes at such high energies require detector effective areas that are impractically large for current space-based instruments. Fortunately, such high-energy photons produce extensive showers of secondary particles in the atmosphere that can be observed on the ground, both directly by radiation counters and optically via the Cherenkov light which the ultra-relativistic shower particles emit. The Imaging Atmospheric Cherenkov Telescope technique currently achieves the highest sensitivity.

Gamma radiation in the TeV range emanating from the Crab Nebula was first detected in 1989 by the Whipple Observatory at Mt. Hopkins, in Arizona in the USA. Modern Cherenkov telescope experiments like H.E.S.S., VERITAS, MAGIC, and CANGAROO III can detect the Crab Nebula in a few minutes. The most energetic photons (up to 16 TeV) observed from an extragalactic object originate from the blazar, Markarian 501 (Mrk 501). These measurements were done by the High-Energy-Gamma-Ray Astronomy (HEGRA) air Cherenkov telescopes.

Gamma-ray astronomy observations are still limited by non-gamma-ray backgrounds at lower energies, and, at higher energy, by the number of photons that can be detected. Larger area detectors and better background suppression are essential for progress in the field. A discovery in 2012 may allow focusing gamma-ray telescopes. At photon energies greater than 700 keV, the index of refraction starts to increase again.

Early History

Long before experiments could detect gamma rays emitted by cosmic sources, scientists had known that the universe should be producing them. Work by Eugene Feenberg and Henry Primakoff in 1948, Sachio Hayakawa and I.B. Hutchinson in 1952, and, especially, Philip Morrison in 1958 had led scientists to believe that a number of different processes which were occurring in the universe would result in gamma-ray emission. These processes included cosmic ray interactions with interstellar gas, supernova explosions, and interactions of energetic electrons with magnetic fields. However, it was not until the 1960s that our ability to actually detect these emissions came to pass.

Most gamma rays coming from space are absorbed by the Earth's atmosphere, so gamma-ray astronomy could not develop until it was possible to get detectors above all or most of the atmosphere using balloons and spacecraft. The first gamma-ray telescope carried into orbit, on the Explorer 11 satellite in 1961, picked up fewer than 100 cosmic gamma-ray photons. They appeared to come from all directions in the Universe, implying some sort of uniform "gamma-ray background". Such a background would be expected from the interaction of cosmic rays (very energetic charged particles in space) with interstellar gas.

The first true astrophysical gamma-ray sources were solar flares, which revealed the strong 2.223 MeV line predicted by Morrison. This line results from the formation of deuterium via the union of a neutron and proton; in a solar flare the neutrons appear as secondaries from interactions of high-energy ions accelerated in the flare process. These first gamma-ray line observations were from OSO-3, OSO-7, and the Solar Maximum Mission, the latter spacecraft launched in 1980. The solar observations inspired theoretical work by Reuven Ramaty and others.

Significant gamma-ray emission from our galaxy was first detected in 1967 by the detector aboard the OSO-3 satellite. It detected 621 events attributable to cosmic gamma rays. However, the field of gamma-ray astronomy took great leaps forward with the SAS-2 (1972) and the COS-B (1975–1982) satellites. These two satellites provided an exciting view into the high-energy universe (sometimes called the 'violent' universe, because the kinds of events in space that produce gamma rays tend to be high-speed collisions and similar processes). They confirmed the earlier findings of the gamma-ray background, produced the first detailed map of the sky at gamma-ray wavelengths, and detected a number of point sources. However the resolution of the instruments was insufficient to identify most of these point sources with specific visible stars or stellar systems.

A discovery in gamma-ray astronomy came in the late 1960s and early 1970s from a constellation of military defense satellites. Detectors on board the Vela satellite series, designed to detect flashes of gamma rays from nuclear bomb blasts, began to record bursts of gamma rays from deep space rather than the vicinity of the Earth. Later detectors determined that these gamma-ray bursts are seen to last for fractions of a second to minutes, appearing suddenly from unexpected directions, flickering, and then fading after briefly dominating the gamma-ray sky. Studied since the mid-1980s with instruments on board a variety of satellites and space probes, including Soviet Venera spacecraft and the Pioneer Venus Orbiter, the sources of these enigmatic high-energy flashes remain a mystery. They appear to come from far away in the Universe, and currently the most likely theory seems to be that at least some of them come from so-called *hypernova* explosions—supernovas creating black holes rather than neutron stars.

Nuclear gamma rays were observed from the solar flares of August 4 and 7, 1972, and November 22, 1977. A solar flare is an explosion in a solar atmosphere and was originally detected visually in our own sun. Solar flares create massive amounts of radiation across the full electromagnetic spectrum from the longest wavelength, radio waves, to high energy gamma rays. The correlations of the high energy electrons energized during the flare and the gamma rays are mostly caused by nuclear combinations of high energy protons and other heavier ions. These gamma rays can be observed and allow scientists to determine the major results of the energy released, which is not provided by the emissions from other wavelengths.

1980s to 1990s

On June 19, 1988, from Birigüi (50° 20' W 21° 20' S) at 10:15 UTC a balloon launch occurred which carried two NaI(Tl) detectors (600 cm^2 total area) to an air pressure altitude of 5.5 mb for a total observation time of 6 hr. The supernova SN1987A in the Large Magellanic Cloud (LMC) was discovered on February 23, 1987, and its progenitor was a blue supergiant, (Sk -69 202), with luminosity of 2-5 x 10^{38} erg/s. The 847 keV and 1238 keV gamma-ray lines from ^{56}Co decay have been detected.

During its High Energy Astronomy Observatory program in 1977, NASA announced plans to build a "great observatory" for gamma-ray astronomy. The Compton Gamma-Ray Observatory (CGRO) was designed to take advantage of the major advances in detector technology during the 1980s, and was launched in 1991. The satellite carried four major instruments which have greatly improved the spatial and temporal resolution of gamma-ray observations. The CGRO provided large amounts of data which are being used to improve our understanding of the high-energy processes in our Universe. CGRO was de-orbited in June 2000 as a result of the failure of one of its stabilizing gyroscopes.

BeppoSAX was launched in 1996 and deorbited in 2003. It predominantly studied X-rays, but also observed gamma-ray bursts. By identifying the first non-gamma ray counterparts to gamma-ray bursts, it opened the way for their precise position determination and optical observation of their fading remnants in distant galaxies.

The High Energy Transient Explorer 2 (HETE-2) was launched in October 2000 (on a nominally 2 year mission) and was still operational (but fading) in March 2007.

Recent Observations

Swift, a NASA spacecraft, was launched in 2004 and carries the BAT instrument for gamma-ray burst observations. Following BeppoSAX and HETE-2, it has observed numerous X-ray and optical counterparts to bursts, leading to distance determinations and detailed optical follow-up. These have established that most bursts originate in the explosions of massive stars (supernovas and hypernovas) in distant galaxies. It is still operational in 2015.

Currently the (other) main space-based gamma-ray observatories are the INTErnational Gamma-Ray Astrophysics Laboratory (INTEGRAL), Fermi, and the Astrorivelatore Gamma ad Immagini Leggero (AGILE).

- INTEGRAL (launched on 17 October 2002) is an ESA mission with additional contributions from the Czech Republic, Poland, US, and Russia.

- AGILE is an all Italian small mission by ASI, INAF and INFN collaboration. It was successfully launched by the Indian PSLV-C8 rocket from the Sriharikota ISRO base on April 23, 2007.

- Fermi was launched by NASA on 11 June 2008. It includes LAT, the Large Area Telescope, and GBM, the GLAST Burst Monitor, for studying gamma-ray bursts.

Two gigantic gamma-ray bubbles at the heart of the Milky Way.

In November 2010, using the Fermi Gamma-ray Space Telescope, two gigantic gamma-ray bubbles, spanning about 25,000 light-years across, were detected at the heart of our galaxy. These bubbles of high-energy radiation are suspected as erupting from a massive black hole or evidence of a burst of star formations from millions of years ago. They were discovered after scientists filtered out the "fog of background gamma-rays suffusing the sky". This discovery confirmed previous clues that a large unknown "structure" was in the center of the Milky Way.

In 2011 the Fermi team released its second catalog of gamma-ray sources detected by the satellite's Large Area Telescope (LAT), which produced an inventory of 1,873 objects shining with the highest-energy form of light. 57% of the sources are Blazars. Over half of the sources are active galaxies, their central black holes created gamma-ray emissions detected by the LAT. One third of the sources have not been detected in other wavelengths.

Top 10 Gamma Ray Sources

The Fermi team created a year 2011 list of "top ten" gamma-ray sources. The top five sources in the Milky Way Galaxy: The Crab Nebula, W44, V407 Cygni, Pulsar PSR J0101-6422, 2FGL J0359.5+5410.

The top five sources outside the Milky Way Galaxy: Centaurus A, The Andromeda Galaxy (M31), The Cigar Galaxy (M82), Blazar PKS 0537-286, 2FGL J1305.0+1152.

A previous year 2009 "top ten" list of gamma-ray sources was created.

Radio Astronomy

The Very Large Array, a radio interferometer in New Mexico, USA

Radio astronomy is a subfield of astronomy that studies celestial objects at radio frequencies. The initial detection of radio waves from an astronomical object was made in the 1930s, when Karl Jansky observed radiation coming from the Milky Way. Subsequent observations have identified a number of different sources of radio emission. These include stars and galaxies, as well as entirely new classes of objects, such as radio galaxies, quasars, pulsars, and masers. The discovery of the cosmic microwave background radiation, regarded as evidence for the Big Bang theory, was made through radio astronomy.

Radio astronomy is conducted using large radio antennas referred to as radio telescopes, that are either used singularly, or with multiple linked telescopes utilizing the techniques of radio interferometry and aperture synthesis. The use of interferometry allows radio astronomy to achieve high angular resolution, as the resolving power of an interferometer is set by the distance between its components, rather than the size of its components.

History

Before Jansky observed the Milky Way in the 1930s, physicists speculated that radio waves could be observed from astronomical sources. In the 1860s, James Clerk Maxwell's equations had shown that electromagnetic radiation is associated with electricity and magnetism, and could exist at any wavelength. Several attempts were made to detect radio emission from the Sun including an experiment by German astrophysicists Johannes Wilsing and Julius Scheiner in 1896 and a centimeter wave radiation apparatus set up by Oliver Lodge between 1897-1900. These attempts were unable to detect any emission due to technical limitations of the instruments. The discovery of the radio reflecting ionosphere in 1902, led physicists to conclude that the layer would bounce any astronomical radio transmission back into space, making them undetectable.

Karl Jansky made the discovery of the first astronomical radio source serendipitously in the early 1930s. As an engineer with Bell Telephone Laboratories, he was investigating static that interfered with short wave transatlantic voice transmissions. Using a large directional antenna, Jansky noticed that his analog pen-and-paper recording system kept recording a repeating signal of unknown origin. Since the signal peaked about every 24 hours, Jansky originally suspected the source of the interference was the Sun crossing the view of his directional antenna. Continued analysis showed that the source was not following the 24-hour daily cycle of the Sun exactly, but instead repeating on a cycle of 23 hours and 56 minutes. Jansky discussed the puzzling phenomena with his friend, astrophysicist and teacher Albert Melvin Skellett, who pointed out that the time between the signal peaks was the exact length of a sidereal day, the time it took for "fixed" astronomical objects, such as a star, to passed in front of the antenna every time the Earth rotated. By comparing his observations with optical astronomical maps, Jansky eventually concluded that the radiation source peaked when his antenna was aimed at the densest part of the Milky Way in the constellation of Sagittarius. He concluded that since the Sun (and therefore other stars) were not large emitters of radio noise, the strange radio interference may be generated by interstellar gas and dust in the galaxy. (Jansky's peak radio source, one of the brightest in the sky, was designated Sagittarius A in the 1950s and, instead of being galactic "gas and dust", was later hypothesized to be emitted by electrons in a strong magnetic field. Current thinking is that these are ions in orbit around a massive Black hole at the center of the galaxy at a point now designated as Sagitarius A*. The asterisk indicates that the particles at Sagitarius A are ionized.) Jansky announced his discovery in 1933. He wanted to investigate the radio waves from the Milky Way in further detail, but Bell Labs reassigned him to another project, so he did no further work in the field of astronomy. His pioneering efforts in the field of radio astronomy have been recognized by the naming of the fundamental unit of flux density, the jansky (Jy), after him.

The Robert C. Byrd Green Bank Telescope (GBT) in West Virginia, United States is the world's largest fully steerable radio telescope.

Grote Reber was inspired by Jansky's work, and built a parabolic radio telescope 9m in diameter in his backyard in 1937. He began by repeating Jansky's observations, and then conducted the first sky survey in the radio frequencies. On February 27, 1942, James Stanley Hey, a British Army research officer, made the first detection of radio waves emitted by the Sun. Later that year George Clark Southworth, at Bell Labs like Jansky, also detected radiowaves from the sun. Both researchers were bound by wartime security surrounding radar, so Reber, who was not, published his 1944 findings first. Several other people independently discovered solar radiowaves, including E. Schott in Denmark and Elizabeth Alexander working on Norfolk Island.

At Cambridge University, where ionospheric research had taken place during World War II, J.A. Ratcliffe along with other members of the Telecommunications Research Establishment that had carried out wartime research into radar, created a radiophysics group at the university where radio wave emissions from the Sun were observed and studied.

This early research soon branched out into the observation of other celestial radio sources and interferometry techniques were pioneered to isolate the angular source of the detected emissions. Martin Ryle and Antony Hewish at the Cavendish Astrophysics Group developed the technique of Earth-rotation aperture synthesis. The radio astronomy group in Cambridge went on to found the Mullard Radio Astronomy Observatory near Cambridge in the 1950s. During the late 1960s and early 1970s, as computers (such as the Titan) became capable of handling the computationally intensive Fourier transform inversions required, they used aperture synthesis to create a 'One-Mile' and later a '5 km' effective aperture using the One-Mile and Ryle telescopes, respectively. They used the Cambridge Interferometer to map the radio sky, producing the famous 2C and 3C surveys of radio sources.

Techniques

First 7-metre ESO/NAOJ/NRAO ALMA Antenna.

Radio astronomers use different techniques to observe objects in the radio spectrum. Instruments may simply be pointed at an energetic radio source to analyze its emission. To "image" a region of the sky in more detail, multiple overlapping scans can be recorded and pieced together in a mosaic image. The type of instrument used depends on the strength of the signal and the amount of detail needed.

Observations from the Earth's surface are limited to wavelengths that can pass through the atmosphere. At low frequencies, or long wavelengths, transmission is limited by the ionosphere, which reflects waves with frequencies less than its characteristic plasma frequency. Water vapor interferes with radio astronomy at higher frequencies, which has led to building radio observatories that conduct observations at millimeter wavelengths at very high and dry sites, in order to minimize the water vapor content in the line of sight. Finally, transmitting devices on earth may cause radio-frequency interference. Because of this, many radio observatories are built at remote places.

Radio Telescopes

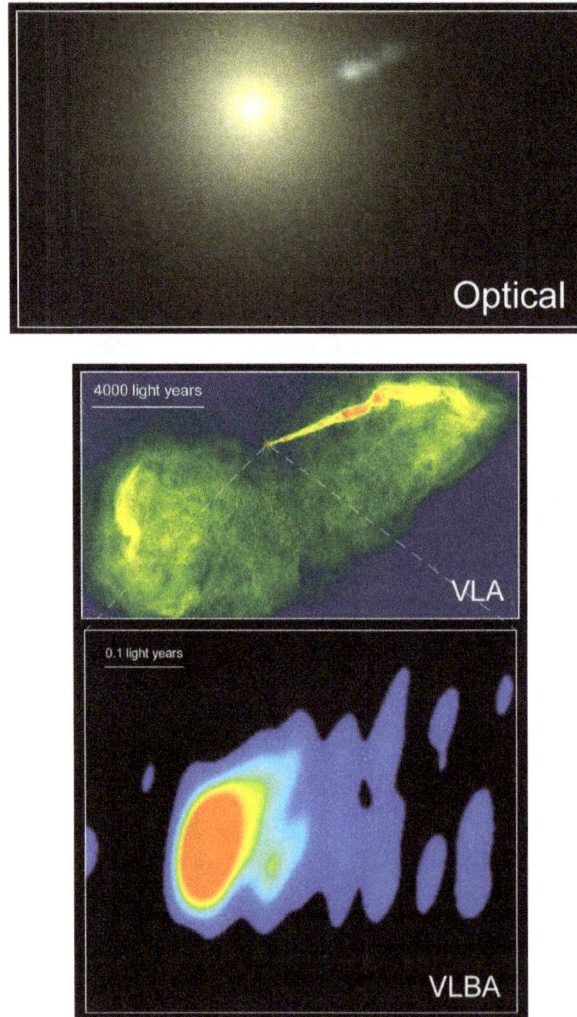

An **optical** image of the galaxy M87 (HST), a radio image of same galaxy using **Interferometry** (Very Large Array-**VLA**), and an image of the center section (**VLBA**) using a *Very Long Baseline Array* (Global VLBI) consisting of antennas in the US, Germany, Italy, Finland, Sweden and Spain. The jet of particles is suspected to be powered by a black hole in the center of the galaxy.

Radio telescopes may need to be extremely large in order to receive signals with high signal-to-noise ratio. Also since angular resolution is a function of the diameter of the "objective" in proportion to the wavelength of the electromagnetic radiation being observed, *radio telescopes* have to be much larger in comparison to their optical counterparts. For example, a 1-meter diameter optical telescope is two million times bigger than the wavelength of light observed giving it a resolution of roughly 0.3 arc seconds, whereas a radio telescope "dish" many times that size may, depending on the wavelength observed, only be able to resolve an object the size of the full moon (30 minutes of arc).

Radio Interferometry

The difficulty in achieving high resolutions with single radio telescopes led to radio interferometry, developed by British radio astronomer Martin Ryle and Australian engineer, radiophysicist, and radio astronomer Joseph Lade Pawsey and Ruby Payne-Scott in 1946. Surprisingly the first use of a radio interferometer for an astronomical observation was carried out by Payne-Scott, Pawsey and Lindsay McCready on 26 January 1946 using a SINGLE converted radar antenna (broadside array) at 200 MHz near Sydney, Australia. This group used the principle of a sea-cliff interferometer in which the antenna (formerly a World War II radar) observed the sun at sunrise with interference arising from the direct radiation from the sun and the reflected radiation from the sea. With this baseline of almost 200 meters, the authors determined that the solar radiation during the burst phase was much smaller than the solar disk and arose from a region associated with a large sunspot group. The Australia group laid out the principles of aperture synthesis in a ground-breaking paper published in 1947. The use of a sea-cliff interferometer had been demonstrated by numerous groups in Australia, Iran and the UK during World War II, who had observed interference fringes (the direct radar return radiation and the reflected signal from the sea) from incoming aircraft.

The Cambridge group of Ryle and Vonberg observed the sun at 175 MHz for the first time in mid July 1946 with a Michelson interferometer consisting of two radio antennas with spacings of some tens of meters up to 240 meters. They showed that the radio radiation was smaller than 10 arc minutes in size and also detected circular polarization in the Type I bursts. Two other groups had also detected circular polarization at about the same time (David Martyn in Australia and Edward Appleton with James Stanley Hey in the UK).

Modern Radio interferometers consist of widely separated radio telescopes observing the same object that are connected together using coaxial cable, waveguide, optical fiber, or other type of transmission line. This not only increases the total signal collected, it can also be used in a process called Aperture synthesis to vastly increase resolution. This technique works by superposing ("interfering") the signal waves from the different telescopes on the principle that waves that coincide with the same phase will add to each other while two waves that have opposite phases will cancel each other out. This creates a combined telescope that is the size of the antennas furthest apart in the array. In order to produce a high quality image, a large number of different separations between different telescopes are required (the projected separation between any two telescopes as seen from the radio source is called a "baseline") - as many different baselines as possible are required in order to get a good quality image. For example, the Very Large Array has 27 telescopes giving 351 independent baselines at once.

Very-long-baseline Interferometry

Beginning in the 1970s, improvements in the stability of radio telescope receivers permitted telescopes from all over the world (and even in Earth orbit) to be combined to perform very-long-baseline interferometry. Instead of physically connecting the antennas, data received at each antenna is paired with timing information, usually from a local atomic clock, and then stored for later analysis on magnetic tape or hard disk. At that later time, the data is correlated with data from other antennas similarly recorded, to produce the resulting image. Using this method it is possible to synthesise an antenna that is effectively the size of the Earth. The large distances between the telescopes enable very high angular resolutions to be achieved, much greater in fact than in any other field of astronomy. At the highest frequencies, synthesised beams less than 1 milliarcsecond are possible.

The Mount Pleasant Radio Telescope is the southern most antenna used in Australia's VLBI network

The pre-eminent VLBI arrays operating today are the Very Long Baseline Array (with telescopes located across North America) and the European VLBI Network (telescopes in Europe, China, South Africa and Puerto Rico). Each array usually operates separately, but occasional projects are observed together producing increased sensitivity. This is referred to as Global VLBI. There are also a VLBI networks, operating in Australia and New Zealand called the LBA (Long Baseline Array), and arrays in Japan, China and South Korea which observe together to form the East-Asian VLBI Network (EAVN).

Since its inception, recording data onto hard media was the only way to bring the data recorded at each telescope together for later correlation. However, the availability today of worldwide, high-bandwidth networks makes it possible to do VLBI in real time. This technique (referred to as e-VLBI) was originally pioneered in Japan, and more recently adopted in Australia and in Europe by the EVN (European VLBI Network) who perform an increasing number of scientific e-VLBI projects per year.

Astronomical Sources

Radio astronomy has led to substantial increases in astronomical knowledge, particularly with the discovery of several classes of new objects, including pulsars, quasars and radio galaxies. This is because radio astronomy allows us to see things that are not detectable in optical astronomy. Such objects represent some of the most extreme and energetic physical processes in the universe.

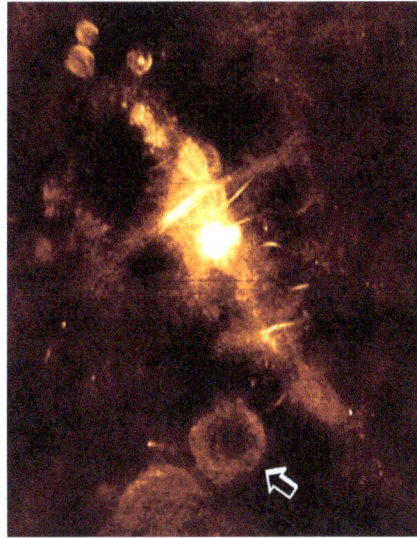

A radio image of the central region of the Milky Way galaxy. The arrow indicates a supernova remnant which is the location of a newly discovered transient, bursting low-frequency radio source GCRT J1745-3009.

The cosmic microwave background radiation was also first detected using radio telescopes. However, radio telescopes have also been used to investigate objects much closer to home, including observations of the Sun and solar activity, and radar mapping of the planets.

Other sources include:

- Sun

- Jupiter

- Sagittarius A, the galactic center of the Milky Way, with one portion Sagittarius A* thought to be a radio wave emitting supermassive black hole

- Active galactic nuclei and pulsars have jets of charged particles which emit synchrotron radiation

- Merging galaxy clusters often show diffuse radio emission

- Supernova remnants can also show diffuse radio emission; pulsars are a type of supernova remant that shows highly synchronous emission.

- The cosmic microwave background is blackbody radio/microwave emission

References

- Longair, Malcolm (2012). Cosmic century: a history of astrophysics and cosmology. Cambridge University Press. ISBN 1107669367.

- Krieg, Uwe (2008). Siegfried Röser, eds. Reviews in Modern Astronomy, Cosmic Matter. 20. WILEY-VCH. p. 191. ISBN 978-3-527-40820-7. Retrieved 2010-11-14.

- Overbye, Dennis (11 February 2016). "Physicists Detect Gravitational Waves, Proving Einstein Right". New York Times. Retrieved 11 February 2016.

- Krauss, Lawrence (11 February 2016). "Finding Beauty in the Darkness". New York Times. Retrieved 11 February 2016.

- "Measuring Intermediate-Mass Black-Hole Binaries with Advanced Gravitational Wave Detectors". Gravitational Physics Group. University of Birmingham. Retrieved 28 November 2015.

- Price, Larry (September 2015). "Looking for the Afterglow: The LIGO Perspective" (PDF). LIGO Magazine (7): 10. Retrieved 28 November 2015.

- Department of Astronautics (2008). "World's first astronautics department celebrates 50 years". Archived from the original on 2012-12-12.

- Morrison, Philip (March 16, 1958). "On gamma-ray astronomy". Il Nuovo Cimento (1955-1965). 7 (6): 858–865. doi:10.1007/BF02745590. Retrieved 2010-11-14.

- "Cosmic Rays Hunted Down: Physicists Closing in on Origin of Mysterious Particles". ScienceDaily. Dec 7, 2009. Retrieved 2010-11-14.

Techniques of Astronomy

Techniques are an important component of any field of study. Techniques used in astronomy are polarimetry, photometry and astronomical spectroscopy. It can be used to accurately position astronomic devices. The following chapter elucidates the techniques used in astronomy in a critical manner providing key analysis to the subject matter.

Polarimetry

Polarimetry is the measurement and interpretation of the polarization of transverse waves, most notably electromagnetic waves, such as radio or light waves. Typically polarimetry is done on electromagnetic waves that have traveled through or have been reflected, refracted, or diffracted by some material in order to characterize that object.

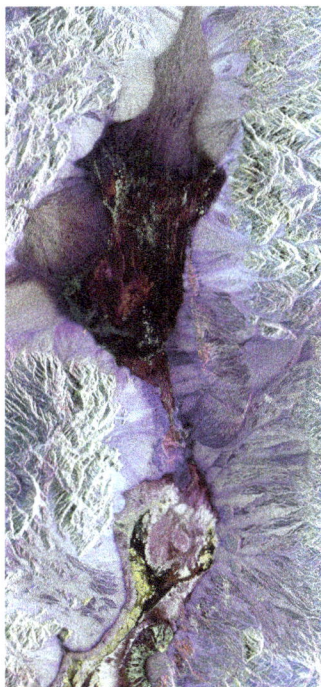

Synthetic aperture radar image of Death Valley colored using polarimetry.

Applications

Polarimetry of thin films and surfaces is commonly known as ellipsometry.

Polarimetry is used in remote sensing applications, such as planetary science, astronomy, and weather radar.

Polarimetry can also be included in computational analysis of waves. For example, radars often consider wave polarization in post-processing to improve the characterization of the targets. In this case, polarimetry can be used to estimate the fine texture of a material, help resolve the orientation of small structures in the target, and, when circularly-polarized antennas are used, resolve the number of bounces of the received signal (the chirality of circularly polarized waves alternates with each reflection).

Equipment

A polarimeter is the basic scientific instrument used to make these measurements, although this term is rarely used to describe a polarimetry process performed by a computer, such as is done in polarimetric synthetic aperture radar.

Polarimetry can be used to measure various optical properties of a material, including linear birefringence, circular birefringence (also known as optical rotation or optical rotary dispersion), linear dichroism, circular dichroism and scattering. To measure these various properties, there have been many designs of polarimeters. Some are archaic and some are in current use. The most sensitive polarimeters are based on interferometers, while more conventional polarimeters are based on arrangements of polarising filters, wave plates or other devices.

Astronomical Polarimetry

Light given off by a star is un-polarized, i.e. the direction of oscillation of the light wave is random. However, when the light is reflected off the atmosphere of a planet, the light waves interact with the molecules in the atmosphere and they are polarized.

By analyzing the polarization in the combined light of an extrasolar planet and its star (about one part in a million), these measurements can in principle be made with very high sensitivity also on ground-based observatories, as polarimetry is not limited by the stability of the Earth's atmosphere. It is akin to transiting of a planet in front of its star.

Measuring Optical Rotation

Optically active samples, such as solutions of chiral molecules, often exhibit circular birefringence. Circular birefringence causes rotation of the polarization of plane polarized light as it passes through the sample.

In ordinary light, the vibrations occur in all planes perpendicular to the direction of propagation. When light passes through a Nicol prism its vibrations in all directions except the direction of axis of the prism are cut off. The light emerging from the prism is said to be plane polarised because its vibration is in one direction. If two Nicol prisms are placed with their polarization planes parallel to each other, then the light rays emerging out of the first prism will enter the second prism. As a result, no loss of light is observed. However, if the second prism is rotated by an angle of 90°, the light emerging from the first prism is stopped by the second prism and no light emerges. The first prism is usually called the polarizer and the second prism is called the analyser.

A simple polarimeter to measure this rotation consists of a long tube with flat glass ends, into which the sample is placed. At each end of the tube is a Nicol prism or other polarizer. Light is

shone through the tube, and the prism at the other end, attached to an eye-piece, is rotated to arrive at the region of complete brightness or that of half-dark, half-bright or that of complete darkness. The angle of rotation is then read from a scale. The same phenomenon is observed after an angle of 180°. The specific rotation of the sample may then be calculated. Temperature can affect the rotation of light, which should be accounted for in the calculations.

$$[\alpha]_\lambda^T = 100\alpha / l\rho$$

where:

- $[\alpha]_\lambda^T$ is the specific rotation.
- T is the temperature.
- λ is the wavelength of light.
- α is the angle of rotation.
- l is the length of the polarimeter tube.
- ρ is the mass concentration of solution.

Photometry (Astronomy)

Photometry is a technique of astronomy concerned with measuring the flux, or intensity of an astronomical object's electromagnetic radiation. When photometry is performed over broad wavelength bands of radiation, where not only the amount of radiation but also its spectral distribution is measured, the term *spectrophotometry* is used.

The word is composed of the Greek affixes *photo-* ("light") and *-metry* ("measure").

Methods

The methods used to perform photometry depend on the wavelength regime under study. At its most basic, photometry is conducted by gathering light in a telescope, sometimes passing it through specialized photometric optical bandpass filters, and then capturing and recording the light energy with a photosensitive instrument. Standard sets of passbands (called a photometric system) are defined to allow accurate comparison of observations.

Historically, photometry in the near-infrared through long-wavelength ultra-violet was done with a photoelectric photometer, an instrument that measured the light intensity of a single object by directing its light onto a photosensitive cell. These have largely been replaced with CCD cameras that can simultaneously image multiple objects, although photoelectric photometers are still used in special situations, such as where fine time resolution is required.

CCD Photometry

A CCD camera is essentially a grid of photometers, simultaneously measuring and recording the photons coming from all the sources in the field of view. Because each CCD image records the photometry of multiple objects at once, various forms of photometric extraction can be performed on the recorded data; typically relative, absolute, and differential. All three will require the extraction of the raw image magnitude of the target object, and a known comparison object. The observed signal from an object will typically cover many pixels according to the point spread function (PSF) of the system. This broadening is due to both the optics in the telescope and the astronomical seeing. When obtaining photometry from a point source, the flux is measured by summing all the light recorded from the object and subtract the light due to the sky. The simplest technique, known as aperture photometry, consists of summing the pixel counts within an aperture centered on the object and subtracting the product of the nearby average sky count per pixel and the number of pixels within the aperture. This will result in the raw flux value of the target object. When doing photometry in a very crowded field, such as a globular cluster, where the profiles of stars overlap significantly, one must use de-blending techniques, such as PSF fitting to determine the individual flux values of the overlapping sources.

Calibrations

After determining the flux of an object in counts, the flux is normally converted into instrumental magnitude. Then, the measurement is calibrated in some way. Which calibrations are used will depend in part on what type of photometry is being done. Typically, observations are processed for relative, or differential photometry. Relative photometry is the measurement of the apparent brightness of multiple objects relative to each other. Absolute photometry is the measurement of the apparent brightness of an object on a standard photometric system; these measurements can be compared with other absolute photometric measurements obtained with different telescopes or instruments. Differential photometry is the measurement of the difference in brightness of two objects. In most cases, differential photometry can be done with the highest precision, while absolute photometry is the most difficult to do with high precision. Also, accurate photometry is usually more difficult when the apparent brightness of the object is fainter.

Absolute Photometry

To perform absolute photometry one must correct for differences between the effective passband through which an object is observed and the passband used to define the standard photometric system. This is often in addition to all of the other corrections discussed above. Typically this correction is done by observing the object(s) of interest through multiple filters and also observing a number of photometric standard stars. If the standard stars cannot be observed simultaneously with the target(s), this correction must be done under photometric conditions, when the sky is cloudless and the extinction is a simple function of the airmass.

Relative Photometry

To perform relative photometry, one compares the instrument magnitude of the object to a known comparison object, and then corrects the measurements for spatial variations in the sensitivity of the instrument and the atmospheric extinction. This is often in addition to correcting for their temporal variations, particularly when the objects being compared are too far apart on the sky to be observed simultaneously. When doing the calibration from an image that contains both the target and comparison objects in close proximity, and using a photometric filter that matches the catalog magnitude of the comparison object most of the measurement variations decrease to null.

Differential Photometry

Differential photometry is the simplest of the calibrations and most useful for time series observations. When using CCD photometry, both the target and comparison objects are observed at the same time, with the same filters, using the same instrument, and viewed through the same optical path. Most of the observational variables drop out and the differential magnitude is simply the difference between the instrument magnitude of the target object and the comparison object (ΔMag = C Mag − T Mag). This is very useful when plotting the change in magnitude over time of a target object, and is usually compiled into a light curve.

Applications

Photometric measurements can be combined with the inverse-square law to determine the luminosity of an object if its distance can be determined, or its distance if its luminosity is known. Other physical properties of an object, such as its temperature or chemical composition, may be determined via broad or narrow-band spectrophotometry. Typically photometric measurements of multiple objects obtained through two filters are plotted on a color-magnitude diagram, which for stars is the observed version of the Hertzsprung-Russell diagram. Photometry is also used to study the light variations of objects such as variable stars, minor planets, active galactic nuclei and supernovae, or to detect transiting extrasolar planets. Measurements of these variations can be used, for example, to determine the orbital period and the radii of the members of an eclipsing binary star system, the rotation period of a minor planet or a star, or the total energy output of a supernova.

Software

A number of free computer programs are available for synthetic aperture photometry and PSF-fitting photometry. SExtractor (www.astromatic.net/software/sextractor) and Aperture Photometry Tool (www.aperturephotometry.org) are popular examples for aperture photometry. The former is geared towards reduction of large scale galaxy-survey data, and the latter has a graphical user interface (GUI) suitable for studying individual images. DAOPHOT is recognized as the best software for PSF-fitting photometry.

Organizations

There are a number of organizations, from professional to amateur, that gather and share photometric data and make it available on-line. Some sites gather the data primarily as a resource for other researchers (e.x. AAVSO) and some solicit contributions of data for their own research (i.e. CBA):

- American Association of Variable Star Observers (AAVSO). www.aavso.org

- Center for Backyard Astrophysics (CBA). www.cbastro.org

- Digital-SF Cataclysmic Variable Database (DSF-Wiki) www.digial-sf.com/dsf-wiki

- Astronomyonlin.org. http://astronomyonline.org/Exoplanets/AmateurDetection.asp

Astronomical Spectroscopy

The Star-Spectroscope of the Lick Observatory in 1898. Designed by James Keeler and constructed by John Brashear.

Astronomical spectroscopy is the study of astronomy using the techniques of spectroscopy to measure the spectrum of electromagnetic radiation, including visible light and radio, which radiates from stars and other hot celestial objects. Spectroscopy can be used to derive many properties of distant stars and galaxies, such as their chemical composition, temperature, density, mass, distance, luminosity, and relative motion using Doppler shift measurements.

Background

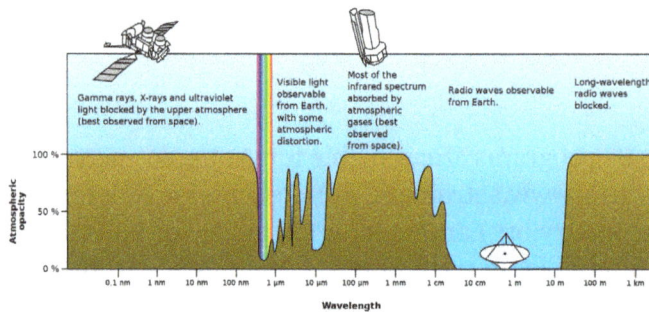

Electromagnetic transmittance, or opacity, of the Earth's atmosphere

Astronomical spectroscopy is used to measure three major bands of radiation: visible spectrum, radio, and X-ray. While all spectroscopy looks at specific areas of the spectrum, different methods are required to acquire the signal depending on the frequency. Ozone (O_3) and molecular oxygen (O_2) absorb light with wavelengths under 300 nm, meaning that X-ray and ultraviolet spectroscopy require the use of a satellite telescope or rocket mounted detectors.[27] Radio signals have much longer wavelengths than optical signals, and require the use of antennas or radio dishes. Infrared light is absorbed by atmospheric water and carbon dioxide, so while the equipment is similar to that used in optical spectroscopy, satellites are required to record much of the infrared spectrum.

Optical Spectroscopy

Physicists have been looking at the solar spectrum since Isaac Newton first used a simple prism to observe the refractive properties of light. In the early 1800s Joseph von Fraunhofer used his skills as a glass maker to create very pure prisms, which allowed him to observe 574 dark lines in a seemingly continuous spectrum. Soon after he combined telescope and prism to observe the spectrum of Venus, the Moon, Mars, and various stars such as Betelgeuse; his company continued to manufacture and sell high-quality refracting telescopes based on his original designs until its closure in 1884.

The resolution of a prism is limited by its size; a larger prism will provide a more detailed spectrum, but the increase in mass makes it unsuitable for highly detailed work. This issue was resolved in the early 1900s with the development of high-quality reflection gratings by J.S. Plaskett at the Dominion Observatory in Ottawa, Canada.[11] Light striking a mirror will reflect at the same angle, however a small portion of the light will be refracted at a different angle; this is dependent upon the indices of refraction of the materials and the wavelength of the light. By creating a "blazed" grating which utilizes a large number of parallel mirrors, the small portion of light can be focused and visualized. These new spectroscopes were more detailed than a prism, required less light, and could be focused on a specific region of the spectrum by tilting the grating.

The limitation to a blazed grating is the width of the mirrors, which can only be ground a finite amount before focus is lost; the maximum is around 1000 lines/mm. In order to overcome this limitation holographic gratings were developed. Volume phase holographic gratings use a thin film of dichromated gelatin on a glass surface, which is subsequently exposed to a wave pattern created by an interferometer. This wave pattern sets up a reflection pattern similar to the blazed gratings but utilizing Bragg diffraction, a process where the angle of reflection is dependent on the arrangement of the atoms in the gelatin. The holographic gratings can have up to 6000 lines/mm and can be up to twice as efficient in collecting light as blazed gratings. Because they are sealed between two sheets of glass, the holographic gratings are very versatile, potentially lasting decades before needing replacement.

Radio Spectroscopy

Radio astronomy was founded with the work of Karl Jansky in the early 1930s, while working for Bell Labs. He built a radio antenna to look at potential sources of interference for transatlantic radio transmissions. One of the sources of noise discovered came not from Earth, but from the center of the Milky Way, in the constellation Sagittarius. In 1942, JS Hey captured the sun's radio frequency using military radar receivers. Radio spectroscopy started with the discovery of the 21-centimeter H I line in 1951.

Radio interferometry was pioneered in 1946, when Joseph Lade Pawsey, Ruby Payne-Scott and Lindsay McCready used a single antenna atop a sea cliff to observe 200 MHz solar radiation. Two incident beams, one directly from the sun and the other reflected from the sea surface, generated the necessary interference. The first multi-receiver interferometer was built in the same year by Martin Ryle and Vonberg. In 1960, Ryle and Antony Hewish published the technique of aperture synthesis to analyze interferometer data. The aperture synthesis process, which involves autocorrelating and discrete Fourier transforming the incoming signal, recovers both the spatial and frequency variation in flux. The result is a 3D image whose third axis is frequency. For this work, Ryle and Hewish were jointly awarded the 1974 Nobel Prize in Physics.

X-ray Spectroscopy

Stars And Their Properties

Chemical Properties

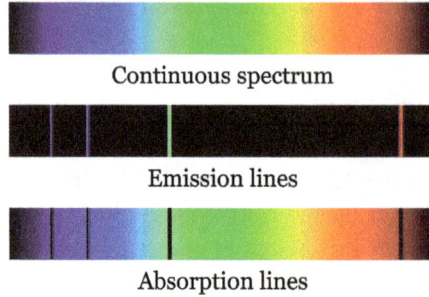

Continuous spectrum

Emission lines

Absorption lines

Newton used a prism to split white light into a spectrum of color, and Fraunhofer's high-quality prisms allowed scientists to see dark lines of an unknown origin. It was not until the 1850s that Gustav Kirchhoff and Robert Bunsen would describe the phenomena behind these dark lines—hot solid objects produce light with a continuous spectrum, hot gasses emit light at specific wavelengths, and hot solid objects surrounded by cooler gasses will show a near-continuous spectrum with dark lines corresponding to the emission lines of the gasses. By comparing the absorption lines of the sun with emission spectra of known gasses, the chemical composition of stars can be determined.

The major Fraunhofer lines, and the elements they are associated with, are shown in the following table. Designations from the early Balmer Series are in parentheses.

Designation	Element	Wavelength (nm)	Designation	Element	Wavelength (nm)
y	O_2	898.765	c	Fe	495.761
Z	O_2	822.696	F (Hβ)	H	486.134
A	O_2	759.370	d	Fe	466.814
B	O_2	686.719	e	Fe	438.355
C (Hα)	H	656.281	G' (Hγ)	H	434.047
a	O_2	627.661	G	Fe	430.790
D1	Na	589.592	G	Ca	430.774
D2	Na	588.995	h (Hδ)	H	410.175
D3 or d	He	587.5618	H	Ca+	396.847
e	Hg	546.073	K	Ca+	393.368
E2	Fe	527.039	L	Fe	382.044
b1	Mg	518.362	N	Fe	358.121
b2	Mg	517.270	P	Ti+	336.112
b3	Fe	516.891	T	Fe	302.108
b4	Mg	516.733	t	Ni	299.444

Not all of the elements in the sun were immediately identified. Two examples are listed below.

- In 1868 Norman Lockyer and Pierre Janssen independently observed a line next to the sodium doublet (D_1 and D_2) which Lockyer determined to be a new element. He named it Helium, but it wasn't until 1895 the element was found on Earth.

- In 1869 the astronomers Charles Augustus Young and William Harkness independently observed a novel green emission line in the Sun's corona during an eclipse. This "new" element was incorrectly named coronium, as it was only found in the corona. It was not until the 1930s that Walter Grotrian and Bengt Edlén discovered that the spectral line at 530.3 nm was due to highly ionized iron (Fe^{13+}). Other unusual lines in the coronal spectrum are also caused by highly charged ions, such as nickel and calcium, the high ionization being due to the extreme temperature of the solar corona.

To date more than 20 000 absorption lines have been listed for the Sun between 293.5 and 877.0 nm, yet only approximately 75% of these lines have been linked to elemental absorption.[69]

By analyzing the width of each spectral line in an emission spectrum, both the elements present in a star and their relative abundances can be determined. Using this information stars can be categorized into stellar populations; Population I stars are the youngest stars and have the highest metal content (our Sun is a Pop I star), while Population III stars are the oldest stars with a very low metal content.

Temperature and Size

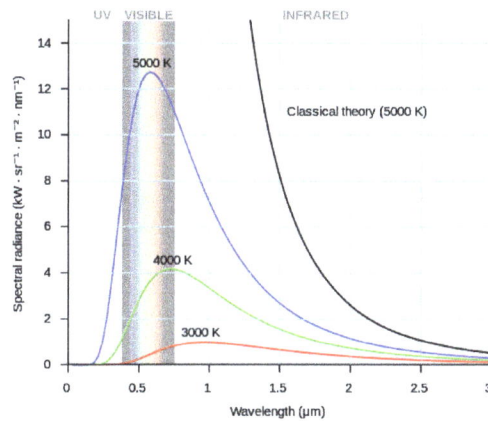

Black body curves for various temperatures.

In 1860 Gustav Kirchhoff proposed the idea of a black body, a material that emits electromagnetic radiation at all wavelengths. In 1894 Wilhelm Wien derived an expression relating the temperature (T) of a black body to its peak emission wavelength (λ_{max}).

$$\lambda_{max} T = b$$

b is a constant of proportionality called *Wien's displacement constant*, equal to $2.8977729(17) \times 10^{-3}$ m K. This equation is called Wien's Law. By measuring the peak wavelength of a star, the surface temperature can be determined. For example, if the peak wavelength of a star is 502 nm the corresponding temperature will be 5778 Kelvin.

The luminosity of a star is a measure of the electromagnetic energy output in a given amount of time. Luminosity (L) can be related to the temperature (T) of a star by

$$L = 4\pi R^2 \sigma T^4 ,$$

where R is the radius of the star and σ is the Stefan–Boltzmann constant, with a value of $5.670367(13) \times 10^{-8}$ W m^{-2} K^{-4}. Thus, when both luminosity and temperature are known (via direct measurement and calculation) the radius of a star can be determined.

Galaxies

The spectra of galaxies look similar to stellar spectra, as they consist of the combined light of millions of stars.

Doppler shift studies of galaxy clusters by Fritz Zwicky in 1937 found that most galaxies were moving much faster than seemed to be possible from what was known about the mass of the cluster. Zwicky hypothesized that there must be a great deal of non-luminous matter in the galaxy clusters, which became known as dark matter. Since his discovery, astronomers have determined that a large portion of galaxies (and most of the universe) is made up of dark matter. In 2003, however, four galaxies (NGC 821, NGC 3379, NGC 4494, and NGC 4697) were found to have little to no dark matter influencing the motion of the stars contained within them; the reason behind the lack of dark matter is unknown.

In the 1950s, strong radio sources were found to be associated with very dim, very red objects. When the first spectrum of one of these objects was taken there were absorption lines at wavelengths where none were expected. It was soon realised that what was observed was a normal galactic spectrum, but highly red shifted. These were named *quasi-stellar radio sources*, or quasars, by Hong-Yee Chiu in 1964. Quasars are now thought to be galaxies formed in the early years of our universe, with their extreme energy output powered by super-massive black holes.

The properties of a galaxy can also be determined by analyzing the stars found within them. NGC 4550, a galaxy in the Virgo Cluster, has a large portion of its stars rotating in the opposite direction as the other portion. It is believed that the galaxy is the combination of two smaller galaxies that were rotating in opposite directions to each other. Bright stars in galaxies can also help determine the distance to a galaxy, which may be a more accurate method than parallax or standard candles.

Interstellar Medium

The interstellar medium is matter that occupies the space between star systems in a galaxy. 99% of this matter is gaseous - hydrogen, helium, and smaller quantities of other ionized elements such as oxygen. The other 1% is dust particles, thought to be mainly graphite, silicates, and ices. Clouds of the dust and gas are referred to as nebulae.

There are three main types of nebula: absorption, reflection, and emission nebulae. Absorption (or dark) nebulae are made of dust and gas in such quantities that they obscure the starlight behind them, making photometry difficult. Reflection nebulae, as their name suggest, reflect the light of nearby stars. Their spectra are the same as the stars surrounding them, though the light is bluer;

shorter wavelengths scatter better than longer wavelengths. Emission nebulae emit light at specific wavelengths depending on their chemical composition.

Gaseous Emission Nebulae

In the early years of astronomical spectroscopy, scientists were puzzled by the spectrum of gaseous nebulae. In 1864 William Huggins noticed that many nebulae showed only emission lines rather than a full spectrum like stars. From the work of Kirchhoff, he concluded that nebulae must contain "enormous masses of luminous gas or vapour." However, there were several emission lines that could not be linked to any terrestrial element, brightest among them lines at 495.9 nm and 500.7 nm. These lines were attributed to a new element, nebulium, until Ira Bowen determined in 1927 that the emission lines were from highly ionised oxygen (O^{+2}). These emission lines could not be replicated in a laboratory because they are forbidden lines; the low density of a nebula (one atom per cubic centimetre) allows for metastable ions to decay via forbidden line emission rather than collisions with other atoms.

Not all emission nebulae are found around or near stars where solar heating causes ionisation. The majority of gaseous emission nebulae are formed of neutral hydrogen. In the ground state neutral hydrogen has two possible spin states: the electron has either the same spin or the opposite spin of the proton. When the atom transitions between these two states, it releases an emission or absorption line of 21 cm. This line is within the radio range and allows for very precise measurements:

- Velocity of the cloud can be measured via Doppler shift

- The intensity of the 21 cm line gives the density and number of atoms in the cloud

- The temperature of the cloud can be calculated

Using this information the shape of the Milky Way has been determined to be a spiral galaxy, though the exact number and position of the spiral arms is the subject of ongoing research.

Complex Molecules

Dust and molecules in the interstellar medium not only obscures photometry, but also causes absorption lines in spectroscopy. Their spectral features are generated by transitions of component electrons between different energy levels, or by rotational or vibrational spectra. Detection usually occurs in radio, microwave, or infrared portions of the spectrum. The chemical reactions that form these molecules can happen in cold, diffuse clouds or in the hot ejecta around a white dwarf star from a nova or supernova. Polycyclic aromatic hydrocarbons such as acetylene (C_2H_2) generally group together to form graphites or other sooty material, but other organic molecules such as acetone ($(CH_3)_2CO$) and buckminsterfullerenes (C_{60} and C_{70}) have been discovered.

Motion in the Universe

Stars and interstellar gas are bound by gravity to form galaxies, and groups of galaxies can be bound by gravity in galaxy clusters. With the exception of stars in the Milky Way and the galaxies in the Local Group, almost all galaxies are moving away from us due to the expansion of the universe.

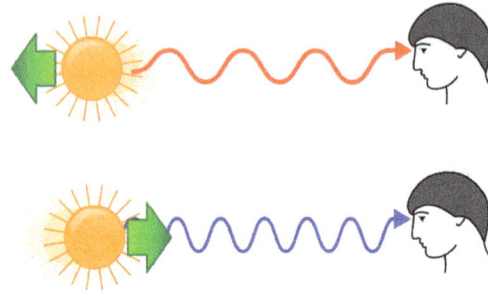

Redshift and blueshift

Doppler Effect and Redshift

The motion of stellar objects can be determined by looking at their spectrum. Because of the Doppler effect, objects moving towards us are blueshifted, and objects moving away are redshifted. The wavelength of redshifted light is longer, appearing redder than the source. Conversely, the wavelength of blueshifted light is shorter, appearing bluer than the source light:

$$\frac{\lambda - \lambda_0}{\lambda_0} = \frac{v_0}{c}$$

where λ_0 is the emitted wavelength, v_0 is the velocity of the object, and λ is the observed wavelength. Note that v<0 corresponds to $\lambda < \lambda_0$, a blueshifted wavelength. A redshifted absorption or emission line will appear more towards the red end of the spectrum than a stationary line. In 1913 Vesto Slipher determined the Andromeda Galaxy was blueshifted, meaning it was moving towards the Milky Way. He recorded the spectra of 20 other galaxies — all but 4 of which were redshifted — and was able to calculate their velocities relative to the Earth. Edwin Hubble would later use this information, as well as his own observations, to define Hubble's law: The further a galaxy is from the Earth, the faster it is moving away from us. Hubble's law can be generalised to

$$v = H_0 d$$

where v is the velocity (or Hubble Flow), H_0 is the Hubble Constant, and d is the distance from Earth.

Redshift (z) can be expressed by the following equations:

Calculation of redshift, z	
Based on wavelength	**Based on frequency**
$z = \dfrac{f_{emit} - f_{obsv}}{f_{obsv}}$	$\dfrac{f_{emit} \quad f_{obsv}}{obsv}$
$1 + z = \dfrac{\lambda_{obsv}}{\lambda_{emit}}$	$1 + z = \dfrac{f_{emit}}{f_{obsv}}$

In these equations, frequency is denoted by f and wavelength by λ. The larger the value of z, the more redshifted the light and the farther away the object is from the Earth. As of January 2013, the largest galaxy redshift of z~12 was found using the Hubble Ultra-Deep Field, corresponding to an age of over 13 billion years (the universe is approximately 13.82 billion years old).

The Doppler effect and Hubble's law can be combined to form the equation $z = \dfrac{v_{Hubble}}{c}$, where c is the speed of light.

Peculiar Motion

Objects that are gravitationally bound will rotate around a common center of mass. For stellar bodies, this motion is known as peculiar velocity, and can alter the Hubble Flow. Thus, an extra term for the peculiar motion needs to be added to Hubble's law:

$$v_{total} = H_0 d + v_{pec}$$

This motion can cause confusion when looking at a solar or galactic spectrum, because the expected redshift based on the simple Hubble law will be obscured by the peculiar motion. For example, the shape and size of the Virgo Cluster has been a matter of great scientific scrutiny due to the very large peculiar velocities of the galaxies in the cluster.

Binary Stars

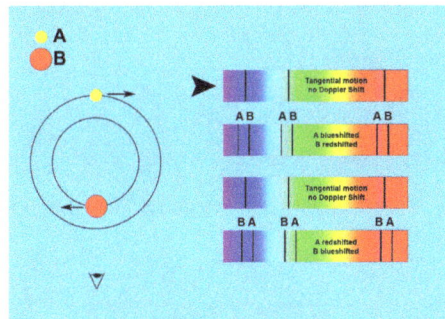

Two stars of different size orbiting the center of mass. The spectrum can be seen to split depending on the position and velocity of the stars.

Just as planets can be gravitationally bound to stars, pairs of stars can orbit each other. Some binary stars are visual binaries, meaning they can be observed orbiting each other through a telescope. Some binary stars, however, are too close together to be resolved. These two stars, when viewed through a spectrometer, will show a composite spectrum: the spectrum of each star will be added together. This composite spectrum becomes easier to detect when the stars are of similar luminosity and of different spectral class.

Spectroscopic binaries can be also detected due to their radial velocity; as they orbit around each other one star may be moving towards the Earth whilst the other moves away, causing a Doppler shift in the composite spectrum. The orbital plane of the system determines the magnitude of the observed shift: if the observer is looking perpendicular to the orbital plane there will be no observed radial velocity. For example, if you look at a carousel from the side, you will see the animals moving toward and away from you, whereas if you look from directly above they will only be moving in the horizontal plane.

Planets, Asteroids, and Comets

Planets and asteroids shine only by the reflected light of their parent star, while comets both absorb and emit light at various wavelengths.

Planets

The reflected light of a planet contains absorption bands due to minerals in the rocks present for rocky bodies, or due to the elements and molecules present in the atmospheres of gas giants. To date almost 1000 exoplanets have been discovered. These include so-called Hot Jupiters, as well as Earth-like planets. Using spectroscopy, compounds such as alkali metals, water vapor, carbon monoxide, carbon dioxide, and methane have all been discovered.

Asteroids

Asteroids can be classified into three major types according to their spectra. The original categories were created by Clark R. Chapman, David Morrison, and Ben Zellner in 1975, and further expanded by David J. Tholen in 1984. In what is now known as the Tholen classification, the C-types are made of carbonaceous material, S-types consist mainly of silicates, and X-types are 'metallic'. There are other classifications for unusual asteroids. C- and S-type asteroids are the most common asteroids. In 2002 the Tholen classification was further "evolved" into the SMASS classification, expanding the number of categories from 14 to 26 to account for more precise spectroscopic analysis of the asteroids.

Comets

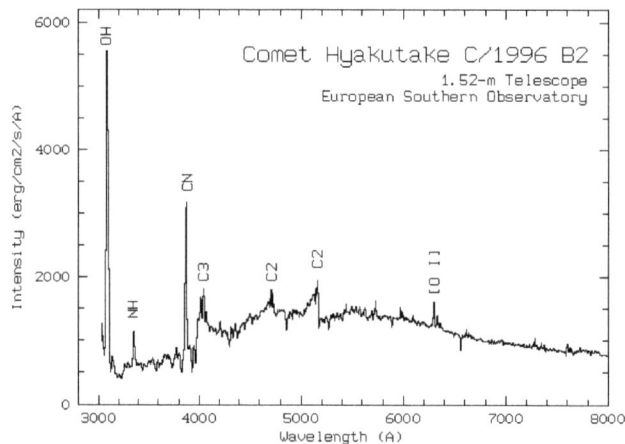

Optical spectrum of Comet Hyakutake.

The spectra of comets consist of a reflected solar spectrum from the dusty clouds surrounding the comet, as well as emission lines from gaseous atoms and molecules excited to fluorescence by sunlight and/or chemical reactions. For example, the chemical composition of Comet ISON was determined by spectroscopy due to the prominent emission lines of cyanogen (CN), as well as two- and three-carbon atoms (C_2 and C_3). Nearby comets can even be seen in X-ray as solar wind ions flying to the coma are neutralized. The cometary X-ray spectra therefore reflect the state of the solar wind rather than that of the comet.

References

- V. Tuchin (2000). Tissue Optics Light Scattering Methods and Instruments for Medical Diagnosis. Society of Photo Optical. ISBN 0-8194-3459-0.

- Sterken, Christiaan; Manfroid, J. (1992), Astronomical photometry: a guide, Astrophysics and space science library, 175, Springer, pp. 1–6, ISBN 0-7923-1653-3

- Hubbell, Gerald R. (2013). Scientific Astrophotography: How Amateurs Can Generate and Use Professional Imaging Data: 264-266. Springer, ISBN 978-1-4614-5173-0

- Ball, David W. (2001). Basics of Spectroscopy. Bellingham, Washington: Society of Photo-Optical Instrumentation Engineers. pp. 24, 28. ISBN 0-8194-4104-X.

- Robertson, Peter (1992). Beyond southern skies: radio astronomy and the Parkes telescope. University of Cambridge. pp. 42, 43. ISBN 0-521-41408-3.

- Jenkins, Francis A.; Harvey E. White (1957). Fundamentals of Optics (4th ed.). New York: McGraw-Hill. pp. 430–437. ISBN 0-07-085346-0.

- Gregory, Stephen A.; Michael Zeilik (1998). Introductory astronomy & astrophysics (4. ed.). Fort Worth [u.a.]: Saunders College Publ. p. 322. ISBN 0-03-006228-4.

- Nahar, Anil K. Pradhan, Sultana N. (2010). Atomic astrophysics and spectroscopy. Cambridge: Cambridge University Press. pp. 7,221. ISBN 978-0-521-82536-8.

- Mahmoud Massoud (2005). "§2.1 Blackbody radiation". Engineering thermofluids: thermodynamics, fluid mechanics, and heat transfer. Springer. p. 568. ISBN 3-540-22292-8.

- Kitchin, C.R. (1987). Stars, nebulae, and the interstellar medium : observational physics and astrophysics. Bristol: A. Hilger. pp. 265–277. ISBN 0-85274-580-X.

- Shu, Frank H. (1982). The physical universe : an introduction to astronomy (12. [Dr.]. ed.). Sausalito, Calif.: Univ. Science Books. pp. 232–234. ISBN 0-935702-05-9.

- Gray, Richard O.; Christopher J. Corbally (2009). Stellar spectral classification. Princeton, N.J.: Princeton University Press. pp. 507–513. ISBN 978-0-691-12510-7.

Major Astronomical Instruments

Telescope, zenith telescope, spectroscopy and astrolabe are the essential instruments used in astronomy. Telescope is a device used to perceive distant objects whereas a zenith telescope is particularly used to measure the distances of stars. The categories of the instruments are dealt with great details in this chapter.

Telescope

A telescope is an optical instrument that aids in the observation of remote objects by collecting electromagnetic radiation (such as visible light). The first known practical telescopes were invented in the Netherlands at the beginning of the 1600s, by using glass lenses. They found use in both terrestrial applications and astronomy.

The 100 inch (2.54 m) Hooker reflecting telescope at Mount Wilson Observatory near Los Angeles, USA.

Within a few decades, the reflecting telescope was invented, which used mirrors to collect and focus the light. In the 20th century many new types of telescopes were invented, including radio telescopes in the 1930s and infrared telescopes in the 1960s. The word *telescope* now refers to a wide range of instruments capable of detecting different regions of the electromagnetic spectrum, and in some cases other types of detectors.

The word *telescope* was coined in 1611 by the Greek mathematician Giovanni Demisiani for one of Galileo Galilei's instruments presented at a banquet at the Accademia dei Lincei. In the *Starry Messenger*, Galileo had used the term *perspicillum*.

History

Modern telescopes typically use CCDs instead of film for recording images.
This is the sensor array in the Kepler spacecraft.

28-inch telescope and 40-foot telescope in Greenwich in 2015.

The earliest recorded working telescopes were the refracting telescopes that appeared in the Netherlands in 1608. Their development is credited to three individuals: Hans Lippershey and Zacharias Janssen, who were spectacle makers in Middelburg, and Jacob Metius of Alkmaar. Galileo heard about the Dutch telescope in June 1609, built his own within a month, and improved upon the design in the following year. In the same year, Galileo became the first person to point a telescope skyward in order to make telescopic observations of a celestial object.

The idea that the objective, or light-gathering element, could be a mirror instead of a lens was being investigated soon after the invention of the refracting telescope. The potential advantages of using parabolic mirrors—reduction of spherical aberration and no chromatic aberration—led to many proposed designs and several attempts to build reflecting telescopes. In 1668, Isaac Newton built the first practical reflecting telescope, of a design which now bears his name, the Newtonian reflector.

The invention of the achromatic lens in 1733 partially corrected color aberrations present in the simple lens and enabled the construction of shorter, more functional refracting telescopes. Reflecting telescopes, though not limited by the color problems seen in refractors, were hampered by the use of fast tarnishing speculum metal mirrors employed during the 18th and early 19th century—a problem alleviated by the introduction of silver coated glass mirrors in 1857, and aluminized mirrors in 1932. The maximum physical size limit for refracting telescopes is about 1 meter (40 inches), dictating that the vast majority of large optical researching telescopes built since the turn of the 20th century have been reflectors. The largest reflecting telescopes currently have objectives larger than 10 m (33 feet), and work is underway on several 30-40m designs.

The 20th century also saw the development of telescopes that worked in a wide range of wavelengths from radio to gamma-rays. The first purpose built radio telescope went into operation in 1937. Since then, a tremendous variety of complex astronomical instruments have been developed.

Types

The name "telescope" covers a wide range of instruments. Most detect electromagnetic radiation, but there are major differences in how astronomers must go about collecting light (electromagnetic radiation) in different frequency bands.

Telescopes may be classified by the wavelengths of light they detect:

- X-ray telescopes, using shorter wavelengths than ultraviolet light
- Ultraviolet telescopes, using shorter wavelengths than visible light
- Optical telescopes, using visible light
- Infrared telescopes, using longer wavelengths than visible light
- Submillimetre telescopes, using longer wavelengths than infrared light
- Fresnel Imager, an optical lens technology
- X-ray optics, optics for certain X-ray wavelengths

Light Comparison				
Name	**Wavelength**	**Frequency (Hz)**	**Photon Energy (eV)**	
Gamma ray	less than 0.01 nm	more than 10 EHZ	100 keV – 300+ GeV	X
X-Ray	0.01 to 10 nm	30 PHz – 30 EHZ	120 eV to 120 keV	X
Ultraviolet	10 nm – 400 nm	30 EHZ – 790 THz	3 eV to 124 eV	
Visible	390 nm – 750 nm	790 THz – 405 THz	1.7 eV – 3.3 eV	X
Infrared	750 nm – 1 mm	405 THz – 300 GHz	1.24 meV – 1.7 eV	X
Microwave	1 mm – 1 meter	300 GHz – 300 MHz	1.24 meV – 1.24 μeV	
Radio	1 mm – km	300 GHz – 3 Hz	1.24 meV – 12.4 feV	X

As wavelengths become longer, it becomes easier to use antenna technology to interact with electromagnetic radiation (although it is possible to make very tiny antenna). The near-infrared can be collected much like visible light, however in the far-infrared and submillimetre range, telescopes can operate more like a radio telescope. For example, the James Clerk Maxwell Telescope observes from wavelengths from 3 μm (0.003 mm) to 2000 μm (2 mm), but uses a parabolic aluminum antenna. On the other hand, the Spitzer Space Telescope, observing from about 3 μm (0.003 mm) to 180 μm (0.18 mm) uses a mirror (reflecting optics). Also using reflecting optics, the Hubble Space Telescope with Wide Field Camera 3 can observe in the frequency range from about 0.2 μm (0.0002 mm) to 1.7 μm (0.0017 mm) (from ultra-violet to infrared light).

At the photon energy of shorter wavelengths (higher frequency), glancing-incident optics, rather than fully reflecting optics are used. Telescopes such as TRACE and SOHO use special mirrors to reflect Extreme ultraviolet, producing higher resolution and brighter images than are otherwise possible. A larger aperture does not just mean that more light is collected, it also enables a finer angular resolution.

Telescopes may also be classified by location: ground telescope, space telescope, or flying telescope. They may also be classified by whether they are operated by professional astronomers or amateur astronomers. A vehicle or permanent campus containing one or more telescopes or other instruments is called an observatory.

Optical Telescopes

An optical telescope gathers and focuses light mainly from the visible part of the electromagnetic spectrum (although some work in the infrared and ultraviolet). Optical telescopes increase the apparent angular size of distant objects as well as their apparent brightness. In order for the image to be observed, photographed, studied, and sent to a computer, telescopes work by employing one or more curved optical elements, usually made from glass lenses and/or mirrors, to gather light and other electromagnetic radiation to bring that light or radiation to a focal point. Optical telescopes are used for astronomy and in many non-astronomical instruments, including: *theodolites* (including *transits*), *spotting scopes, monoculars, binoculars, camera lenses*, and *spyglasses*. There are three main optical types:

- The refracting telescope which uses lenses to form an image.

- The reflecting telescope which uses an arrangement of mirrors to form an image.

- The catadioptric telescope which uses mirrors combined with lenses to form an image.

Beyond these basic optical types there are many sub-types of varying optical design classified by the task they perform such as astrographs, comet seekers, solar telescope, etc.

Radio Telescopes

Radio telescopes are directional radio antennas used for radio astronomy. The dishes are sometimes constructed of a conductive wire mesh whose openings are smaller than the wavelength being observed. Multi-element Radio telescopes are constructed from pairs or larger groups of these dishes to synthesize large 'virtual' apertures that are similar in size to the separation between the telescopes; this process is known as aperture synthesis. As of 2005, the current record array size is many times the width of the Earth—utilizing space-based Very Long Baseline Interferometry (VLBI) telescopes such as the Japanese HALCA (Highly Advanced Laboratory for Communications and Astronomy) VSOP (VLBI Space Observatory Program) satellite. Aperture synthesis is now also being applied to optical telescopes using optical interferometers (arrays of optical telescopes) and aperture masking interferometry at single reflecting telescopes. Radio telescopes are also used to collect microwave radiation, which is used to collect radiation when any visible light is obstructed or faint, such as from quasars. Some radio telescopes are used by programs such as SETI and the Arecibo Observatory to search for extraterrestrial life.

X-ray Telescopes

X-ray telescopes can use X-ray optics, such as a Wolter telescopes composed of ring-shaped 'glancing' mirrors made of heavy metals that are able to reflect the rays just a few degrees. The mirrors are usually a section of a rotated parabola and a hyperbola, or ellipse. In 1952, Hans Wolter outlined 3 ways a telescope could be built using only this kind of mirror. Examples of an observatory using this type of telescope are the Einstein Observatory, ROSAT, and the Chandra X-Ray Observatory. By 2010, Wolter focusing X-ray telescopes are possible up to 79 keV.

Einstein Observatory was a space-based focusing optical X-ray telescope from 1978.

Gamma-ray Telescopes

Higher energy X-ray and Gamma-ray telescopes refrain from focusing completely and use coded aperture masks: the patterns of the shadow the mask creates can be reconstructed to form an image.

X-ray and Gamma-ray telescopes are usually on Earth-orbiting satellites or high-flying balloons since the Earth's atmosphere is opaque to this part of the electromagnetic spectrum. However, high energy X-rays and gamma-rays do not form an image in the same way as telescopes at visible wavelengths. An example of this type of telescope is the Fermi Gamma-ray Space Telescope.

The detection of very high energy gamma rays, with shorter wavelength and higher frequency than regular gamma rays, requires further specialization. An example of this type of observatory is VERITAS. Very high energy gamma-rays are still photons, like visible light, whereas cosmic rays includes particles like electrons, protons, and heavier nuclei.

A discovery in 2012 may allow focusing gamma-ray telescopes. At photon energies greater than 700 keV, the index of refraction starts to increase again.

High-energy Particle Telescopes

High-energy astronomy requires specialized telescopes to make observations since most of these particles go through most metals and glasses.

In other types of high energy particle telescopes there is no image-forming optical system. Cosmic-ray telescopes usually consist of an array of different detector types spread out over a large area. A Neutrino telescope consists of a large mass of water or ice, surrounded by an array of sensitive light detectors known as photomultiplier tubes. Originating direction of the neutrinos is determined by reconstructing the path of secondary particles scattered by neutrino impacts, from their interaction with multiple detectors. Energetic neutral atom observatories like Interstellar Boundary Explorer detect particles traveling at certain energies.

Other Types of Telescopes

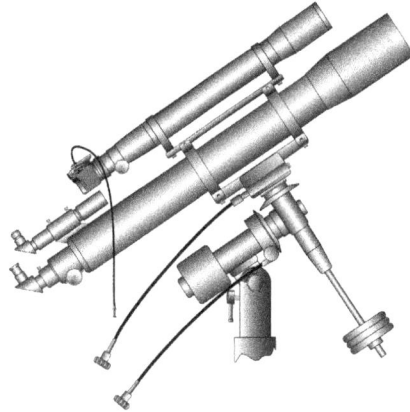

Equatorial-mounted Keplerian telescope

Astronomy is not limited to using electromagnetic radiation. Additional information can be obtained using other media. The detectors used to observe the Universe are analogous to telescopes, these are:

- Gravitational-wave detector, the equivalent of a gravitational wave telescope, used for gravitational-wave astronomy.

- Neutrino detector, the equivalent of a neutrino telescope, used for neutrino astronomy.

Types of Mount

A telescope mount is a mechanical structure which supports a telescope. Telescope mounts are designed to support the mass of the telescope and allow for accurate pointing of the instrument. Many sorts of mounts have been developed over the years, with the majority of effort being put into systems that can track the motion of the stars as the Earth rotates. The two main types of tracking mount are:

- Altazimuth mount

- Equatorial mount

Atmospheric Electromagnetic Opacity

Since the atmosphere is opaque for most of the electromagnetic spectrum, only a few bands can be observed from the Earth's surface. These bands are visible – near-infrared and a portion of the radio-wave part of the spectrum. For this reason there are no X-ray or far-infrared ground-based telescopes as these have to be observed from orbit. Even if a wavelength is observable from the ground, it might still be advantageous to place a telescope on a satellite due to astronomical seeing.

Zenith Telescope

A zenith telescope is a type of telescope that is designed to point straight up at or near the zenith. They are used for precision measurement of star positions, to simplify telescope construction, or both.

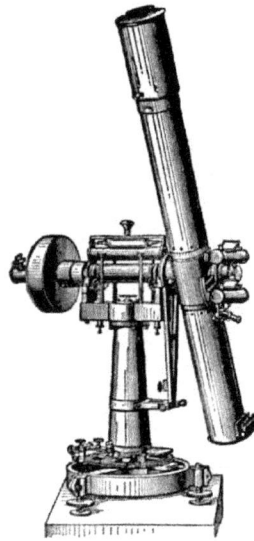

Zenith Telescope

Astronomical transit and zenith telescope, 1898

A classic zenith telescope, also known as a **zenith sector** employs a strong altazimuth mount, fitted with levelling screws. Extremely sensitive levels are attached to the telescope mount to make angle measurements and the telescope has an eyepiece fitted with a micrometer. They are used for the measurement of small differences of zenith distance, and used in the determination of astronomic latitude.

Other types of zenith telescopes include the Monument to the Great Fire of London, which includes a central shaft meant for use as a zenith telescope. High-precision (and fixed building) zenith telescopes were also used until the early 1980s to track Earth's north pole position e.g. Earth's rotation axis position. Since then radio astronomical quasar measurements have also measured Earth's rotation axis several orders of magnitude more accurately than optical tracking. The NASA Orbital Debris Observatory, which used a 3 m diameter aperture liquid mirror, and the Large Zenith Telescope, which uses a 6 m diameter aperture liquid mirror, are both zenith telescopes, as the use of liquid mirror meant these telescopes could only point straight up.

Spectroscopy

Spectroscopy is the study of the interaction between matter and electromagnetic radiation. Historically, spectroscopy originated through the study of visible light dispersed according to its wavelength, by a prism. Later the concept was expanded greatly to include any interaction with radiative energy as a function of its wavelength or frequency. Spectroscopic data is often represented by an emission spectrum, a plot of the response of interest as a function of wavelength or frequency.

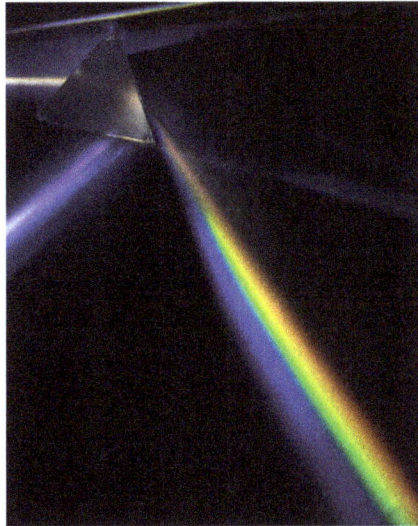

Analysis of white light by dispersing it with a prism is an example of spectroscopy.

Introduction

Spectroscopy and spectrography are terms used to refer to the measurement of radiation intensity as a function of wavelength and are often used to describe experimental spectroscopic methods. Spectral measurement devices are referred to as spectrometers, spectrophotometers, spectrographs or spectral analyzers.

Daily observations of color can be related to spectroscopy. Neon lighting is a direct application of atomic spectroscopy. Neon and other noble gases have characteristic emission frequencies (colors). Neon lamps use collision of electrons with the gas to excite these emissions. Inks, dyes and paints include chemical compounds selected for their spectral characteristics in order to generate specific colors and hues. A commonly encountered molecular spectrum is that of nitrogen dioxide. Gaseous nitrogen dioxide has a characteristic red absorption feature, and this gives air polluted with nitrogen dioxide a reddish brown color. Rayleigh scattering is a spectroscopic scattering phenomenon that accounts for the color of the sky.

Spectroscopic studies were central to the development of quantum mechanics and included Max Planck's explanation of blackbody radiation, Albert Einstein's explanation of the photoelectric effect and Niels Bohr's explanation of atomic structure and spectra. Spectroscopy is used in physical and analytical chemistry because atoms and molecules have unique spectra. As a result, these spectra can be used to detect, identify and quantify information about the atoms and molecules.

Spectroscopy is also used in astronomy and remote sensing on earth. Most research telescopes have spectrographs. The measured spectra are used to determine the chemical composition and physical properties of astronomical objects (such as their temperature and velocity).

Theory

One of the central concepts in spectroscopy is a resonance and its corresponding resonant frequency. Resonances were first characterized in mechanical systems such as pendulums. Mechanical systems that vibrate or oscillate will experience large amplitude oscillations when they are driven at their resonant frequency. A plot of amplitude vs. excitation frequency will have a peak centered at the resonance frequency. This plot is one type of spectrum, with the peak often referred to as a spectral line, and most spectral lines have a similar appearance.

In quantum mechanical systems, the analogous resonance is a coupling of two quantum mechanical stationary states of one system, such as an atom, via an oscillatory source of energy such as a photon. The coupling of the two states is strongest when the energy of the source matches the energy difference between the two states. The energy (E) of a photon is related to its frequency (v) by $E = hv$ where h is Planck's constant, and so a spectrum of the system response vs. photon frequency will peak at the resonant frequency or energy. Particles such as electrons and neutrons have a comparable relationship, the de Broglie relations, between their kinetic energy and their wavelength and frequency and therefore can also excite resonant interactions.

Spectra of atoms and molecules often consist of a series of spectral lines, each one representing a resonance between two different quantum states. The explanation of these series, and the spectral patterns associated with them, were one of the experimental enigmas that drove the development and acceptance of quantum mechanics. The hydrogen spectral series in particular was first successfully explained by the Rutherford-Bohr quantum model of the hydrogen atom. In some cases spectral lines are well separated and distinguishable, but spectral lines can also overlap and appear to be a single transition if the density of energy states is high enough. Named series of lines include the principal, sharp, diffuse and fundamental series.

Classification of Methods

Spectroscopy is a sufficiently broad field that many sub-disciplines exist, each with numerous implementations of specific spectroscopic techniques. The various implementations and techniques can be classified in several ways.

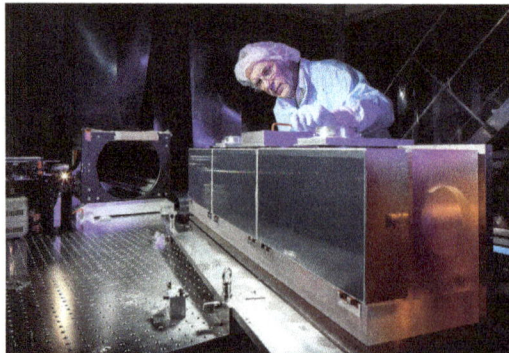

A huge diffraction grating at the heart of the ultra-precise ESPRESSO spectrograph.

Type of Radiative Energy

Types of spectroscopy are distinguished by the type of radiative energy involved in the interaction. In many applications, the spectrum is determined by measuring changes in the intensity or frequency of this energy. The types of radiative energy studied include:

- Electromagnetic radiation was the first source of energy used for spectroscopic studies. Techniques that employ electromagnetic radiation are typically classified by the wavelength region of the spectrum and include microwave, terahertz, infrared, near infrared, visible and ultraviolet, x-ray and gamma spectroscopy.

- Particles, due to their de Broglie wavelength, can also be a source of radiative energy and both electrons and neutrons are commonly used. For a particle, its kinetic energy determines its wavelength.

- Acoustic spectroscopy involves radiated pressure waves.

- Mechanical methods can be employed to impart radiating energy, similar to acoustic waves, to solid materials.

Nature of the Interaction

Types of spectroscopy can also be distinguished by the nature of the interaction between the energy and the material. These interactions include:

- Absorption occurs when energy from the radiative source is absorbed by the material. Absorption is often determined by measuring the fraction of energy transmitted through the material; absorption will decrease the transmitted portion.

- Emission indicates that radiative energy is released by the material. A material's blackbody spectrum is a spontaneous emission spectrum determined by its temperature; this feature can be measured in the infrared by instruments such as the Atmospheric Emitted Radiance Interferometer (AERI). Emission can also be induced by other sources of energy such as flames or sparks or electromagnetic radiation in the case of fluorescence.

- Elastic scattering and reflection spectroscopy determine how incident radiation is reflected or scattered by a material. Crystallography employs the scattering of high energy radiation, such as x-rays and electrons, to examine the arrangement of atoms in proteins and solid crystals.

- Impedance spectroscopy studies the ability of a medium to impede or slow the transmittance of energy. For optical applications, this is characterized by the index of refraction.

- Inelastic scattering phenomena involve an exchange of energy between the radiation and the matter that shifts the wavelength of the scattered radiation. These include Raman and Compton scattering.

- Coherent or resonance spectroscopy are techniques where the radiative energy couples two quantum states of the material in a coherent interaction that is sustained by the radiating field. The coherence can be disrupted by other interactions, such as particle collisions and energy transfer, and so often require high intensity radiation to be sustained. Nuclear magnetic resonance (NMR) spectroscopy is a widely used resonance method and ultrafast laser methods are also now possible in the infrared and visible spectral regions.

Type of Material

Spectroscopic studies are designed so that the radiant energy interacts with specific types of matter.

Atoms

Atomic spectroscopy was the first application of spectroscopy developed. Atomic absorption spectroscopy (AAS) and atomic emission spectroscopy (AES) involve visible and ultraviolet light. These absorptions and emissions, often referred to as atomic spectral lines, are due to electronic transitions of outer shell electrons as they rise and fall from one electron orbit to another. Atoms also have distinct x-ray spectra that are attributable to the excitation of inner shell electrons to excited states.

Atoms of different elements have distinct spectra and therefore atomic spectroscopy allows for the identification and quantitation of a sample's elemental composition. Robert Bunsen and Gustav Kirchhoff discovered new elements by observing their emission spectra. Atomic absorption lines are observed in the solar spectrum and referred to as Fraunhofer lines after their discoverer. A comprehensive explanation of the hydrogen spectrum was an early success of quantum mechanics and explained the Lamb shift observed in the hydrogen spectrum, which further led to the development of quantum electrodynamics.

Modern implementations of atomic spectroscopy for studying visible and ultraviolet transitions include flame emission spectroscopy, inductively coupled plasma atomic emission spectroscopy, glow discharge spectroscopy, microwave induced plasma spectroscopy, and spark or arc emission spectroscopy. Techniques for studying x-ray spectra include X-ray spectroscopy and X-ray fluorescence (XRF).

Molecules

The combination of atoms into molecules leads to the creation of unique types of energetic states and therefore unique spectra of the transitions between these states. Molecular spectra can be obtained due to electron spin states (electron paramagnetic resonance), molecular rotations, molecular vibration and electronic states. Rotations are collective motions of the atomic nuclei and typically lead to spectra in the microwave and millimeter-wave spectral regions; rotational spectroscopy and microwave spectroscopy are synonymous. Vibrations are relative motions of the atomic nuclei and are studied by both infrared and Raman spectroscopy. Electronic excitations are studied using visible and ultraviolet spectroscopy as well as fluorescence spectroscopy.

Studies in molecular spectroscopy led to the development of the first maser and contributed to the subsequent development of the laser.

Crystals and Extended Materials

The combination of atoms or molecules into crystals or other extended forms leads to the creation of additional energetic states. These states are numerous and therefore have a high density of states. This high density often makes the spectra weaker and less distinct, i.e., broader.

For instance, blackbody radiation is due to the thermal motions of atoms and molecules within a material. Acoustic and mechanical responses are due to collective motions as well. Pure crystals, though, can have distinct spectral transitions, and the crystal arrangement also has an effect on the observed molecular spectra. The regular lattice structure of crystals also scatters x-rays, electrons or neutrons allowing for crystallographic studies.

Nuclei

Nuclei also have distinct energy states that are widely separated and lead to gamma ray spectra. Distinct nuclear spin states can have their energy separated by a magnetic field, and this allows for NMR spectroscopy.

Other Types

Other types of spectroscopy are distinguished by specific applications or implementations:

- Acoustic resonance spectroscopy is based on sound waves primarily in the audible and ultrasonic regions

- Auger spectroscopy is a method used to study surfaces of materials on a micro-scale. It is often used in connection with electron microscopy.

- Cavity ring down spectroscopy

- Circular Dichroism spectroscopy

- Coherent anti-Stokes Raman spectroscopy (CARS) is a recent technique that has high sensitivity and powerful applications for *in vivo* spectroscopy and imaging.

- Cold vapour atomic fluorescence spectroscopy

- Correlation spectroscopy encompasses several types of two-dimensional NMR spectroscopy.

- Deep-level transient spectroscopy measures concentration and analyzes parameters of electrically active defects in semiconducting materials

- Dual polarisation interferometry measures the real and imaginary components of the complex refractive index

- Electron phenomenological spectroscopy measures physicochemical properties and characteristics of electronic structure of multicomponent and complex molecular systems.

- EPR spectroscopy

- Force spectroscopy

- Fourier transform spectroscopy is an efficient method for processing spectra data obtained using interferometers. Fourier transform infrared spectroscopy (FTIR) is a common implementation of infrared spectroscopy. NMR also employs Fourier transforms.

- Hadron spectroscopy studies the energy/mass spectrum of hadrons according to spin, parity, and other particle properties. Baryon spectroscopy and meson spectroscopy are both types of hadron spectroscopy.

- Hyperspectral imaging is a method to create a complete picture of the environment or various objects, each pixel containing a full visible, VNIR, NIR, or infrared spectrum.

- Inelastic electron tunneling spectroscopy (IETS) uses the changes in current due to inelastic electron-vibration interaction at specific energies that can also measure optically forbidden transitions.

- Inelastic neutron scattering is similar to Raman spectroscopy, but uses neutrons instead of photons.

- Laser-Induced Breakdown Spectroscopy (LIBS), also called Laser-induced plasma spectrometry (LIPS)

- Laser spectroscopy uses tunable lasers and other types of coherent emission sources, such as optical parametric oscillators, for selective excitation of atomic or molecular species.

- Mass spectroscopy is an historical term used to refer to mass spectrometry. Current recommendations are to use the latter term. Use of the term mass spectroscopy originated in the use of phosphor screens to detect ions.

- Mössbauer spectroscopy probes the properties of specific isotopic nuclei in different atomic environments by analyzing the resonant absorption of gamma-rays.

- Neutron spin echo spectroscopy measures internal dynamics in proteins and other soft matter systems

- Photoacoustic spectroscopy measures the sound waves produced upon the absorption of radiation.

- Photoemission spectroscopy

- Photothermal spectroscopy measures heat evolved upon absorption of radiation.

- Pump-probe spectroscopy can use ultrafast laser pulses to measure reaction intermediates in the femtosecond timescale.

- Raman optical activity spectroscopy exploits Raman scattering and optical activity effects to reveal detailed information on chiral centers in molecules.

- Raman spectroscopy

- Saturated spectroscopy

- Scanning tunneling spectroscopy

- Spectrophotometry

- Time-resolved spectroscopy measures the decay rate(s) of excited states using various spectroscopic methods.

- Time-Stretch Spectroscopy

- Thermal infrared spectroscopy measures thermal radiation emitted from materials and surfaces and is used to determine the type of bonds present in a sample as well as their lattice environment. The techniques are widely used by organic chemists, mineralogists, and planetary scientists.

- Ultraviolet photoelectron spectroscopy (UPS)

- Ultraviolet–visible spectroscopy

- Vibrational circular dichroism spectroscopy

- Video spectroscopy

- X-ray photoelectron spectroscopy (XPS)

Applications

UVES is a high-resolution spectrograph on the Very Large Telescope.

- Cure monitoring of composites using optical fibers.

- Estimate weathered wood exposure times using near infrared spectroscopy.

- Measurement of different compounds in food samples by absorption spectroscopy both in visible and infrared spectrum.

History

The history of spectroscopy began with Isaac Newton's optics experiments (1666–1672). Newton applied the word "spectrum" to describe the rainbow of colors that combine to form white light and that are revealed when the white light is passed through a prism. During the early 1800s, Joseph von Fraunhofer made experimental advances with dispersive spectrometers that enabled spectroscopy to become a more precise and quantitative scientific technique. Since then, spectroscopy has played and continues to play a significant role in chemistry, physics and astronomy.

Astrolabe

An astrolabe (Greek: *astrolabos*, «star-taker») is an elaborate inclinometer, historically used by astronomers, navigators, and astrologers. Its many uses include locating and predicting the positions of the Sun, Moon, planets, and stars, determining local time given local latitude and vice versa, surveying, and triangulation. It was used in classical antiquity, the Islamic Golden Age, the European Middle Ages and Renaissance for all these purposes.

A modern astrolabe, made in Tabriz, Iran in 2013.

Astrolabe quadrant, England, 1388

A 16th-century astrolabe, showing a tulip rete and rule

In the Islamic world, it was also used to calculate the direction to Mecca.

There is often confusion between the astrolabe and the mariner's astrolabe. While the astrolabe could be useful for determining latitude on land, it was an awkward instrument for use on the heaving deck of a ship or in wind. The mariner's astrolabe was developed to solve these problems.

Etymology

OED gives the translation "star-taker" for the English word "astrolabe" and traces it, through medieval Latin, to the Greek word *astrolabos* from *astron* "star" and *lambanein* "to take". In the medieval Islamic world the word "asturlab" (i.e. astrolabe) was given various etymologies. In Arabic texts, the word is translated as "akhdh al-kawakib" (lit. "taking the stars") which corresponds to an interpretation of the Greek word. Al-Biruni quotes and criticizes the medieval scientist Hamzah al-Isfahani who had stated: "asturlab is an arabization of this Persian phrase" (*sitara yab*, meaning "taker of the stars"). In medieval Islamic sources, there is also a "fictional" and popular etymology of the words as "lines of lab". In this popular etymology, "Lab" is a certain son of Idris (=Enoch). This etymology is mentioned by a 10th-century scientist named al-Qummi but rejected by al-Khwarizmi. "Lab" in Arabic also means "sun" and "black stony places" (cf. Dictionary).

History

Ancient World

An early astrolabe was invented in the Hellenistic world by Apollonius of Perga, around 220 BCE or in 150 BC and is often attributed to Hipparchus. A marriage of the planisphere and dioptra, the astrolabe was effectively an analog calculator capable of working out several different kinds of problems in spherical astronomy. Theon of Alexandria wrote a detailed treatise on the astrolabe, and Lewis argues that Ptolemy used an astrolabe to make the astronomical observations recorded in the *Tetrabiblos*. It is generally accepted that Greek astrologers, in either the first or second centuries BC, invented the *astrolabe*, an instrument that measures the altitude of stars and planets above the horizon. Some historians attribute its invention to Hypatia, the daughter of the mathematician Theon Alexandricus (c. 335 – c. 405), and others note that Synesius, a student of Hypatia, credits her for the invention in his letters.

Astrolabes continued in use in the Greek-speaking world throughout the Byzantine period. About 550 AD the Christian philosopher John Philoponus wrote a treatise on the astrolabe in Greek, which is the earliest extant Greek treatise on the instrument. In addition, Severus Sebokht, a bishop who lived in Mesopotamia, also wrote a treatise on the astrolabe in Syriac in the mid-7th century. Severus Sebokht refers in the introduction of his treatise to the astrolabe as being made of brass, indicating that metal astrolabes were known in the Christian East well before they were developed in the Islamic world or the Latin West.

Medieval Era

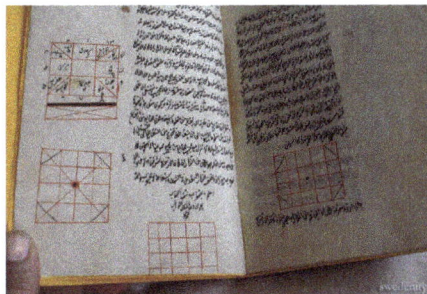

A treatise explaining the importance of the astrolabe by Nasir al-Din al-Tusi, Persian scientist.

Astrolabes were further developed in the medieval Islamic world, where Muslim astronomers introduced angular scales to the astrolabe, adding circles indicating azimuths on the horizon. It was widely used throughout the Muslim world, chiefly as an aid to navigation and as a way of finding the Qibla, the direction of Mecca. The first person credited with building the astrolabe in the Islamic world is reportedly the 8th-century mathematician Muhammad al-Fazari. The mathematical background was established by the Muslim astronomer Albatenius in his treatise *Kitab az-Zij* (ca. 920 AD), which was translated into Latin by Plato Tiburtinus (*De Motu Stellarum*). The earliest surviving dated astrolabe is dated AH 315 (927/8 AD). In the Islamic world, astrolabes were used to find the times of sunrise and the rising of fixed stars, to help schedule morning prayers (salat). In the 10th century, al-Sufi first described over 1,000 different uses of an astrolabe, in areas as diverse as astronomy, astrology, navigation, surveying, timekeeping, prayer, Salat, Qibla, etc.

Astrolabe of Jean Fusoris (fr), made in Paris, 1400

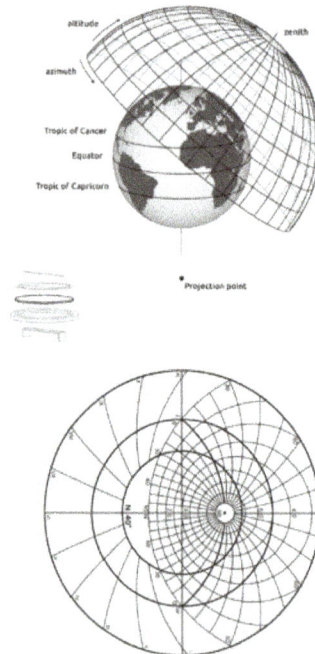

Animation showing how celestial and geographic coordinates are mapped on an astrolabe's tympan through a stereographic projection. Hypothetical tympan (40 degrees North Latitude) of a 16th-century European planispheric astrolabe.

Astrolabium Masha'Allah Public Library Bruges (nl) Ms. 522

The spherical astrolabe, a variation of both the astrolabe and the armillary sphere, was invented during the Middle Ages by astronomers and inventors in the Islamic world. The earliest description of the spherical astrolabe dates back to Al-Nayrizi (fl. 892–902). In the 12th century, Sharaf al-Dīn al-Tūsī invented the *linear astrolabe*, sometimes called the "staff of al-Tusi," which was "a simple wooden rod with graduated markings but without sights. It was furnished with a plumb line and a double chord for making angular measurements and bore a perforated pointer." The first geared mechanical astrolabe was later invented by Abi Bakr of Isfahan in 1235.

Peter of Maricourt, in the last half of the 13th century, also wrote a treatise on the construction and use of a universal astrolabe (*Nova compositio astrolabii particularis*). Universal astrolabes can be found at the History of Science Museum in Oxford.

The English author Geoffrey Chaucer (ca. 1343–1400) compiled a treatise on the astrolabe for his son, mainly based on Messahalla. The same source was translated by the French astronomer and astrologer Pélerin de Prusse and others. The first printed book on the astrolabe was *Composition and Use of Astrolabe* by Christian of Prachatice, also using Messahalla, but relatively original.

In 1370, the first Indian treatise on the astrolabe was written by the Jain astronomer Mahendra Suri.

The first known metal astrolabe in Western Europe is the Destombes astrolabe made from brass in tenth-century Spain. Metal astrolabes avoided the warping that large wooden astrolabes were prone to, allowing the construction of larger and therefore more accurate instruments; however, metal astrolabes were also heavier than wooden instruments of the same size, making it difficult to use them as navigational instruments. The astrolabe was almost certainly first brought north of the Pyrenees by Gerbert of Aurillac (future Pope Sylvester II), where it was integrated into the quadrivium at the school in Reims, France, sometime before the turn of the 11th century. In the 15th century, the French instrument-maker Jean Fusoris (fr) (ca. 1365–1436) also started remaking and selling astrolabes in his shop in Paris, along with portable sundials and other popular scientific gadgets of the day. Thirteen of his astrolabes survive to this day. Finally, one more special example of craftsmanship in the early 15th-century Europe is the astrolabe dated 1420, designed by Antonius de Pacento and made by Dominicus de Lanzano.

In the 16th century, Johannes Stöffler published *Elucidatio fabricae ususque astrolabii*, a manual of the construction and use of the astrolabe. Four identical 16th-century astrolabes made by Georg Hartmann provide some of the earliest evidence for batch production by division of labor.

Astrolabes and Clocks

At first mechanical astronomical clocks were influenced by the astrolabe; in many ways they could be seen as clockwork astrolabes designed to produce a continual display of the current position of the sun, stars, and planets. For example, Richard of Wallingford's clock (c. 1330) consisted essentially of a star map rotating behind a fixed rete, similar to that of an astrolabe.

Many astronomical clocks, such as the famous clock at Prague, use an astrolabe-style display, adopting a stereographic projection of the ecliptic plane.

In recent times, astrolabe watches have become a feature of haute horologie. For example, in 1985 Swiss watchmaker Dr. Ludwig Oechslin designed and built an astrolabe wristwatch in conjunction with Ulysse Nardin. Dutch watchmaker Christaan van der Klauuw also manufactures astrolabe watches today.

Construction

An astrolabe consists of a disk, called the *mater* (mother), which is deep enough to hold one or more flat plates called *tympans*, or *climates*. A tympan is made for a specific latitude and is engraved with a stereographic projection of circles denoting azimuth and altitude and representing the portion of the celestial sphere above the local horizon. The rim of the mater is typically graduated into hours of time, degrees of arc, or both. Above the mater and tympan, the *rete*, a framework bearing a projection of the ecliptic plane and several pointers indicating the positions of the brightest stars, is free to rotate. These pointers are often just simple points, but depending on the skill of the craftsman can be very elaborate and artistic. There are examples of astrolabes with artistic pointers in the shape of balls, stars, snakes, hands, dogs' heads, and leaves, among others. Some astrolabes have a narrow *rule* or *label* which rotates over the rete, and may be marked with a scale of declinations.

The Hartmann astrolabe in Yale collection. This instrument shows its rete and rule.

Celestial Globe, Isfahan (?), Iran 1144. Shown at the Louvre Museum, this globe is the 3rd oldest surviving in the world

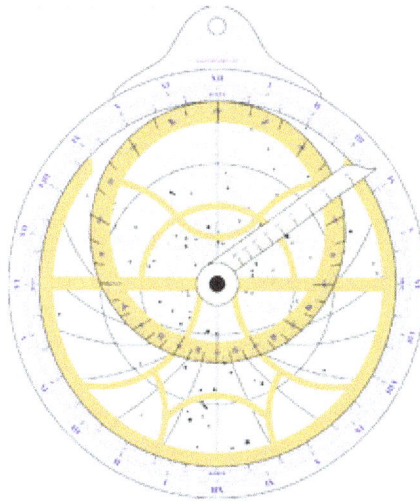

Computer-generated planispheric astrolabe

The rete, representing the sky, functions as a star chart. When it is rotated, the stars and the ecliptic move over the projection of the coordinates on the tympan. One complete rotation corresponds to the passage of a day. The astrolabe is therefore a predecessor of the modern planisphere.

On the back of the mater there is often engraved a number of scales that are useful in the astrolabe's various applications; these vary from designer to designer, but might include curves for time conversions, a calendar for converting the day of the month to the sun's position on the ecliptic, trigonometric scales, and a graduation of 360 degrees around the back edge. The *alidade* is attached to the back face. An alidade can be seen in the lower right illustration of the Persian astrolabe above. When the astrolabe is held vertically, the alidade can be rotated and the sun or a star sighted along its length, so that its altitude in degrees can be read ("taken") from the graduated edge of the astrolabe; hence the word's Greek roots: "astron" = star + "lab-" = to take.

Models of Astronomy

Some of the models of astronomy are theoretical astronomy, geocentric model, heliocentrism and cosmogony. The geocentric model of astronomy considers the Earth to be the center of the universe, under this model, the sun, the moon, the planets all circle around the Earth. The following content explains all the models of astronomy, providing the reader with a detailed understanding on the topic.

Theoretical Astronomy

Theoretical astronomy is the use of the analytical models of physics and chemistry to describe astronomical objects and astronomical phenomena.

Ptolemy's Almagest, although a brilliant treatise on theoretical astronomy combined with a practical handbook for computation, nevertheless includes many compromises to reconcile discordant observations. Theoretical astronomy is usually assumed to have begun with Johannes Kepler (1571–1630), and Kepler's laws. It is co-equal with observation. The general history of astronomy deals with the history of the descriptive and theoretical astronomy of the Solar System, from the late sixteenth century to the end of the nineteenth century. The major categories of works on the history of modern astronomy include general histories, national and institutional histories, instrumentation, descriptive astronomy, theoretical astronomy, positional astronomy, and astrophysics. Astronomy was early to adopt computational techniques to model stellar and galactic formation and celestial mechanics. From the point of view of theoretical astronomy, not only must the mathematical expression be reasonably accurate but it should preferably exist in a form which is amenable to further mathematical analysis when used in specific problems. Most of theoretical astronomy uses Newtonian theory of gravitation, considering that the effects of general relativity are weak for most celestial objects. The obvious fact is that theoretical astronomy cannot (and does not try) to predict the position, size and temperature of every star in the heavens. Theoretical astronomy by and large has concentrated upon analyzing the apparently complex but periodic motions of celestial objects.

Integrating Astronomy and Physics

"Contrary to the belief generally held by laboratory physicists, astronomy has contributed to the growth of our understanding of physics." Physics has helped in the elucidation of astronomical phenomena, and astronomy has helped in the elucidation of physical phenomena:

1. discovery of the law of gravitation came from the information provided by the motion of the Moon and the planets,

2. viability of nuclear fusion as demonstrated in the Sun and stars and yet to be reproduced on earth in a controlled form.

Integrating astronomy with physics involves

Physical interaction	Astronomical phenomena
Electromagnetism:	observation using the electromagnetic spectrum
black body radiation	stellar radiation
synchrotron radiation	radio and X-ray sources
inverse-Compton scattering	astronomical X-ray sources
acceleration of charged particles	pulsars and cosmic rays
absorption/scattering	interstellar dust
Strong and weak interaction:	nucleosynthesis in stars
	cosmic rays
	supernovae
	primeval universe
Gravity:	motion of planets, satellites and binary stars, stellar structure and evolution, N-body motions in clusters of stars and galaxies, black holes, and the expanding universe.

The aim of astronomy is to understand the physics and chemistry from the laboratory that is behind cosmic events so as to enrich our understanding of the cosmos and of these sciences as well.

Integrating Astronomy and Chemistry

Astrochemistry, the overlap of the disciplines of astronomy and chemistry, is the study of the abundance and reactions of chemical elements and molecules in space, and their interaction with radiation. The formation, atomic and chemical composition, evolution and fate of molecular gas clouds, is of special interest because it is from these clouds that solar systems form.

Infrared astronomy, for example, has revealed that the interstellar medium contains a suite of complex gas-phase carbon compounds called aromatic hydrocarbons, often abbreviated (PAHs or PACs). These molecules composed primarily of fused rings of carbon (either neutral or in an ionized state) are said to be the most common class of carbon compound in the galaxy. They are also the most common class of carbon molecule in meteorites and in cometary and asteroidal dust (cosmic dust). These compounds, as well as the amino acids, nucleobases, and many other compounds in meteorites, carry deuterium and isotopes of carbon, nitrogen, and oxygen that are very rare on earth, attesting to their extraterrestrial origin. The PAHs are thought to form in hot circumstellar environments (around dying carbon rich red giant stars).

The sparseness of interstellar and interplanetary space results in some unusual chemistry, since symmetry-forbidden reactions cannot occur except on the longest of timescales. For this reason, molecules and molecular ions which are unstable on earth can be highly abundant in space, for example the H_3^+ ion. Astrochemistry overlaps with astrophysics and nuclear physics in characterizing the nuclear reactions which occur in stars, the consequences for stellar evolution, as well as stellar 'generations'. Indeed, the nuclear reactions in stars produce every naturally occurring chemical element. As the stellar 'generations' advance, the mass of the newly formed elements increases. A first-generation star uses elemental hydrogen (H) as a fuel source and produces helium (He). Hydrogen is the most abundant element, and it is the basic building block for all other elements as its nucleus has only one proton. Gravitational pull toward the center of a star creates massive amounts of heat and pressure, which cause nuclear fusion. Through this process of merging nuclear mass, heavier elements are formed. Lithium, carbon, nitrogen and oxygen are examples of elements that form in stellar fusion. After many stellar generations, very heavy elements are formed (e.g. iron and lead).

Tools of Theoretical Astronomy

Theoretical astronomers use a wide variety of tools which include analytical models (for example, polytropes to approximate the behaviors of a star) and computational numerical simulations. Each has some advantages. Analytical models of a process are generally better for giving insight into the heart of what is going on. Numerical models can reveal the existence of phenomena and effects that would otherwise not be seen.

Astronomy theorists endeavor to create theoretical models and figure out the observational consequences of those models. This helps observers look for data that can refute a model or help in choosing between several alternate or conflicting models.

Theorists also try to generate or modify models to take into account new data. Consistent with the general scientific approach, in the case of an inconsistency, the general tendency is to try to make minimal modifications to the model to fit the data. In some cases, a large amount of inconsistent data over time may lead to total abandonment of a model.

Topics of Theoretical Astronomy

Topics studied by theoretical astronomers include:

1. stellar dynamics and evolution;

2. galaxy formation;

3. large-scale structure of matter in the Universe;

4. origin of cosmic rays;

5. general relativity and physical cosmology, including string cosmology and astroparticle physics.

Astrophysical relativity serves as a tool to gauge the properties of large scale structures for which gravitation plays a significant role in physical phenomena investigated and as the basis for black hole (*astro*)physics and the study of gravitational waves.

Astronomical Models

Some widely accepted and studied theories and models in astronomy, now included in the Lambda-CDM model are the Big Bang, Cosmic inflation, dark matter, and fundamental theories of physics.

A few examples of this process:

Physical process	Experimental tool	Theoretical model	Explains/predicts
Gravitation	Radio telescopes	Self-gravitating system	Emergence of a star system
Nuclear fusion	Spectroscopy	Stellar evolution	How the stars shine and how metals formed
The Big Bang	Hubble Space Telescope, COBE	Expanding universe	Age of the Universe
Quantum fluctuations		Cosmic inflation	Flatness problem
Gravitational collapse	X-ray astronomy	General relativity	Black holes at the center of Andromeda galaxy
CNO cycle in stars			

Leading Topics in Theoretical Astronomy

Dark matter and dark energy are the current leading topics in astronomy, as their discovery and controversy originated during the study of the galaxies.

Theoretical Astrophysics

Of the topics approached with the tools of theoretical physics, particular consideration is often given to stellar photospheres, stellar atmospheres, the solar atmosphere, planetary atmospheres, gaseous nebulae, nonstationary stars, and the interstellar medium. Special attention is given to the internal structure of stars.

Weak Equivalence Principle

The observation of a neutrino burst within 3 h of the associated optical burst from Supernova 1987A in the Large Magellanic Cloud (LMC) gave theoretical astrophysicists an opportunity to test that neutrinos and photons follow the same trajectories in the gravitational field of the galaxy.

Thermodynamics for Stationary Black Holes

A general form of the first law of thermodynamics for stationary black holes can be derived from the microcanonical functional integral for the gravitational field. The boundary data

1. the gravitational field as described with a micocanonical system in a spatially finite region and

2. the density of states expressed formally as a functional integral over Lorentzian metrics and as a functional of the geometrical boundary data that are fixed in the corresponding action,

are the thermodynamical extensive variables, including the energy and angular momentum of the system. For the simpler case of nonrelativistic mechanics as is often observed in astrophysical phenomena associated with a black hole event horizon, the density of states can be expressed as a real-time functional integral and subsequently used to deduce Feynman's imaginary-time functional integral for the canonical partition function.

Theoretical Astrochemistry

Reaction equations and large reaction networks are an important tool in theoretical astrochemistry, especially as applied to the gas-grain chemistry of the interstellar medium. Theoretical astrochemistry offers the prospect of being able to place constraints on the inventory of organics for exogenous delivery to the early Earth.

Interstellar Organics

"An important goal for theoretical astrochemistry is to elucidate which organics are of true interstellar origin, and to identify possible interstellar precursors and reaction pathways for those molecules which are the result of aqueous alterations." One of the ways this goal can be achieved is through the study of carbonaceous material as found in some meteorites. Carbonaceous chondrites (such as C_1 and C_2) include organic compounds such as amines and amides; alcohols, aldehydes, and ketones; aliphatic and aromatic hydrocarbons; sulfonic and phosphonic acids; amino, hydroxycarboxylic, and carboxylic acids; purines and pyrimidines; and kerogen-type material. The organic inventories of primitive meteorites display large and variable enrichments in deuterium, ^{13}C and ^{15}N which is indicative of their retention of an interstellar heritage.

Chemistry in Cometary Comae

The chemical composition of comets should reflect both the conditions in the outer solar nebula some 4.5×10^9 ayr, and the nature of the natal interstellar cloud from which the Solar system was formed. While comets retain a strong signature of their ultimate interstellar origins, significant processing must have occurred in the protosolar nebula. Early models of coma chemistry showed that reactions can occur rapidly in the inner coma, where the most important reactions are proton transfer reactions. Such reactions can potentially cycle deuterium between the different coma molecules, altering the initial D/H ratios released from the nuclear ice, and necessitating the construction of accurate models of cometary deuterium chemistry, so that gas-phase coma observations can be safely extrapolated to give nuclear D/H ratios.

Theoretical Chemical Astronomy

While the lines of conceptual understanding between theoretical astrochemistry and theoretical chemical astronomy often become blurred so that the goals and tools are the same, there are subtle differences between the two sciences. Theoretical chemistry as applied to astronomy seeks to find new ways to observe chemicals in celestial objects, for example. This often leads to theoretical astrochemistry having to seek new ways to describe or explain those same observations.

Astronomical Spectroscopy

The new era of chemical astronomy had to await the clear enunciation of the chemical principles of spectroscopy and the applicable theory.

Chemistry of Dust Condensation

Supernova radioactivity dominates light curves and the chemistry of dust condensation is also dominated by radioactivity. Dust is usually either carbon or oxides depending on which is more abundant, but Compton electrons dissociate the CO molecule in about one month. The new chemical astronomy of supernova solids depends on the supernova radioactivity:

1. the radiogenesis of ^{44}Ca from ^{44}Ti decay after carbon condensation establishes their supernova source,

2. their opacity suffices to shift emission lines blueward after 500 d and emits significant infrared luminosity,

3. parallel kinetic rates determine trace isotopes in meteoritic supernova graphites,

4. the chemistry is kinetic rather than due to thermal equilibrium and

5. is made possible by radiodeactivation of the CO trap for carbon.

Theoretical Physical Astronomy

Like theoretical chemical astronomy, the lines of conceptual understanding between theoretical astrophysics and theoretical physical astronomy are often blurred, but, again, there are subtle differences between these two sciences. Theoretical physics as applied to astronomy seeks to find new ways to observe physical phenomena in celestial objects and what to look for, for example. This often leads to theoretical astrophysics having to seek new ways to describe or explain those same observations, with hopefully a convergence to improve our understanding of the local environment of Earth and the physical Universe.

Weak interaction and Nuclear Double Beta Decay

Nuclear matrix elements of relevant operators as extracted from data and from a shell-model and theoretical approximations both for the two-neutrino and neutrinoless modes of decay are used to explain the weak interaction and nuclear structure aspects of nuclear double beta decay.

Neutron-rich Isotopes

New neutron-rich isotopes, ^{34}Ne, ^{37}Na, and ^{43}Si have been produced unambiguously for the first time, and convincing evidence for the particle instability of three others, ^{33}Ne, ^{36}Na, and ^{39}Mg has been obtained. These experimental findings compare with recent theoretical predictions.

Theory of Astronomical Time Keeping

Until recently all the time units that appear natural to us are caused by astronomical phenomena:

1. Earth's orbit around the Sun => the year, and the seasons,

2. Moon's orbit around the Earth => the month,

3. Earth's rotation and the succession of brightness and darkness => the day (and night).

High precision appears problematic:

1. amibiguities arise in the exact definition of a rotation or revolution,

2. some astronomical processes are uneven and irregular, such as the noncommensurability of year, month, and day,

3. there are a multitude of time scales and calendars to solve the first two problems.

Some of these time scales are sidereal time, solar time, and universal time.

Atomic Time

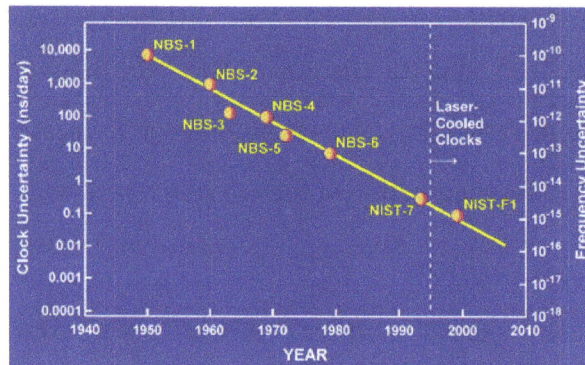

Historical accuracy of atomic clocks from NIST.

From the Systeme Internationale (SI) comes the second as defined by the duration of 9 192 631 770 cycles of a particular hyperfine structure transition in the ground state of [133]Cesium. For practical usability a device is required that attempts to produce the SI second (s) such as an atomic clock. But not all such clocks agree. The weighted mean of many clocks distributed over the whole Earth defines the Temps Atomique International; i.e., the Atomic Time TAI. From the General theory of relativity the time measured depends on the altitude on earth and the spatial velocity of the clock so that TAI refers to a location on sea level that rotates with the Earth.

Ephemeris Time

Since the Earth's rotation is irregular, any time scale derived from it such as Greenwich Mean Time led to recurring problems in predicting the Ephemerides for the positions of the Moon, Sun, planets and their natural satellites. In 1976 the International Astronomical Union (IAU) resolved that the theoretical basis for ephemeris time (ET) was wholly non-relativistic, and therefore, beginning in 1984 ephemeris time would be replaced by two further time scales with allowance for relativistic corrections. Their names, assigned in 1979, emphasized their dynamical nature or origin, Barycentric Dynamical Time (TDB) and Terrestrial Dynamical Time (TDT). Both were defined for continuity with ET and were based on what had become the standard SI second, which in turn had been derived from the measured second of ET.

During the period 1991–2006, the TDB and TDT time scales were both redefined and replaced, owing to difficulties or inconsistencies in their original definitions. The current fundamental relativistic time scales are Geocentric Coordinate Time (TCG) and Barycentric Coordinate Time (TCB). Both of these have rates that are based on the SI second in respective reference frames (and hypothetically outside the relevant gravity well), but due to relativistic effects, their rates would appear slightly faster when observed at the Earth's surface, and therefore diverge from local Earth-based time scales using the SI second at the Earth's surface.

The currently defined IAU time scales also include Terrestrial Time (TT) (replacing TDT, and now defined as a re-scaling of TCG, chosen to give TT a rate that matches the SI second when observed at the Earth's surface), and a redefined Barycentric Dynamical Time (TDB), a re-scaling of TCB to give TDB a rate that matches the SI second at the Earth's surface.

Stellar Dynamical Time Scale

For a star, the dynamical time scale is defined as the time that would be taken for a test particle released at the surface to fall under the star's potential to the centre point, if pressure forces were negligible. In other words, the dynamical time scale measures the amount of time it would take a certain star to collapse in the absence of any internal pressure. By appropriate manipulation of the equations of stellar structure this can be found to be

$$\tau_{dynamical} \simeq \frac{R}{v} = \sqrt{\frac{R^3}{2GM}} \sim 1/\sqrt{G\rho}$$

where R is the radius of the star, G is the gravitational constant, M is the mass of the star and v is the escape velocity. As an example, the Sun dynamical time scale is approximately 1133 seconds. Note that the actual time it would take a star like the Sun to collapse is greater because internal pressure is present.

The 'fundamental' oscillatory mode of a star will be at approximately the dynamical time scale. Oscillations at this frequency are seen in Cepheid variables.

Theory of Astronomical Navigation

On Earth

The basic characteristics of applied astronomical navigation are

1. usable in all areas of sailing around the earth,

2. applicable autonomously (does not depend on others – persons or states) and passively (does not emit energy),

3. conditional usage via optical visibility (of horizon and celestial bodies), or state of cloudiness,

4. precisional measurement, sextant is 0.1', altitude and position is between 1.5' and 3.0'.

5. temporal determination takes a couple of minutes (using the most modern equipment) and ≤ 30 min (using classical equipment).

The superiority of satellite navigation systems to astronomical navigation are currently undeniable, especially with the development and use of GPS/NAVSTAR. This global satellite system

1. enables automated three-dimensional positioning at any moment,

2. automatically determines position continuously (every second or even more often),

3. determines position independent of weather conditions (visibility and cloudiness),

4. determines position in real time to a few meters (two carrying frequencies) and 100 m (modest commercial receivers), which is two to three orders of magnitude better than by astronomical observation,

5. is simple even without expert knowledge,

6. is relatively cheap, comparable to equipment for astronomical navigation, and

7. allows incorporation into integrated and automated systems of control and ship steering. The use of astronomical or celestial navigation is disappearing from the surface and beneath or above the surface of the earth.

Geodetic astronomy is the application of astronomical methods into networks and technical projects of geodesy for

- apparent places of stars, and their proper motions

- precise astronomical navigation

- astro-geodetic geoid determination and

- modelling the rock densities of the topography and of geological layers in the subsurface

- Satellite geodesy using the stellar background.

- Monitoring of the Earth rotation and polar wandering

- Contribution to the time system of physics and geosciences

Astronomical algorithms are the algorithms used to calculate ephemerides, calendars, and positions (as in celestial navigation or satellite navigation).

Many astronomical and navigational computations use the Figure of the Earth as a surface representing the earth.

The International Earth Rotation and Reference Systems Service (IERS), formerly the International Earth Rotation Service, is the body responsible for maintaining global time and reference frame standards, notably through its Earth Orientation Parameter (EOP) and International Celestial Reference System (ICRS) groups.

Deep Space

The Deep Space Network, or DSN, is an international network of large antennas and communication facilities that supports interplanetary spacecraft missions, and radio and radar astronomy observations for the exploration of the solar system and the universe. The network also supports selected Earth-orbiting missions. DSN is part of the NASA Jet Propulsion Laboratory (JPL).

Aboard an Exploratory Vehicle

An observer becomes a deep space explorer upon escaping Earth's orbit. While the Deep Space Network maintains communication and enables data download from an exploratory vessel, any local probing performed by sensors or active systems aboard usually require astronomical navigation, since the enclosing network of satellites to ensure accurate positioning is absent.

Geocentric Model

In astronomy, the Geocentric model (also known as Geocentrism, or the Ptolemaic system) is a superseded description of the universe with the Earth at the center. Under the geocentric model, the Sun, Moon, stars, and naked eye planets all circled Earth. This model served as the predominant cosmological system in many ancient civilizations, such as those of Aristotle and Ptolemy.

Figure of the heavenly bodies — An illustration of the Ptolemaic Geocentric system by Portuguese cosmographer and cartographer Bartolomeu Velho, 1568 (Bibliothèque Nationale, Paris)

Two observations supported the idea that the Earth was the center of the Universe. First, the Sun appears to revolve around the Earth once per day. While the Moon and the planets have their own motions, they also appear to revolve around the Earth about once per day. The stars appeared to be on a celestial sphere, rotating once each day along an axis through the north and south geographic poles of the Earth. Second, the Earth does not seem to move from the perspective of an Earth-bound observer; it appears to be solid, stable, and unmoving.

Ancient Greek, ancient Roman and medieval philosophers usually combined the Geocentric model with a spherical Earth. It is not the same as the older flat Earth model implied in some mythology. [n 1][n 2] The ancient Jewish Babylonian uranography pictured a flat Earth with a dome-shaped rigid canopy named firmament placed over it. (רקיע- rāqîa'). [n 3][n 4][n 5][n 6][n 7][n 8] However, the ancient Greeks believed that the motions of the planets were circular and not elliptical, a view that was not challenged in Western culture until the 17th century through the synthesis of theories by Copernicus and Kepler.

The astronomical predictions of Ptolemy's Geocentric model were used to prepare astrological and astronomical charts for over 1500 years. The Geocentric model held sway into the early modern age, but from the late 16th century onward was gradually superseded by the Heliocentric model of Copernicus, Galileo and Kepler. There was much resistance to the transition between these two theories. Christian theologians were reluctant to reject a theory that agreed with Bible passages (e.g. "Sun, stand you still upon Gibeon", Joshua 10:12 – King James 2000 Bible). Others felt a new, unknown theory could not subvert an accepted consensus for Geocentrism.

Ancient Greece

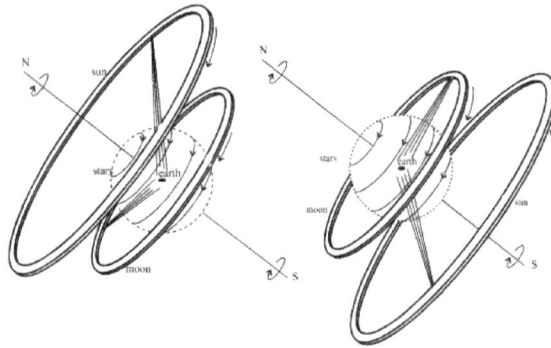

Illustration of Anaximander's models of the universe. On the left, daytime in summer; on the right, nighttime in winter.

The Geocentric model entered Greek astronomy and philosophy at an early point; it can be found in Pre-Socratic philosophy. In the 6th century BC, Anaximander proposed a cosmology with the Earth shaped like a section of a pillar (a cylinder), held aloft at the center of everything. The Sun, Moon, and planets were holes in invisible wheels surrounding the Earth; through the holes, humans could see concealed fire. About the same time, the Pythagoreans thought that the Earth was a sphere (in accordance with observations of eclipses), but not at the center; they believed that it was in motion around an unseen fire. Later these views were combined, so most educated Greeks from the 4th century BC on thought that the Earth was a sphere at the center of the universe.

In the 4th century BC, two influential Greek philosophers, Plato and his student Aristotle, wrote works based on the Geocentric model. According to Plato, the Earth was a sphere, stationary at the center of the universe. The stars and planets were carried around the Earth on spheres or circles, arranged in the order (outwards from the center): Moon, Sun, Venus, Mercury, Mars, Jupiter, Saturn, fixed stars, with the fixed stars located on the celestial sphere. In his "Myth of Er", a section of the *Republic*, Plato describes the cosmos as the Spindle of Necessity, attended by the Sirens and turned by the three Fates. Eudoxus of Cnidus, who worked with Plato, developed a less mythical, more mathematical explanation of the planets' motion based on Plato's dictum stating that all phenomena in the heavens can be explained with uniform circular motion. Aristotle elaborated on Eudoxus' system.

In the fully developed Aristotelian system, the spherical Earth is at the center of the universe, and all other heavenly bodies are attached to 47–55 transparent, rotating spheres surrounding the Earth, all concentric with it. (The number is so high because several spheres are needed for each planet.) These spheres, known as crystalline spheres, all moved at different uniform speeds to create the revolution of bodies around the Earth. They were composed of an incorruptible substance called aether. Aristotle believed that the moon was in the innermost sphere and therefore touches the realm of Earth, causing the dark spots (macula) and the ability to go through lunar phases. He further described his system by explaining the natural tendencies of the terrestrial elements: Earth, water, fire, air, as well as celestial aether. His system held that Earth was the heaviest element, with the strongest movement towards the center, thus water formed a layer surrounding the sphere of Earth. The tendency of air and fire, on the other hand, was to move upwards, away from the center, with fire being lighter than air. Beyond the layer of fire, were the solid spheres of aether in which the celestial bodies were embedded. They, themselves, were also entirely composed of aether.

Adherence to the Geocentric model stemmed largely from several important observations. First of all, if the Earth did move, then one ought to be able to observe the shifting of the fixed stars due to stellar parallax. In short, if the Earth was moving, the shapes of the constellations should change considerably over the course of a year. If they did not appear to move, the stars are either much farther away than the Sun and the planets than previously conceived, making their motion undetectable, or in reality they are not moving at all. Because the stars were actually much further away than Greek astronomers postulated (making movement extremely subtle), stellar parallax was not detected until the 19th century. Therefore, the Greeks chose the simpler of the two explanations. The lack of any observable parallax was considered a fatal flaw in any non-Geocentric theory. Another observation used in favor of the Geocentric model at the time was the apparent consistency of Venus' luminosity, which implies that it is usually about the same distance from Earth, which in turn is more consistent with geocentrism than heliocentrism. In reality, that is because the loss of light caused by Venus' phases compensates for the increase in apparent size caused by its varying distance from Earth. Objectors to heliocentrism noted that terrestrial bodies naturally tend to come to rest as near as possible to the center of the Earth. Further barring the opportunity to fall closer the center, terrestrial bodies tend not to move unless forced by an outside object, or transformed to a different element by heat or moisture.

Atmospheric explanations for many phenomena were preferred because the Eudoxan–Aristotelian model based on perfectly concentric spheres was not intended to explain changes in the brightness of the planets due to a change in distance. Eventually, perfectly concentric spheres were abandoned as it was impossible to develop a sufficiently accurate model under that ideal. However, while providing for similar explanations, the later deferent and epicycle model was flexible enough to accommodate observations for many centuries.

Ptolemaic Model

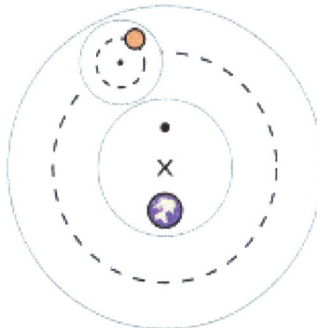

The basic elements of Ptolemaic astronomy, showing a planet on an epicycle with an eccentric deferent and an equant point. The Green shaded area is the celestial sphere which the planet occupies.

Although the basic tenets of Greek Geocentrism were established by the time of Aristotle, the details of his system did not become standard. The Ptolemaic system, developed by the Hellenistic astronomer Claudius Ptolemaeus in the 2nd century AD finally standardised Geocentrism. His main astronomical work, the *Almagest*, was the culmination of centuries of work by Hellenic, Hellenistic and Babylonian astronomers. For over a millennium European and Islamic astronomers assumed it was the correct cosmological model. Because of its influence, people sometimes wrongly think the Ptolemaic system is identical with the Geocentric model.

Ptolemy argued that the Earth was a sphere in the center of the universe, from the simple observation that half the stars were above the horizon and half were below the horizon at any time (stars on rotating stellar sphere), and the assumption that the stars were all at some modest distance from the center of the universe. If the Earth was substantially displaced from the center, this division into visible and invisible stars would not be equal.[n 9]

Ptolemaic System

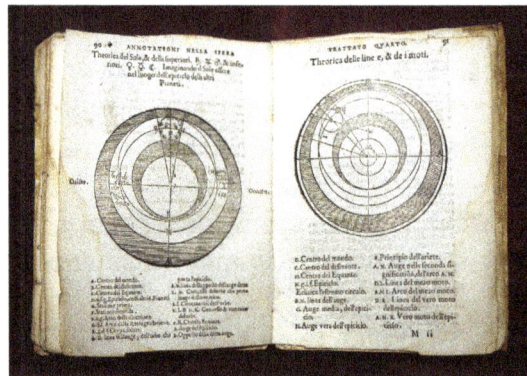

Pages from 1550 *Annotazione* on Sacrobosco's *Tractatus de Sphaera*, showing the Ptolemaic system.

In the Ptolemaic system, each planet is moved by a system of two spheres: one called its deferent; the other, its epicycle. The deferent is a circle whose center point, called the eccentric and marked in the diagram with an X, is removed from the Earth. The original purpose of the eccentric was to account for the differences of the lengths of the seasons (autumn is the shortest by a week or so), by placing the Earth away from the center of rotation of the rest of the universe. Another sphere, the epicycle, is embedded inside the deferent sphere and is represented by the smaller dotted line to the right. A given planet then moves around the epicycle at the same time the epicycle moves along the path marked by the deferent. These combined movements cause the given planet to move closer to and further away from the Earth at different points in its orbit, and explained the observation that planets slowed down, stopped, and moved backward in retrograde motion, and then again reversed to resume normal, or prograde, motion.

The deferent-and-epicycle model had been used by Greek astronomers for centuries along with the idea of the eccentric (a deferent which is slightly off-center from the Earth), which was even older. In the illustration, the center of the deferent is not the Earth but the spot marked X, making it eccentric (from the Greek *ec-* meaning "from," and *kentron* meaning "center"), from which the spot takes its name. Unfortunately, the system that was available in Ptolemy's time did not quite match observations, even though it was considerably improved over Hipparchus' system. Most noticeably the size of a planet's retrograde loop (especially that of Mars) would be smaller, and sometimes larger, than expected, resulting in positional errors of as much as 30 degrees. To alleviate the problem, Ptolemy developed the equant. The equant was a point near the center of a planet's orbit which, if you were to stand there and watch, the center of the planet's epicycle would always appear to move at uniform speed; all other locations would see non-uniform speed, like on the Earth. By using an equant, Ptolemy claimed to keep motion which was uniform and circular, although it departed from the Platonic ideal of uniform circular motion. The resultant system, which eventually came to be widely accepted in the west, seems unwieldy to modern astronomers; each planet required an epicycle revolving on a deferent, offset by an equant which was different for each planet. It predicted various celestial motions, including the beginning and end of retrograde motion, to within a maximum error of 10 degrees, considerably better than without the equant.

The model with epicycles is in fact a very good model of an elliptical orbit with low eccentricity. The well known ellipse shape does not appear to a noticeable extent when the eccentricity is less than 5%, but the offset distance of the 'center' (in fact the focus occupied by the sun) is very noticeable even with low eccentricities as possessed by the planets.

To summarize, Ptolemy devised a system that was compatible with Aristotelian philosophy and managed to track actual observations and predict future movement mostly to within the limits of the next 1000 years of observations. The observed motions and his mechanisms for explaining them include:

The Geocentric model was eventually replaced by the heliocentric model. The earliest heliocentric model, Copernican heliocentrism, could remove Ptolemy's epicycles because the retrograde motion could be seen to be the result of the combination of Earth and planet movement and speeds. Copernicus felt strongly that equants were a violation of Aristotelian purity, and proved that replacement of the equant with a pair of new epicycles was entirely equivalent. Astronomers often continued using the equants instead of the epicycles because the former was easier to calculate, and gave the same result.

It has been determined, in fact, that the Copernican, Ptolemaic and even the Tychonic models provided identical results to identical inputs. They are computationally equivalent. It wasn't until Kepler demonstrated a physical observation that could show that the physical sun is directly involved in determining an orbit that a new model was required.

The Ptolemaic order of spheres from Earth outward is:

1. Moon

2. Mercury

3. Venus

4. Sun

5. Mars

6. Jupiter

7. Saturn

8. Fixed Stars

9. *Primum Mobile* (First Moved).

Ptolemy did not invent or work out this order, which aligns with the ancient Seven Heavens religious cosmology common to the major Eurasian religious traditions. It also follows the decreasing orbital periods of the moon, sun, planets and stars.

Geocentrism and Islamic Astronomy

Muslim astronomers accepted the Ptolemaic system and the Geocentric model, but by the 10th century texts appeared regularly whose subject matter was doubts concerning Ptolemy (*shukūk*). Several Muslim scholars questioned the Earth's apparent immobility and centrality within the universe. Some Muslim astronomers believed that the Earth rotates around its axis, such as Abu Sa'id al-Sijzi (d. circa 1020). According to al-Biruni, Sijzi invented an astrolabe called *al-zūraqī* based on a belief held by some of his contemporaries "that the motion we see is due to the Earth's movement and not to that of the sky." The prevalence of this view is further confirmed by a reference from the 13th century which states:

According to the geometers [or engineers] (*muhandisīn*), the Earth is in constant circular motion, and what appears to be the motion of the heavens is actually due to the motion of the Earth and not the stars.

Early in the 11th century Alhazen wrote a scathing critique of Ptolemy's model in his *Doubts on Ptolemy* (c. 1028), which some have interpreted to imply he was criticizing Ptolemy's geocentrism, but most agree that he was actually criticizing the details of Ptolemy's model rather than his geocentrism.

In the 12th century, Arzachel departed from the ancient Greek idea of uniform circular motions by hypothesizing that the planet Mercury moves in an elliptic orbit, while Alpetragius proposed a planetary model that abandoned the equant, epicycle and eccentric mechanisms, though this resulted in a system that was mathematically less accurate. Fakhr al-Din al-Razi (1149–1209), in dealing with his conception of physics and the physical world in his *Matalib*, rejects the Aristotelian and Avicennian notion of the Earth's centrality within the universe, but instead argues that there are "a thousand thousand worlds (*alfa alfi 'awalim*) beyond this world such that each one of those worlds be bigger and more massive than this world as well as having the like of what this world has." To support his theological argument, he cites the Qur'anic verse, "All praise belongs to God, Lord of the Worlds," emphasizing the term "Worlds."

The "Maragha Revolution" refers to the Maragha school's revolution against Ptolemaic astronomy. The "Maragha school" was an astronomical tradition beginning in the Maragha observatory and continuing with astronomers from the Damascus mosque and Samarkand observatory. Like their Andalusian predecessors, the Maragha astronomers attempted to solve the equant problem (the circle around whose circumference a planet or the center of an epicycle was conceived to move uniformly) and produce alternative configurations to the Ptolemaic model without abandoning geocentrism. They were more successful than their Andalusian predecessors in producing non-Ptolemaic configurations which eliminated the equant and eccentrics, were more accurate than the Ptolemaic model in numerically predicting planetary positions, and were in better agreement with empirical observations. The most important of the Maragha astronomers included Mo'ayyeduddin Urdi (d. 1266), Nasīr al-Dīn al-Tūsī (1201–1274), Qutb al-Din al-Shirazi (1236–1311), Ibn al-Shatir (1304–1375), Ali Qushji (c. 1474), Al-Birjandi (d. 1525), and Shams al-Din al-Khafri (d. 1550). Ibn al-Shatir, the Damascene astronomer (1304–1375 AD) working at the Umayyad Mosque, wrote a major book entitled *Kitab Nihayat al-Sul fi Tashih al-Usul* (*A Final Inquiry Concerning the Rectification of Planetary Theory*) on a theory which departs largely from the Ptolemaic system known at that time. In his book, *Ibn al-Shatir, an Arab astronomer of the fourteenth century*, E. S. Kennedy wrote "what is of most interest, however, is that Ibn al-Shatir's lunar theory, except for trivial differences in parameters, is identical with that of Copernicus (1473–1543 AD)." The discovery that the models of Ibn al-Shatir are mathematically identical to those of Copernicus suggests the possible transmission of these models to Europe. At the Maragha and Samarkand observatories, the Earth's rotation was discussed by al-Tusi and Ali Qushji (b. 1403); the arguments and evidence they used resemble those used by Copernicus to support the Earth's motion.

However, the Maragha school never made the paradigm shift to heliocentrism. The influence of the Maragha school on Copernicus remains speculative, since there is no documentary evidence to prove it. The possibility that Copernicus independently developed the Tusi couple remains open, since no researcher has yet demonstrated that he knew about Tusi's work or that of the Maragha school.

Geocentrism and Rival Systems

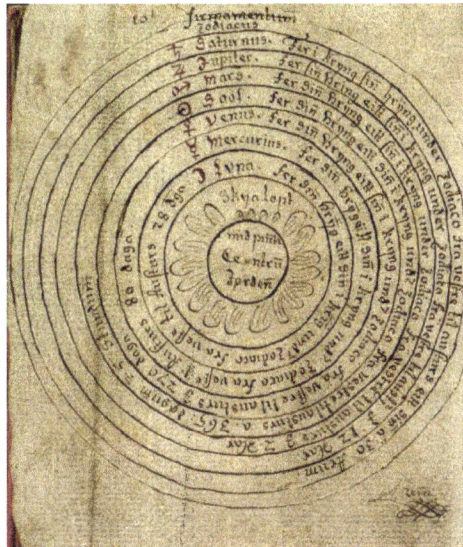

This drawing from an Icelandic manuscript dated around 1750 illustrates the Geocentric model.

Not all Greeks agreed with the Geocentric model. The Pythagorean system has already been mentioned; some Pythagoreans believed the Earth to be one of several planets going around a central fire. Hicetas and Ecphantus, two Pythagoreans of the 5th century BC, and Heraclides Ponticus in the 4th century BC, believed that the Earth rotated on its axis but remained at the center of the universe. Such a system still qualifies as Geocentric. It was revived in the Middle Ages by Jean Buridan. Heraclides Ponticus was once thought to have proposed that both Venus and Mercury went around the Sun rather than the Earth, but this is no longer accepted. Martianus Capella definitely put Mercury and Venus in orbit around the Sun. Aristarchus of Samos was the most radical. He wrote a work, which has not survived, on heliocentrism, saying that the Sun was at the center of the universe, while the Earth and other planets revolved around it. His theory was not popular, and he had one named follower, Seleucus of Seleucia.

Copernican System

In 1543, the Geocentric system met its first serious challenge with the publication of Copernicus' *De revolutionibus orbium coelestium* (*On the Revolutions of the Heavenly Spheres*), which posited that the Earth and the other planets instead revolved around the Sun. The Geocentric system was still held for many years afterwards, as at the time the Copernican system did not offer better predictions than the Geocentric system, and it posed problems for both natural philosophy and scripture. The Copernican system was no more accurate than Ptolemy's system, because it still used circular orbits. This was not altered until Johannes Kepler postulated that they were elliptical (Kepler's first law of planetary motion).

With the invention of the telescope in 1609, observations made by Galileo Galilei (such as that Jupiter has moons) called into question some of the tenets of geocentrism but did not seriously threaten it. Because he observed dark "spots" on the moon, craters, he remarked that the moon was not a perfect celestial body as had been previously conceived. This was the first time someone could see imperfections on a celestial body that was supposed to be composed of perfect aether. As such, because the moon's imperfections could now be related to those seen on Earth, one could argue that neither was unique: rather, they were both just celestial bodies made from Earth-like material. Galileo could also see the moons of Jupiter, which he dedicated to Cosimo II de' Medici, and stated that they orbited around Jupiter, not Earth. This was a significant claim as it would mean not only that not everything revolved around Earth as stated in the Ptolemaic model, but also showed a secondary celestial body could orbit a moving celestial body, strengthening the heliocentric argument that a moving Earth could retain the Moon. Galileo's observations were verified by other astronomers of the time period who quickly adopted use of the telescope, including Christoph Scheiner, Johannes Kepler, and Giovan Paulo Lembo.

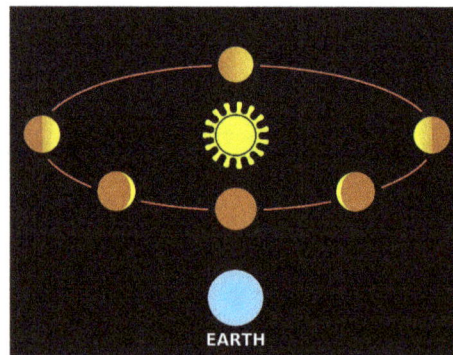

Phases of Venus

In December 1610, Galileo Galilei used his telescope to observe that Venus showed all phases, just like the Moon. He thought that while this observation was incompatible with the Ptolemaic system, it was a natural consequence of the heliocentric system.

However, Ptolemy placed Venus' deferent and epicycle entirely inside the sphere of the Sun (between the Sun and Mercury), but this was arbitrary; he could just as easily have swapped Venus and Mercury and put them on the other side of the Sun, or made any other arrangement of Venus and Mercury, as long as they were always near a line running from the Earth through the Sun, such as placing the center of the Venus epicycle near the Sun. In this case, if the Sun is the source of all the light, under the Ptolemaic system:

If Venus is between Earth and the Sun, the phase of Venus must always be crescent or all dark.

If Venus is beyond the Sun, the phase of Venus must always be gibbous or full.

But Galileo saw Venus at first small and full, and later large and crescent.

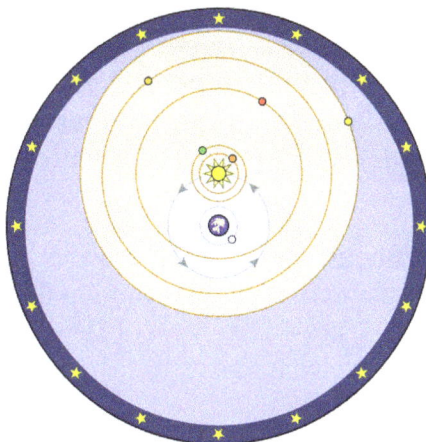

In this depiction of the Tychonic system, the objects on blue orbits (the moon and the sun) revolve around the Earth. The objects on orange orbits (Mercury, Venus, Mars, Jupiter, and Saturn) revolve around the sun. Around all is a sphere of stars, which rotates.

This showed that with a Ptolemaic cosmology, the Venus epicycle can be neither completely inside nor completely outside of the orbit of the Sun. As a result, Ptolemaics abandoned the idea that the epicycle of Venus was completely inside the Sun, and later 17th century competition between astronomical cosmologies focused on variations of Tycho Brahe's Tychonic system (in which the Earth was still at the center of the universe, and around it revolved the Sun, but all other planets revolved around the Sun in one massive set of epicycles), or variations on the Copernican system.

Gravitation

Johannes Kepler analysed Tycho Brahe's famously accurate observations and afterwards constructed his three laws in 1609 and 1619, based on a heliocentric view where the planets move in elliptical paths. Using these laws, he was the first astronomer to successfully predict a transit of Venus (for the year 1631). The change from circular orbits to elliptical planetary paths dramatically improved the accuracy of celestial observations and predictions. Because the heliocentric model by Copernicus was no more accurate than Ptolemy's system, new observations were needed to persuade those who still held on to the Geocentric model. However, Kepler's laws based on Brahe's data became a problem which geocentrists could not easily overcome.

In 1687, Isaac Newton stated the law of universal gravitation, described earlier as a hypothesis by Robert Hooke and others. His main achievement was to mathematically derive Kepler's laws of planetary motion from the law of gravitation, thus helping to prove the latter. This introduced gravitation as the force that both kept the Earth and planets moving through the heavens and also kept the air from flying away. The theory of gravity allowed scientists to construct a plausible heliocentric model for the solar system quickly. In his *Principia*, Newton explained his system of how gravity, (previously thought of as an occult force) directed the movements of celestial bodies, and kept our solar system in its working order. His descriptions of centripetal force were a break-through in scientific thought which used the newly developed differential calculus, and finally replaced the previous schools of scientific thought, i.e. those of Aristotle and Ptolemy. However, the process was gradual.

Several empirical tests of Newton's theory, explaining the longer period of oscillation of a pendulum at the equator and the differing size of a degree of latitude, gradually became available over the period 1673–1738. In addition, stellar aberration was observed by Robert Hooke in 1674 and tested in a series of observations by Jean Picard over ten years finishing in 1680. However, it was not explained until 1729 when James Bradley provided an approximate explanation in terms of the Earth's revolution about the sun.

In 1838, astronomer Friedrich Wilhelm Bessel measured the parallax of the star 61 Cygni successfully, and disproved Ptolemy's claim that parallax motion did not exist. This finally confirmed the assumptions made by Copernicus, provided accurate, dependable scientific observations, and displayed truly how far away stars were from Earth.

A Geocentric frame is useful for many everyday activities and most laboratory experiments, but is a less appropriate choice for solar-system mechanics and space travel. While a heliocentric frame is most useful in those cases, galactic and extra-galactic astronomy is easier if the sun is treated as neither stationary nor the center of the universe, but rotating around the center of our galaxy, and in turn our galaxy is also not at rest in the cosmic background.

Relativity

Albert Einstein and Leopold Infeld wrote in The Evolution of Physics (1938): "Can we formulate physical laws so that they are valid for all CS (=coordinate systems), not only those moving uniformly, but also those moving quite arbitrarily, relative to each other? If this can be done, our difficulties will be over. We shall then be able to apply the laws of nature to any CS. The struggle, so violent in the early days of science, between the views of Ptolemy and Copernicus would then be quite meaningless. Either CS could be used with equal justification. The two sentences, "the sun is at rest and the Earth moves", or "the sun moves and the Earth is at rest", would simply mean two different conventions concerning two different CS. Could we build a real relativistic physics valid in all CS; a physics in which there would be no place for absolute, but only for relative, motion? This is indeed possible!"

Religious and Contemporary Adherence to Geocentrism

The Ptolemaic model of the solar system held sway into the early modern age; from the late 16th century onward it was gradually replaced as the consensus description by the heliocentric model. Geocentrism as a separate religious belief, however, never completely died out. In the United States between 1870 and 1920, for example, various members of the Lutheran Church – Missouri Synod published articles disparaging Copernican astronomy, and geocentrism was widely taught within the synod during that period. However, in the 1902 Theological Quarterly, A. L. Graebner claimed that the synod had no doctrinal position on geocentrism, heliocentrism, or any scientific model, unless it were to contradict Scripture. He stated that any possible declarations of geocentrists within the synod did not set the position of the church body as a whole.

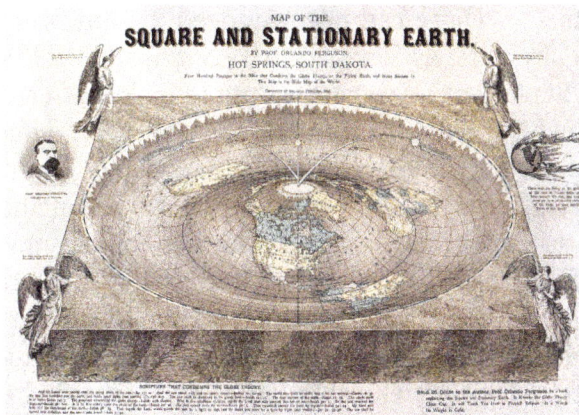

Map of the Square and Stationary Earth, by Orlando Ferguson (1893)

Articles arguing that geocentrism was the biblical perspective appeared in some early creation science newsletters associated with the Creation Research Society pointing to some passages in the Bible, which, when taken literally, indicate that the daily apparent motions of the Sun and the Moon are due to their actual motions around the Earth rather than due to the rotation of the Earth about its axis. For example, in Joshua 10:12, the Sun and Moon are said to stop in the sky, and in Psalms 93:1 the world is described as immobile. (Psalms 9:1 says in part "the world is established, firm and secure".) Contemporary advocates for such religious beliefs include Robert Sungenis (president of Bellarmine Theological Forum and author of the 2006 book *Galileo Was Wrong*). These people subscribe to the view that a plain reading of the Bible contains an accurate account of the manner in which the universe was created and requires a Geocentric worldview. Most contemporary creationist organizations reject such perspectives.

After all, Copernicanism was the first major victory of science over religion, so it's inevitable that some folks would think that everything that's wrong with the world began there. (Steven Dutch of the University of Wisconsin–Madison)

Morris Berman quotes a 2006 survey that show currently some 20% of the U.S. population believe that the sun goes around the Earth (Geocentricism) rather than the Earth goes around the sun (heliocentricism), while a further 9% claimed not to know. Polls conducted by Gallup in the 1990s found that 16% of Germans, 18% of Americans and 19% of Britons hold that the Sun revolves around the Earth. A study conducted in 2005 by Jon D. Miller of Northwestern University, an expert in the public understanding of science and technology, found that about 20%, or one in five, of American adults believe that the Sun orbits the Earth. According to 2011 VTSIOM poll, 32% of Russians believe that the Sun orbits the Earth.

Historical Positions of the Roman Catholic Hierarchy

The famous Galileo affair pitted the Geocentric model against the claims of Galileo. In regards to the theological basis for such an argument, two Popes addressed the question of whether the use of phenomenological language would compel one to admit an error in Scripture. Both taught that it would not. Pope Leo XIII (1878–1903) wrote:

> we have to contend against those who, making an evil use of physical science, minutely scrutinize the Sacred Book in order to detect the writers in a mistake, and to take occasion

to vilify its contents. . . . There can never, indeed, be any real discrepancy between the theologian and the physicist, as long as each confines himself within his own lines, and both are careful, as St. Augustine warns us, "not to make rash assertions, or to assert what is not known as known." If dissension should arise between them, here is the rule also laid down by St. Augustine, for the theologian: "Whatever they can really demonstrate to be true of physical nature, we must show to be capable of reconciliation with our Scriptures; and whatever they assert in their treatises which is contrary to these Scriptures of ours, that is to Catholic faith, we must either prove it as well as we can to be entirely false, or at all events we must, without the smallest hesitation, believe it to be so." To understand how just is the rule here formulated we must remember, first, that the sacred writers, or to speak more accurately, the Holy Ghost "Who spoke by them, did not intend to teach men these things (that is to say, the essential nature of the things of the visible universe), things in no way profitable unto salvation." Hence they did not seek to penetrate the secrets of nature, but rather described and dealt with things in more or less figurative language, or in terms which were commonly used at the time, and which in many instances are in daily use at this day, even by the most eminent men of science. Ordinary speech primarily and properly describes what comes under the senses; and somewhat in the same way the sacred writers-as the Angelic Doctor also reminds us – `went by what sensibly appeared," or put down what God, speaking to men, signified, in the way men could understand and were accustomed to. (Providentissimus Deus 18).

Maurice Finocchiaro, author of a book on the Galileo affair, notes that this is "a view of the relationship between biblical interpretation and scientific investigation that corresponds to the one advanced by Galileo in the "Letter to the Grand Duchess Christina". Pope Pius XII (1939–1958) repeated his predecessor's teaching:

The first and greatest care of Leo XIII was to set forth the teaching on the truth of the Sacred Books and to defend it from attack. Hence with grave words did he proclaim that there is no error whatsoever if the sacred writer, speaking of things of the physical order "went by what sensibly appeared" as the Angelic Doctor says, speaking either "in figurative language, or in terms which were commonly used at the time, and which in many instances are in daily use at this day, even among the most eminent men of science." For "the sacred writers, or to speak more accurately – the words are St. Augustine's – the Holy Spirit, Who spoke by them, did not intend to teach men these things – that is the essential nature of the things of the universe – things in no way profitable to salvation"; which principle "will apply to cognate sciences, and especially to history," that is, by refuting, "in a somewhat similar way the fallacies of the adversaries and defending the historical truth of Sacred Scripture from their attacks (*Divino afflante Spiritu*, 3).

In 1664 Pope Alexander VII republished the *Index Librorum Prohibitorum* (*List of Prohibited Books*) and attached the various decrees connected with those books, including those concerned with heliocentrism. He stated in a Papal Bull that his purpose in doing so was that "the succession of things done from the beginning might be made known [*quo rei ab initio gestae series innotescat*]."

The position of the curia evolved slowly over the centuries towards permitting the heliocentric view. In 1757, during the papacy of Benedict XIV, the Congregation of the Index withdrew the decree which prohibited *all* books teaching the Earth's motion, although the *Dialogue* and a few other books continued to be explicitly included. In 1820, the Congregation of the Holy Office, with the pope's approval, decreed that Catholic astronomer Giuseppe Settele was allowed to treat the Earth's motion as an established fact and removed any obstacle for Catholics to hold to the motion of the Earth:

> The Assessor of the Holy Office has referred the request of Giuseppe Settele, Professor of Optics and Astronomy at La Sapienza University, regarding permission to publish his work Elements of Astronomy in which he espouses the common opinion of the astronomers of our time regarding the Earth's daily and yearly motions, to His Holiness through Divine Providence, Pope Pius VII. Previously, His Holiness had referred this request to the Supreme Sacred Congregation and concurrently to the consideration of the Most Eminent and Most Reverend General Cardinal Inquisitor. His Holiness has decreed that no obstacles exist for those who sustain Copernicus' affirmation regarding the Earth's movement in the manner in which it is affirmed today, even by Catholic authors. He has, moreover, suggested the insertion of several notations into this work, aimed at demonstrating that the above mentioned affirmation [of Copernicus], as it is has come to be understood, does not present any difficulties; difficulties that existed in times past, prior to the subsequent astronomical observations that have now occurred. [Pope Pius VII] has also recommended that the implementation [of these decisions] be given to the Cardinal Secretary of the Supreme Sacred Congregation and Master of the Sacred Apostolic Palace. He is now appointed the task of bringing to an end any concerns and criticisms regarding the printing of this book, and, at the same time, ensuring that in the future, regarding the publication of such works, permission is sought from the Cardinal Vicar whose signature will not be given without the authorization of the Superior of his Order.

In 1822, the Congregation of the Holy Office removed the prohibition on the publication of books treating of the Earth's motion in accordance with modern astronomy and Pope Pius VII ratified the decision:

> The most excellent [cardinals] have decreed that there must be no denial, by the present or by future Masters of the Sacred Apostolic Palace, of permission to print and to publish works which treat of the mobility of the Earth and of the immobility of the sun, according to the common opinion of modern astronomers, as long as there are no other contrary indications, on the basis of the decrees of the Sacred Congregation of the Index of 1757 and of this Supreme [Holy Office] of 1820; and that those who would show themselves to be reluctant or would disobey, should be forced under punishments at the choice of [this] Sacred Congregation, with derogation of [their] claimed privileges, where necessary.

The 1835 edition of the Catholic Index of Prohibited Books for the first time omits the *Dialogue* from the list. In his 1921 papal encyclical, *In praeclara summorum*, Pope Benedict XV stated that, "though this Earth on which we live may not be the center of the universe as at one time was thought, it was the scene of the original happiness of our first ancestors, witness of their unhappy fall, as too of the Redemption of mankind through the Passion and Death of Jesus Christ." In 1965 the Second Vatican Council stated that, "Consequently, we cannot but deplore certain habits of mind, which are sometimes found too among Christians, which do not sufficiently attend to the rightful independence of science and which, from the arguments and controversies they spark, lead many minds to conclude that faith and science are mutually opposed." The footnote on this statement is to Msgr. Pio Paschini's, *Vita e opere di Galileo Galilei*, 2 volumes, Vatican Press (1964). Pope John Paul II regretted the treatment which Galileo received, in a speech to the Pontifical Academy of Sciences in 1992. The Pope declared the incident to be based on a "tragic mutual miscomprehension". He further stated:

> Cardinal Poupard has also reminded us that the sentence of 1633 was not irreformable, and that the debate which had not ceased to evolve thereafter, was closed in 1820 with the imprimatur given to the work of Canon Settele. . . . The error of the theologians of the time, when they maintained the centrality of the Earth, was to think that our understanding of the physical world's structure was, in some way, imposed by the literal sense of Sacred Scripture. Let us recall the celebrated saying attributed to Baronius "Spiritui Sancto mentem fuisse nos docere quomodo ad coelum eatur, non quomodo coelum gradiatur". In fact, the Bible does not concern itself with the details of the physical world, the understanding of which is the competence of human experience and reasoning. There exist two realms of knowledge, one which has its source in Revelation and one which reason can discover by its own power. To the latter belong especially the experimental sciences and philosophy. The distinction between the two realms of knowledge ought not to be understood as opposition.

Orthodox Judaism

Some Orthodox Jewish leaders, particularly the Lubavitcher Rebbe, maintain a Geocentric model of the universe based on the aforementioned Biblical verses and an interpretation of Maimonides to the effect that he ruled that the Earth is orbited by the sun. The Lubavitcher Rebbe also explained that geocentrism is defensible based on the theory of Relativity, which establishes that "when two bodies in space are in motion relative to one another, ... science declares with absolute certainty that from the scientific point of view both possibilities are equally valid, namely that the Earth revolves around the sun, or the sun revolves around the Earth."

While geocentrism is important in Maimonides' calendar calculations, the great majority of Jewish religious scholars, who accept the divinity of the Bible and accept many of his rulings as legally binding, do not believe that the Bible or Maimonides command a belief in geocentrism.

However, there is some evidence that geocentrist beliefs are becoming increasingly common among Orthodox Jews.

The Zohar implies: "The entire world and those upon it, spin round in a circle like a ball,' both those at the bottom of the ball and those at the top. All God's creatures, wherever they live on the different parts of the ball, look different (in color, in their features) because the air is different in each place, but they stand erect as all other human beings.

Therefore there are places in the world where, when some have light, others have darkness; when some have day, others have night.

Islam

Prominent cases of modern geocentrism in Islam are very isolated. Very few individuals promoted a Geocentric view of the universe. One of them was Ahmed Raza Khan Barelvi, a Sunni scholar of Indian subcontinent. He rejected the heliocentric model and wrote a book that explains the movement of sun, moon and other planets around the Earth. The Grand Mufti of Saudi Arabia from 1993 to 1999, Ibn Baz also promoted the Geocentric view between 1966 and 1985.

Planetariums

The Geocentric (Ptolemaic) model of the solar system is still of interest to planetarium makers, as, for technical reasons, a Ptolemaic-type motion for the planet light apparatus has some advantages over a Copernican-type motion. The celestial sphere, still used for teaching purposes and sometimes for navigation, is also based on a Geocentric system which in effect ignores parallax. However this effect is negligible at the scale of accuracy that applies to a planetarium.

Geocentric Models in Fiction

Alternate history science fiction has produced some literature of interest on the proposition that some alternate universes and Earths might indeed have laws of physics and cosmologies that are Ptolemaic and Aristotelian in design. This subcategory began with Philip Jose Farmer's short story, "Sail On! Sail On!" (1952), where Columbus has access to radio technology, and where his Spanish-financed exploratory and trade fleet sail off the edge of the (flat) world in his Geocentric alternate universe in 1492, instead of discovering North America and South America.

Sir Terry Pratchett's Discworld, a flat disc balanced on the backs of four elephants which in turn stand on the back of a giant turtle, is Geocentric (or, perhaps, turtle-centric!) with a small sun and moon in orbit around the main mass.

Richard Garfinkle's *Celestial Matters* (1996) is set in a more elaborated Geocentric cosmos, where Earth is divided by two contending factions, the Classical Greece-dominated Delian League and the Chinese Middle Kingdom, both of which are capable of flight within an alternate universe based on Ptolemaic astronomy, Aristotle's physics and Taoist thought. Unfortunately, both superpowers have been fighting a thousand-year war since the time of Alexander the Great.

In the C.S. Lewis novel, *The Voyage of the Dawn Treader*, one of the Chronicles of Narnia series, the characters involved set out on a naval voyage to discover the edge of the world. The events of the book follow their journey across a flat, Geocentric "world" and beyond its fringes.

1. There is also recognition of the ability of humans to change the environment in which they lived. This same understanding occurred also in the great creation stories of Mesopotamia; these stories formed the basis for the Jewish theological reflections of the Hebrew Scriptures concerning the creation of the world. The Jewish priests and theologians who constructed the narrative took accepted ideas about the structure of the world and reflected theologically on them in the light of their experience and faith. There was never any clash between Jewish and Babylonian people about the structure of the world, but only about who was responsible for it and its ultimate theological meaning. The envisaged structure is simple: Earth was seen as being situated in the middle of a great volume of water, with water both above and below Earth. A great dome was thought to be set above Earth (like an inverted glass bowl), maintaining the water above Earth in its place. Earth was pictured as resting on foundations that go down into the deep. These foundations secured the stability of the land as something that is not floating on the water and so

could not be tossed about by wind and wave. The waters surrounding Earth were thought to have been gathered together in their place. The stars, sun, moon, and planets moved in their allotted paths across the great dome above Earth, with their movements defining the months, seasons, and year.

2. From Myth to Cosmos: The earliest speculations about the origin and nature of the world took the form of religious myths. Almost all ancient cultures developed cosmological stories to explain the basic features of the cosmos: Earth and its inhabitants, sky, sea, sun, moon, and stars. For example, for the Babylonians, the creation of the universe was seen as born from a primeval pair of human-like gods. In early Egyptian cosmology, eclipses were explained as the moon being swallowed temporarily by a sow or as the sun being attacked by a serpent. For the early Hebrews, whose account is preserved in the biblical book of Genesis, a single God created the universe in stages within the relatively recent past. Such pre-scientific cosmologies tended to assume a flat Earth, a finite past, ongoing active interference by deities or spirits in the cosmic order, and stars and planets (visible to the naked eye only as points of light) that were different in nature from Earth.

3. This argument is given in Book I, Chapter 5, of the *Almagest*.

4. Donald B. DeYoung, for example, states that "Similar terminology is often used today when we speak of the sun's rising and setting, even though the earth, not the sun, is doing the moving. Bible writers used the 'language of appearance,' just as people always have. Without it, the intended message would be awkward at best and probably not understood clearly. When the Bible touches on scientific subjects, it is entirely accurate."

Heliocentrism

Andreas Cellarius's illustration of the Copernican system, from the *Harmonia Macrocosmica* (1708).

Heliocentrism, or heliocentricism, is the astronomical model in which the Earth and planets revolve around the Sun at the center of the Solar System. The word comes from the Greek (*helios* "sun" and *kentron* "center"). Historically, Heliocentrism was opposed to geocentrism, which placed the Earth at the center. The notion that the Earth revolves around the Sun had been proposed as early as the 3rd century BC by Aristarchus of Samos, but at least in the medieval world, Aristarchus's Heliocentrism attracted little attention—possibly because of the loss of scientific works of the Hellenistic Era.

It was not until the 16th century that a geometric mathematical model of a heliocentric system was presented, by the Renaissance mathematician, astronomer, and Catholic cleric Nicolaus Copernicus, leading to the Copernican Revolution. In the following century, Johannes Kepler elaborated upon and expanded this model to include elliptical orbits, and Galileo Galilei presented supporting observations made using a telescope.

With the observations of William Herschel, Friedrich Bessel, and others, astronomers realized that the sun, although the center of Earth's solar system, was not the center of the universe. And even more recent thinking is that there is no specific location that is the center of the universe, per Albert Einstein's principle of relativity.

Early Developments

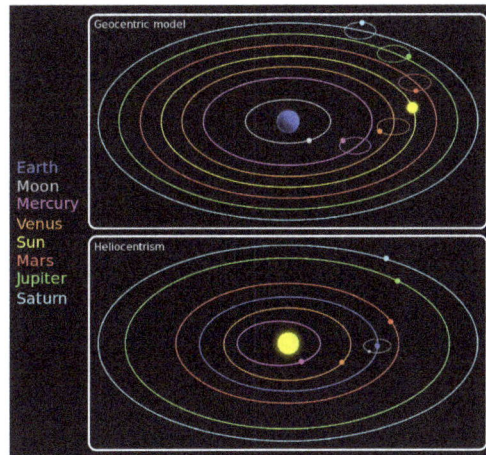

A hypothetical geocentric model of the solar system (upper panel) in comparison to the heliocentric model (lower panel).

To anyone who stands and looks up at the sky, it seems that the Earth stays in one place, while everything in the sky rises in the east and sets in the west once a day. However, with more scrutiny one will observe more complicated movements. The positions at which the Sun and moon rise change over the course of a year, some planets and stars do not appear at all for many months, and planets sometimes appear to have moved in the reverse direction for a while, relative to the background stars.

As these motions became better understood, more elaborate descriptions were required, the most famous of which was the geocentric Ptolemaic system, which achieved its full expression in the 2nd century. The Ptolemaic system was a sophisticated astronomical system that managed to calculate the positions for the planets to a fair degree of accuracy. Ptolemy himself, in his *Almagest*, points out that any model for describing the motions of the planets is merely a mathematical device, and since there is no actual way to know which is true, the simplest model that gets the right numbers should be used. However, he rejected the idea of a spinning earth as absurd as he believed it would create huge winds. His planetary hypotheses were sufficiently real that the distances of moon, sun, planets and stars could be determined by treating orbits' celestial spheres as contiguous realities.

This made the stars' distance less than 20 Astronomical Units, a regression, since Aristarchus of Samos's heliocentric scheme had centuries earlier necessarily placed the stars at least two orders of magnitude more distant.

Greek and Hellenistic World

Pythagoreans

The non-geocentric model of the Universe was proposed by the Pythagorean philosopher Philolaus (d. 390 BC), who taught that at the center of the Universe was a "central fire", around which the Earth, Sun, Moon and Planets revolved in uniform circular motion. This system postulated the existence of a counter-earth collinear with the Earth and central fire, with the same period of revolution around the central fire as the Earth. The Sun revolved around the central fire once a year, and the stars were stationary. The Earth maintained the same hidden face towards the central fire, rendering both it and the "counter-earth" invisible from Earth. The Pythagorean concept of uniform circular motion remained unchallenged for approximately the next 2000 years, and it was to the Pythagoreans that Copernicus referred to show that the notion of a moving Earth was neither new nor revolutionary. Kepler gave an alternative explanation of the Pythagoreans' "central fire" as the Sun, "as most sects purposely hid[e] their teachings".

Heraclides of Pontus (4th century BC) said that the rotation of the Earth explained the apparent daily motion of the celestial sphere. It used to be thought that he believed Mercury and Venus to revolve around the Sun, which in turn (along with the other planets) revolves around the Earth. Macrobius Ambrosius Theodosius (AD 395–423) later described this as the "Egyptian System," stating that "it did not escape the skill of the Egyptians," though there is no other evidence it was known in ancient Egypt.

Aristarchus of Samos

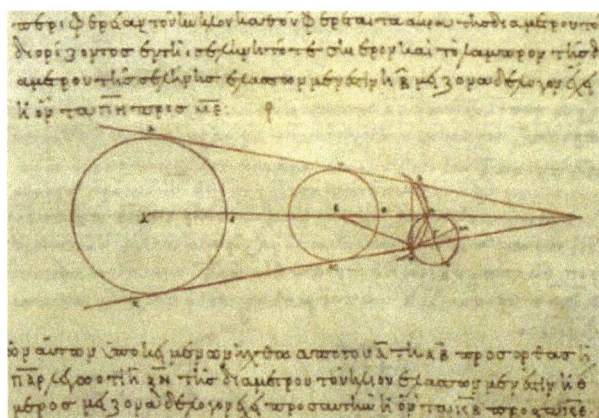

Aristarchus's 3rd century BC calculations on the relative sizes of the Earth, Sun and Moon, from a 10th-century CE Greek copy

The first person known to have proposed a heliocentric system, however, was Aristarchus of Samos (c. 270 BC). Like Eratosthenes, Aristarchus calculated the size of the Earth, and measured the size and distance of the Moon and Sun, in a treatise which has survived. From his estimates, he concluded that the Sun was six to seven times wider than the Earth and thus hundreds of times more voluminous. His writings on the heliocentric system are lost, but some information is known from surviving descriptions and critical commentary by his contemporaries, such as Archimedes. Some have suggested that his calculation of the relative size of the Earth and Sun led Aristarchus to conclude that it made more sense for the Earth to be moving than for the huge Sun to be moving around it. Though the original text has been lost, a reference in Archimedes' book *The Sand Reckoner* describes another work by Aristarchus in which he advanced an alternative hypothesis of the heliocentric model. Archimedes wrote:

You King Gelon are aware the 'universe' is the name given by most astronomers to the sphere the center of which is the center of the Earth, while its radius is equal to the straight line between the center of the Sun and the center of the Earth. This is the common account as you have heard from astronomers. But Aristarchus has brought out a book consisting of certain hypotheses, wherein it appears, as a consequence of the assumptions made, that the universe is many times greater than the 'universe' just mentioned. His hypotheses are that the fixed stars and the Sun remain unmoved, that the Earth revolves about the Sun on the circumference of a circle, the Sun lying in the middle of the orbit, and that the sphere of fixed stars, situated about the same center as the Sun, is so great that the circle in which he supposes the Earth to revolve bears such a proportion to the distance of the fixed stars as the center of the sphere bears to its surface.

Aristarchus believed the stars to be very far away, and saw this as the reason why there was no visible parallax, that is, an observed movement of the stars relative to each other as the Earth moved around the Sun. The stars are in fact much farther away than the distance that was generally assumed in ancient times, which is why stellar parallax is only detectable with telescopes.

Archimedes says that Aristarchus made the stars' distance larger, suggesting that he was answering the natural objection that Heliocentrism requires stellar parallactic oscillations. He apparently agreed to the point but placed the stars so distant as to make the parallactic motion invisibly minuscule. Thus Heliocentrism opened the way for realization that the universe was larger than the geocentrists taught.

Seleucus of Seleucia

Since Plutarch mentions the "followers of Aristarchus" in passing, it is likely that there were other astronomers in the Classical period who also espoused Heliocentrism, but whose work was lost. The only other astronomer from antiquity known by name who is known to have supported Aristarchus' heliocentric model was Seleucus of Seleucia (b. 190 BC), a Hellenistic astronomer who flourished a century after Aristarchus in the Seleucid empire. Seleucus adopted the heliocentric system of Aristarchus and is said to have proved the heliocentric theory. According to Bartel Leendert van der Waerden, Seleucus may have proved the heliocentric theory by determining the constants of a geometric model for the heliocentric theory and by developing methods to compute planetary positions using this model. He may have used early trigonometric methods that were available in his time, as he was a contemporary of Hipparchus. A fragment of a work by Seleucus has survived in Arabic translation, which was referred to by Rhazes (b. 865).

Alternatively, his explanation may have involved the phenomenon of tides, which he supposedly theorized to be caused by the attraction to the Moon and by the revolution of the Earth around the Earth-Moon 'center of mass'.

Medieval Europe

Nicholas of Cusa, 15th century, asked whether there was any reason to assert that any point was the center of the universe.

There were occasional speculations about Heliocentrism in Europe before Copernicus. In Roman Carthage, the pagan Martianus Capella (5th century A.D.) expressed the opinion that the planets Venus and Mercury did not go about the Earth but instead circled the Sun. Capella's model was discussed in the Early Middle Ages by various anonymous 9th-century commentators and Copernicus mentions him as an influence on his own work.

During the Late Middle Ages, Bishop Nicole Oresme discussed the possibility that the Earth rotated on its axis, while Cardinal Nicholas of Cusa in his *Learned Ignorance* asked whether there was any reason to assert that the Sun (or any other point) was the center of the universe. In parallel to a mystical definition of God, Cusa wrote that "Thus the fabric of the world (*machina mundi*) will *quasi* have its center everywhere and circumference nowhere."

India

According to theosophists, the earliest traces of heliocentrism is found in several Vedic Sanskrit texts written in ancient India. Aitareya Mahidasa (c. 9th–8th century BC), a Vedic Rishi and author of Aitareya Upanishad, wrote that "The Sun never sets nor rises. When people think the sun is setting, it is not so; they are mistaken. It only changes about after reaching the end of the day and makes night below and day to what is on the other side" as described in the Vedas at the time in his astronomical text Aitareya Brahmana. The earliest traces of a counter-intuitive idea that it is the Earth that is actually moving and the Sun that is at the centre of the solar system (hence the concept of Heliocentrism) is found in several Vedic Sanskrit texts written in ancient India. Yajnavalkya (c. 9th– 8th century BC) recognized that the Earth is spherical and believed that the Sun was "the centre of the spheres" as described in the Vedas at the time. In his astronomical text Shatapatha Brahmana (8.7.3.10) he states: "The sun strings these worlds - the earth, the planets, the atmosphere - to himself on a thread." He recognized that the Sun was much larger than the Earth, which would have influenced this early heliocentric concept. He also accurately measured the relative distances of the Sun and the Moon from the Earth as 108 times the diameters of these heavenly bodies, close to the modern measurements of 107.6 for the Sun and 110.6 for the Moon. He also described a solar calendar in the Shatapatha Brahmana.

The Vedic Sanskrit text Aitareya Brahmana (2.7) (c. 9th–8th century BC) also states: "The Sun never sets nor rises thats right. When people think the sun is setting, it is not so; they are mistaken." This indicates that the Sun is stationary (hence the Earth is moving around it), which is elaborated in a later commentary Vishnu Purana (2.8) (c. 1st century), which states: "The sun is stationed for all time, in the middle of the day. Of the sun, which is always in one and the same place, there is neither setting nor rising."

Aryabhata (476–550), in his magnum opus *Aryabhatiya* (499), propounded a planetary model in which the Earth was taken to be spinning on its axis and the periods of the planets were given with respect to the Sun. He accurately calculated many astronomical constants, such as the periods of the planets, times of the solar and lunar eclipses, and the instantaneous motion of the Moon. Early followers of Aryabhata's model included Varahamihira, Brahmagupta, and Bhaskara II.

Nilakantha Somayaji (1444–1544), in his *Aryabhatiyabhasya*, a commentary on Aryabhata's *Aryabhatiya*, developed a computational system for a partially heliocentric planetary model, in which the planets orbit the Sun, which in turn orbits the Earth, similar to the Tychonic system later proposed by Tycho Brahe in the late 16th century. In the *Tantrasangraha* (1500), he further revised his planetary system, which was mathematically more accurate at predicting the heliocentric orbits of the interior planets than both the Tychonic and Copernican models, but like Indian astronomy in general fell short of proposing models of the universe. Nilakantha's planetary system also incorporated the Earth's rotation on its axis. Most astronomers of the Kerala school of astronomy and mathematics seem to have accepted his planetary model.

Medieval Islamic World

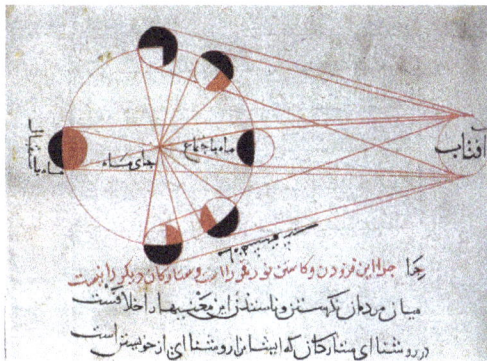

An illustration from al-Biruni's astronomical works, explains the different phases of the moon, with respect to the position of the sun. Al-Biruni suggested that if the Earth rotated on its axis this would be consistent with astronomical theory. He discussed Heliocentrism but considered it a problem of natural philosophy.

Muslim astronomers generally accepted the Ptolemaic system and the geocentric model, but by the 10th century texts appeared regularly whose subject matter was doubts concerning Ptolemy (*shukūk*). Several Muslim scholars questioned the Earth's apparent immobility and centrality within the universe. Some accepted that the Earth rotates around its axis, such as the 10th-century astronomer Abu Sa'id al-Sijzi (d. circa 1020). According to Al-Biruni, Sijzi invented an astrolabe called *al-zūraqī* based on a belief held by some of his contemporaries "That the motion we see is due to the Earth's movement and not to that of the sky." The prevalence of this view is further confirmed by a reference from the 13th century which states:

According to the Geometers [or engineers] (*muhandisīn*), the earth is in constant circular motion, and what appears to be the motion of the heavens is actually due to the motion of the earth and not the stars.

Early in the 11th century Alhazen wrote a scathing critique of Ptolemy's model in his *Doubts on Ptolemy* (c. 1028), which some have interpreted to imply he was criticizing Ptolemy's geocentrism, but most agree that he was actually criticizing the details of Ptolemy's model rather than his geocentrism. Abu Rayhan Biruni (b. 973) discussed the possibility of whether the Earth rotated about its own axis and around the Sun, but in his *Masudic Canon*, he set forth the principles that the Earth is at the center of the universe and that it has no motion of its own. He was aware that if the Earth rotated on its axis, this would be consistent with his astronomical parameters, but he considered it a problem of natural philosophy rather than mathematics.

In the 12th century, some Islamic astronomers developed complete alternatives to the Ptolemaic system (although not heliocentric), such as Nur ad-Din al-Bitruji, who considered the Ptolemaic model as mathematical, and not physical. Al-Bitruji's alternative system spread through most of Europe in the 13th century, with debates and refutations of his ideas continued up to the 16th century.

At the Maragha and Samarkand observatories, the Earth's rotation was discussed by al-Kātibī (d. 1277), Tusi (b. 1201) and Qushji (b. 1403). The arguments and evidence used by Tusi and Qushji resemble those used by Copernicus to support the Earth's motion. However, it remains a fact that the Maragha school never made the big leap to Heliocentrism. Some historians maintain that the thought of the Maragha school influenced Copernicus, in particular the mathematical devices known as the Urdi lemma and the Tusi couple. Copernicus used such devices in the same planetary models as found in Arabic sources. Furthermore, the exact replacement of the equant by two epicycles used by Copernicus in the *Commentariolus* was found in an earlier work by Ibn al-Shatir (d. c. 1375) of Damascus. Ibn al-Shatir's lunar and Mercury models are also identical to those of Copernicus. However, this remains speculative as no researcher has yet proven that Copernicus knew about Ibn al-Shatir's work or the Maragha school. It has been argued that Copernicus could have independently discovered the Tusi couple or took the idea from Proclus's *Commentary on the First Book of Euclid*, which Copernicus cited. Another possible source for Copernicus's knowledge of this mathematical device is the *Questiones de Spera* of Nicole Oresme, who described how a reciprocating linear motion of a celestial body could be produced by a combination of circular motions similar to those proposed by al-Tusi. Nevertheless, Copernicus cited some of the Islamic astronomers whose theories and observations he used in *De Revolutionibus*, namely al-Battani, Thabit ibn Qurra, al-Zarqali, Ibn Rushd, and al-Bitruji.

Copernican Revolution

Astronomical Model

In the 16th century, Nicolaus Copernicus's *De revolutionibus* presented a discussion of a heliocentric model of the universe in much the same way as Ptolemy's *Almagest* had presented his geocentric model in the 2nd century. Copernicus discussed the philosophical implications of his proposed system, elaborated it in geometrical detail, used selected astronomical observations to derive the parameters of his model, and wrote astronomical tables which enabled one to compute the past and future positions of the stars and planets. In doing so, Copernicus moved Heliocentrism from philosophical speculation to predictive geometrical astronomy. In reality, Copernicus's system did not predict the planets' positions any better than the Ptolemaic system. This theory resolved the issue of planetary retrograde motion by arguing that such motion was only perceived and apparent, rather than real: it was a parallax effect, as an object that one is passing seems to move backwards against the horizon. This issue was also resolved in the geocentric Tychonic system; the latter, however, while eliminating the major epicycles, retained as a physical reality the irregular back-and-forth motion of the planets, which Kepler characterized as a "pretzel".

Nicolaus Copernicus, 16th century, described the first computational system explicitly tied to a heliocentric model

Copernicus cited Aristarchus in an early (unpublished) manuscript of *De Revolutionibus* (which still survives), stating: "Philolaus believed in the mobility of the earth, and some even say that Aristarchus of Samos was of that opinion." However, in the published version he restricts himself to noting that in works by Cicero he had found an account of the theories of Hicetas and that Plutarch had provided him with an account of the Pythagoreans Heraclides Ponticus, Philolaus, and Ecphantus. These authors had proposed a moving earth, which did not, however, revolve around a central sun.

Religious Attitudes to Heliocentrism

Heliocentrism had been in conflict with religion before Copernicus. One of the few pieces of information we have about the reception of Aristarchus's heliocentric system comes from a passage in Plutarch's dialogue, *Concerning the Face which Appears in the Orb of the Moon*. According to one of Plutarch's characters in the dialogue, the philosopher Cleanthes had held that Aristarchus should be charged with impiety for "moving the hearth of the world".

Circulation of Commentariolus (Before 1515)

The first information about the heliocentric views of Nicolaus Copernicus was circulated in manuscript completed some time before May 1, 1514. Although only in manuscript, Copernicus' ideas were well known among astronomers and others. His ideas contradicted the then-prevailing understanding of the Bible. In the King James Bible First Chronicles 16:30 state that "the world also shall be stable, that it be not moved." Psalm 104:5 says, "[the Lord] Who laid the foundations of the earth, that it should not be removed for ever." Ecclesiastes 1:5 states that "The sun also ariseth, and the sun goeth down, and hasteth to his place where he arose."

Nonetheless, in 1533, Johann Albrecht Widmannstetter delivered in Rome a series of lectures outlining Copernicus' theory. The lectures were heard with interest by Pope Clement VII and several Catholic cardinals. On November 1, 1536, Archbishop of Capua Nikolaus von Schönberg wrote a letter to Copernicus from Rome encouraging him to publish a full version of his theory.

However, in 1539, Martin Luther said:

"There is talk of a new astrologer who wants to prove that the earth moves and goes around instead of the sky, the sun, the moon, just as if somebody were moving in a carriage or ship might hold that he was sitting still and at rest while the earth and the trees walked and moved. But that is how things are nowadays: when a man wishes to be clever he must . . . invent something special, and the way he does it must needs be the best! The fool wants to turn the whole art of astronomy upside-down. However, as Holy Scripture tells us, so did Joshua bid the sun to stand still and not the earth."

This was reported in the context of a conversation at the dinner table and not a formal statement of faith. Melanchthon, however, opposed the doctrine over a period of years.

Publication of De Revolutionibus (1543)

Nicolaus Copernicus published the definitive statement of his system in *De Revolutionibus* in 1543. Copernicus began to write it in 1506 and finished it in 1530, but did not publish it until the year of his death. Although he was in good standing with the Church and had dedicated the book to Pope Paul III, the published form contained an unsigned preface by Osiander defending the system and arguing that it was useful for computation even if its hypotheses were not necessarily true. Possibly because of that preface, the work of Copernicus inspired very little debate on whether it might be heretical during the next 60 years. There was an early suggestion among Dominicans that the teaching of Heliocentrism should be banned, but nothing came of it at the time.

Some years after the publication of *De Revolutionibus* John Calvin preached a sermon in which he denounced those who "pervert the order of nature" by saying that "the sun does not move and that it is the earth that revolves and that it turns".

On the other hand, Calvin is *not* responsible for another famous quotation which has often been misattributed to him: "Who will venture to place the authority of Copernicus above that of the Holy Spirit?" It has long been established that this line cannot be found in any of Calvin's works. It has been suggested that the quotation was originally sourced from the works of Lutheran theologian Abraham Calovius.

Tycho Brahe's Geo-heliocentric System c. 1587

Prior to the publication of *De Revolutionibus*, the most widely accepted system had been proposed by Ptolemy, in which the Earth was the center of the universe and all celestial bodies orbited it. Tycho Brahe, arguably the most accomplished astronomer of his time, advocated against Copernicus's heliocentric system and for an alternative to the Ptolemaic geocentric system: a geo-heliocentric system now known as the Tychonic system in which the five then known planets orbit the sun, while the sun and the moon orbit the earth.

Tycho appreciated the Copernican system, but objected to the idea of a moving Earth on the basis of physics, astronomy, and religion. The Aristotelian physics of the time (modern Newtonian physics was still a century away) offered no physical explanation for the motion of a massive body like Earth, whereas it could easily explain the motion of heavenly bodies by postulating that they were made of a different sort substance called aether that moved naturally. So Tycho said that the Copernican system "... expertly and completely circumvents all that is superfluous or discordant in the system of Ptolemy. On no point does it offend the principle of mathematics. Yet it ascribes to the Earth, that hulking, lazy body, unfit for motion, a motion as quick as that of the aethereal torches, and a triple motion at that." Likewise, Tycho took issue with the vast distances to the stars that Aristarchus and Copernicus had assumed in order to explain the lack of any visible parallax. Tycho had measured the apparent sizes of stars (now known to be illusory – see stellar magnitude), and used geometry to calculate that in order to both have those apparent sizes and be as far away as Heliocentrism required, stars would have to be huge (much larger than the sun; the size of Earth's orbit or larger). Regarding this Tycho wrote, "Deduce these things geometrically if you like, and you will see how many absurdities (not to mention others) accompany this assumption [of the motion of the earth] by inference." He also cited the Copernican system's "opposition to the authority of Sacred Scripture in more than one place" as a reason why one might wish to reject it, and observed that his own geoheliocentric alternative "offended neither the principles of physics nor Holy Scripture".

The Jesuit astronomers in Rome were at first unreceptive to Tycho's system; the most prominent, Clavius, commented that Tycho was "confusing all of astronomy, because he wants to have Mars lower than the Sun." However, after the advent of the telescope showed problems with some geocentric models (by demonstrating that Venus circles the sun, for example), the Tychonic system and variations on that system became very popular among geocentrists, and the Jesuit astronomer Giovanni Battista Riccioli would continue Tycho's use of physics, stellar astronomy (now with a telescope), and religion to argue against Heliocentrism and for Tycho's system well into the seventeenth century.

Publication of Starry Messenger (1610)

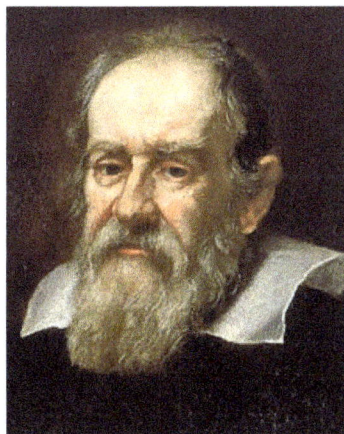

In the 17th century AD Galileo Galilei opposed the Roman Catholic Church by his strong support for Heliocentrism

Galileo was able to look at the night sky with the newly invented telescope. Then he published his discoveries in Sidereus Nuncius including (among other things) the moons of Jupiter and that Venus exhibited a full range of phases. These discoveries were not consistent with the Ptolemeic model of the solar system. As the Jesuit astronomers confirmed Galileo's observations, the Jesuits moved toward Tycho's teachings.

Publication of Letter to the Grand Duchess (1615)

In a Letter to the Grand Duchess Christina, Galileo defended Heliocentrism, and claimed it was not contrary to Scriptures. He took Augustine's position on Scripture: not to take every passage literally when the scripture in question is a book of poetry and songs, not a book of instructions or history. The writers of the Scripture wrote from the perspective of the terrestrial world, and from that vantage point the sun does rise and set. In fact, it is the Earth's rotation which gives the impression of the sun in motion across the sky.

1616 Ban Against Copernicanism

In February 1615, prominent Dominicans including Thomaso Caccini and Niccolò Lorini brought Galileo's writings on Heliocentrism to the attention of the Inquisition, because they appeared to violate Holy Scripture and the decrees of the Council of Trent. Cardinal and Inquisitor Robert Bellarmine was called upon to adjudicate, and wrote in April that treating Heliocentrism as a real phenomenon would be "a very dangerous thing," irritating philosophers and theologians, and harming "the Holy Faith by rendering Holy Scripture as false."

In January 1616 Msgr. Francesco Ingoli addressed an essay to Galileo disputing the Copernican system. Galileo later stated that he believed this essay to have been instrumental in the ban against Copernicanism that followed in February. According to Maurice Finocchiaro, Ingoli had probably been commissioned by the Inquisition to write an expert opinion on the controversy, and the essay provided the "chief direct basis" for the ban. The essay focused on eighteen physical and mathematical arguments against Heliocentrism. It borrowed primarily from the arguments of Tycho Brahe, and it notably mentioned the problem that Heliocentrism requires the stars to be much larger than the sun. Ingoli wrote that the great distance to the stars in the heliocentric theory "clearly proves ... the fixed stars to be of such size, as they may surpass or equal the size of the orbit circle of the Earth itself." Ingoli included four theological arguments in the essay, but suggested to Galileo that he focus on the physical and mathematical arguments. Galileo did not write a response to Ingoli until 1624.

In February 1616, the Inquisition assembled a committee of theologians, known as qualifiers, who delivered their unanimous report condemning Heliocentrism as "foolish and absurd in philosophy, and formally heretical since it explicitly contradicts in many places the sense of Holy Scripture." The Inquisition also determined that the Earth's motion "receives the same judgement in philosophy and ... in regard to theological truth it is at least erroneous in faith."

–Bellarmine personally ordered Galileo

"to abstain completely from teaching or defending this doctrine and opinion or from discussing it... to abandon completely... the opinion that the sun stands still at the center of the world and the earth moves, and henceforth not to hold, teach, or defend it in any way whatever, either orally or in writing."

— Bellarmine and the Inquisition's injunction against Galileo, 1616

In March, after the Inquisition's injunction against Galileo, the papal Master of the Sacred Palace, Congregation of the Index, and Pope banned all books and letters advocating the Copernican system, which they called "the false Pythagorean doctrine, altogether contrary to Holy Scripture." In 1618 the Holy Office recommended that a modified version of Copernicus' *De Revolutionibus* be allowed for use in calendric calculations, though the original publication remained forbidden until 1758.

Publication of Epitome Astronomia Copernicanae (1617–1621)

In *Astronomia nova* (1609), Johannes Kepler had used an elliptical orbit to explain the motion of Mars. In *Epitome astronomiae Copernicanae* he developed a heliocentric model of the solar system in which all the planets have elliptical orbits. This provided significantly increased accuracy in predicting the position of the planets. Kepler's ideas were not immediately accepted. Galileo for example completely ignored Kepler's work. Kepler proposed Heliocentrism as a physical description of the solar system and *Epitome astronomia Copernicanae* was placed on the index of prohibited books despite Kepler being a Protestant.

Publication of Dialogue Concerning the Two Chief World Systems

Pope Urban VIII encouraged Galileo to publish the pros and cons of Heliocentrism. Galileo's response, *Dialogue concerning the two chief world systems,* clearly advocated Heliocentrism, despite his declaration in the preface that

I will endeavour to show that all experiments that can be made upon the Earth are insufficient means to conclude for its mobility but are indifferently applicable to the Earth, movable or immovable...

and his straightforward statement,

I might very rationally put it in dispute, whether there be any such centre in nature, or no; being that neither you nor any one else hath ever proved, whether the World be finite and figurate, or else infinite and interminate; yet nevertheless granting you, for the present, that it is finite, and of a terminate Spherical Figure, and that thereupon it hath its centre...

Some ecclesiastics also interpreted the book as characterizing the Pope as a simpleton, since his viewpoint in the dialogue was advocated by the character Simplicio. Urban VIII became hostile to Galileo and he was again summoned to Rome. Galileo's trial in 1633 involved making fine distinctions between "teaching" and "holding and defending as true". For advancing heliocentric theory Galileo was forced to recant Copernicanism and was put under house arrest for the last few years of his life.

According to J. L. Heilbron, informed contemporaries of Galileo's:

"appreciated that the reference to heresy in connection with Galileo or Copernicus had no general or theological significance."

Subsequent Developments

René Descartes postponed, and ultimately never finished, his treatise *The World*, which included a heliocentric model, but the Galileo affair did little to slow the spread of Heliocentrism across Europe, as Kepler's *Epitome of Copernican Astronomy* became increasingly influential in the coming decades. By 1686 the model was well enough established that the general public was reading about it in *Conversations on the Plurality of Worlds*, published in France by Bernard le Bovier de Fontenelle and translated into English and other languages in the coming years. It has been called "one of the first great popularizations of science."

In 1687, Isaac Newton published *Philosophiæ Naturalis Principia Mathematica*, which provided an explanation for Kepler's laws in terms of universal gravitation and what came to be known as Newton's laws of motion. This placed Heliocentrism on a firm theoretical foundation, although Newton's Heliocentrism was of a somewhat modern kind. Already in the mid-1680s he recognized the "deviation of the Sun" from the centre of gravity of the solar system. For Newton it was not precisely the centre of the Sun or any other body that could be considered at rest, but "the common centre of gravity of the Earth, the Sun and all the Planets is to be esteem'd the Centre of the World", and this centre of gravity "either is at rest or moves uniformly forward in a right line". Newton adopted the "at rest" alternative in view of common consent that the centre, wherever it was, was at rest.

Meanwhile, the Church remained opposed to Heliocentrism as a literal description, but this did not by any means imply opposition to all astronomy; indeed, it needed observational data to maintain its calendar. In support of this effort it allowed the cathedrals themselves to be used as solar observatories called *meridiane*; i.e., they were turned into "reverse sundials", or gigantic pinhole cameras, where the Sun's image was projected from a hole in a window in the cathedral's lantern onto a meridian line.

In 1664, Pope Alexander VII published his *Index Librorum Prohibitorum Alexandri VII Pontificis Maximi jussu editus* (Index of Prohibited Books, published by order of Alexander VII, P.M.) which included all previous condemnations of heliocentric books.

In the mid-eighteenth century the Church's opposition began to fade. An annotated copy of Newton's *Principia* was published in 1742 by Fathers le Seur and Jacquier of the Franciscan Minims, two Catholic mathematicians, with a preface stating that the author's work assumed Heliocentrism and could not be explained without the theory. In 1758 the Catholic Church dropped the general prohibition of books advocating Heliocentrism from the *Index of Forbidden Books*. Pope Pius VII approved a decree in 1822 by the Sacred Congregation of the Inquisition to allow the printing of heliocentric books in Rome.

The Roman Catholic Church currently operates the Vatican Observatory, home to multiple award winning scholars, showing their ultimate acceptance of Heliocentrism by furthering the field.

Heliocentrism and Judaism

Already in the Talmud, Greek philosophy and science under general name "Greek wisdom" were considered dangerous. They were put under ban then and later for some periods. For example, in 13-5 a *beit din* (rabbinical court) in Barcelona forbade men younger than 25 from studying secular philosophy or the natural sciences (although an exception was made for those who studied medicine). Possibly due to this the system of Nicolaus Copernicus did not cause furious resistance, although it was found to be contradicting verses of Tanakh (Jewish Bible).

The first to mention the new system was Maharal of Prague, although he did not mention Copernicus, the author of the system. In his book "Be'er ha-Golah", in 1593 Maharal used the appearance of the new system to show that scientific theories are not reliable enough – even astronomy was turned upside-down.

Copernicus is mentioned for the first time in Hebrew in the books of David Gans (1541–1613), who worked with Tycho Brahe and Johannes Kepler. Gans wrote two books on astronomy: a short one "Magen David" (1612) and a full one "Nehmad veNaim" (published only in 1743). He described objectively three systems: Ptolemy, Copernicus and of Tycho Brahe without taking sides.

In 1629 a new Hebrew book "Elim" by Joseph Solomon Delmedigo (1591–1655) appeared. The author says that the arguments of Copernicus are so strong, that only an imbecile will not accept them. Delmedigo studied at Padua and was acquainted with Galileo.

The following wave of Hebrew literature on the subject is from the 18th century. Most of its authors were for Copernicus, although David Nieto and Tobias Cohn were exceptions. These two authors gave the same reason for opposing Heliocentrism—namely, contradiction of the Bible—although Nieto merely rejected the new system on those grounds without much passion, whereas Hacohen went so far as to call Copernicus "a first-born of Satan". Hacohen also mentions the fact that the Sages of Talmud derived the Hebrew name of Earth from the verb "run".

In later periods there were no explicit attacks on Heliocentrism, although some Rabbis were not sure about the point.

In the 20th century R. M.M. Schneerson suggested that the theory of relativity makes the question obsolete, as he writes; "on the basis of the presently accepted scientific view (in accordance with the theory of Relativity) that where two bodies in space are in motion relative to one another, it is impossible scientifically to ascertain which revolves around which, or which is stationary and the other in motion. Therefore, to say that there is, or can be, "scientific proof" that the earth revolves around the sun is quite an unscientific and uncritical statement."

The View of Modern Science

Kepler's laws of planetary motion were used as arguments in favor of the heliocentric hypothesis. Three apparent proofs of the heliocentric hypothesis were provided in 1727 by James Bradley, in 1838 by Friedrich Wilhelm Bessel and in 1851 by Foucault. Bessel proved that the parallax of a star was greater than zero by measuring the parallax of 0.314 arcseconds of a star named 61 Cygni. In the same year Friedrich Georg Wilhelm Struve and Thomas Henderson measured the parallaxes of other stars, Vega and Alpha Centauri.

The thinking that the heliocentric view was also not true in a strict sense was achieved in steps. That the Sun was not the center of the universe, but one of innumerable stars, was strongly advocated by the mystic Giordano Bruno. Over the course of the 18th and 19th centuries, the status of the Sun as merely one star among many became increasingly obvious. By the 20th century, even before the discovery that there are many galaxies, it was no longer an issue.

The concept of an absolute velocity, including being "at rest" as a particular case, is ruled out by the principle of relativity, also eliminating any obvious "center" of the universe as a natural origin of coordinates. Some forms of Mach's principle consider the frame at rest with respect to the distant masses in the universe to have special properties.

Even if the discussion is limited to the solar system, the Sun is not at the geometric center of any planet's orbit, but rather approximately at one focus of the elliptical orbit. Furthermore, to the extent that a planet's mass cannot be neglected in comparison to the Sun's mass, the center of gravity of the solar system is displaced slightly away from the center of the Sun. (The masses of the planets, mostly Jupiter, amount to 0.14% of that of the Sun.) Therefore, a hypothetical astronomer on an extrasolar planet would observe a small "wobble" in the Sun's motion.

Modern use of *Geocentric* and *Heliocentric*

In modern calculations the terms "geocentric" and "heliocentric" are often used to refer to reference frames. In such systems the origin in the center of mass of the Earth, of the Earth–Moon system, of the Sun, of the Sun plus the major planets, or of the entire solar system can be selected; see center-of-mass frame. Right Ascension and Declination are examples of geocentric coordinates, used in Earth-based observations, while the heliocentric latitude and longitude are used for orbital calculations. This leads to such terms as "heliocentric velocity" and "heliocentric angular momentum". In this heliocentric picture, any planet of the Solar System can be used as a source of mechanical energy because it moves relatively to the Sun. A smaller body (either artificial or natural) may gain heliocentric velocity due to gravity assist – this effect can change the body›s mechanical energy in heliocentric reference frame (although it will not changed in the planetary one). However, such selection of «geocentric» or «heliocentric» frames is merely a matter of computation. It does not have philosophical implications and does not constitute a distinct physical or scientific model. From the point of view of General Relativity, inertial reference frames do not exist at all, and any practical reference frame is only an approximation to the actual space-time, which can have higher or lower precision.

Cosmogony

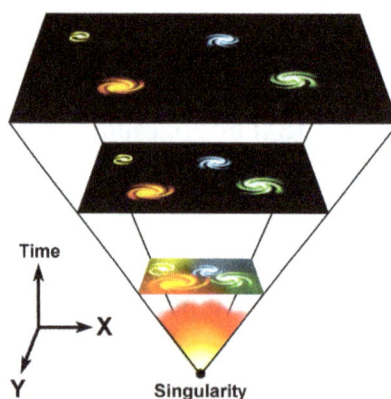

The Big Bang theory, which states that the universe expanded from and was a singularity whose radius was zero, is widely accepted by physicists.

Cosmogony (or **cosmogeny**) is any model concerning the origin of either the cosmos or universe. Developing a complete theoretical model has implications in both the philosophy of science and epistemology.

Etymology

The word comes from the Koine Greek (from "cosmos, the world") and the root of ("come into a new state of being"). In astronomy, cosmogony refers to the study of the origin of particular astrophysical objects or systems, and is most commonly used in reference to the origin of the universe, the solar system, or the earth-moon system.

The Big Bang theory is the prevailing cosmological model of the early development of the universe. The most commonly held view is that the universe was once a gravitational singularity, which expanded extremely rapidly from its hot and dense state. However, while this expansion is well-modeled by the Big Bang theory, the origins of the singularity remain as one of the unsolved problems in physics.

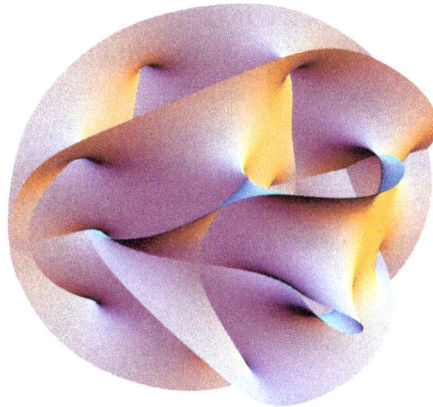

Projection of a Calabi–Yau manifold from string theory. In quantum physics, there remain different, plausible theories regarding what combination of "stuff", space, or time emerged along with the singularity (and therefore this universe). The main disagreement among scientists is whether time existed "before" the emergence of our universe or not.

Cosmologist and science communicator Sean M. Carroll explains two competing types of explanations for the origins of the singularity which is the main disagreement between the scientists who study cosmogony and centers on the question of whether time existed "before" the emergence of our universe or not. One cosmogonical view sees time as fundamental and even eternal: The universe could have contained the singularity because the universe evolved or changed from a prior state (the prior state was "empty space", or maybe a state that could not be called "space" at all). The other view, held by proponents like Stephen Hawking, says that there was no change through time because "time" itself emerged along with this universe (in other words, there can be no "prior" to the universe). Thus, it remains unclear what combination of "stuff", space, or time emerged with the singularity and this universe.

One problem in cosmogony is that there is currently no theoretical model that explains the earliest moments of the universe's existence (during the Planck time) because of a lack of a testable theory of quantum gravity. Researchers in string theory and its extensions (for example, M theory), and of loop quantum cosmology, have nevertheless proposed solutions of the type just discussed.

Another issue facing the field of particle physics is a need for more expensive and technologically advanced particle accelerators to test proposed theories (for example, that the universe was caused by colliding membranes).

Compared with Cosmology

Cosmology is the study of the structure and changes in the present universe, while the scientific field of cosmogony is concerned with the origin of the universe. Observations about our present universe may not only allow predictions to be made about the future, but they also provide clues to events that happened long ago when ... the cosmos began. So the work of cosmologists and cosmogonists overlaps.

National Aeronautics and Space Administration (NASA)

Cosmogony can be distinguished from cosmology, which studies the universe at large and throughout its existence, and which technically does not inquire directly into the source of its origins. There is some ambiguity between the two terms. For example, the cosmological argument from theology regarding the existence of God is technically an appeal to cosmogonical rather than cosmological ideas. In practice, there is a scientific distinction between cosmological and cosmogonical ideas. Physical cosmology is the science that attempts to explain all observations relevant to the development and characteristics of the universe as a whole. Questions regarding why the universe behaves in such a way have been described by physicists and cosmologists as being extra-scientific (i.e., metaphysical), though speculations are made from a variety of perspectives that include extrapolation of scientific theories to untested regimes (i.e., at Planck scales), and philosophical or religious ideas.

Theoretical Scenarios

Cosmogonists have only tentative theories for the early stages of the universe and its beginning. As of 2011, no accelerator experiments probe energies of sufficient magnitude to provide any experimental insight into the behavior of matter at the energy levels that prevailed shortly after the Big Bang. Furthermore, since astronomical observations imply a singularity at the origin of the universe, experiments at any given high energy level will always be dwarfed by the infinite energy level predicted by Big Bang Theory. Therefore, significant technological and conceptual advances would be needed to propose a scientific test for cosmogonical theories.

Proposed theoretical scenarios differ radically, and include string theory and M-theory, the Hartle–Hawking initial state, string landscape, brane inflation, the Big Bang, and the ekpyrotic universe. Some of these models are mutually compatible, whereas others are not.

References

- Hoskin, Michael (1999-03-18). The Cambridge Concise History of Astronomy. Cambridge University Press. p. 60. ISBN 9780521576000.

- Young, M. J. L., ed. (2006-11-02). Religion, Learning and Science in the 'Abbasid Period. Cambridge University Press. p. 413. ISBN 9780521028875.

- Nasr, Seyyed Hossein (1993-01-01). An Introduction to Islamic Cosmological Doctrines. SUNY Press. p. 135. ISBN 9781438414195.

- Saliba, George (1994). A History of Arabic Astronomy: Planetary Theories During the Golden Age of Islam. New York University Press. pp. 233–234, 240. ISBN 0814780237.

- Huff, Toby E. (2003). The Rise of Early Modern Science: Islam, China and the West. Cambridge University Press. p. 58. ISBN 9780521529945.

- Numbers, Ronald L. (1993). The Creationists: The Evolution of Scientific Creationism. University of California Press. p. 237. ISBN 0520083938.

- Berman, Morris (2006). Dark Ages America: The Final Phase of Empire. W.W. Norton & Company. ISBN 9780393058666.

- Finocchiaro, Maurice A. (1989). The Galileo Affair: A Documentary History. Berkeley: University of California Press. p. 307. ISBN 9780520066625.

- Schneersohn, Menachem Mendel; Gotfryd, Arnie (2003). Mind over Matter: The Lubavitcher Rebbe on Science, Technology and Medicine. Shamir. pp. 76ff.; cf. xvi-xvii, 69, 100–1, 171–2, 408ff. ISBN 9789652930804.

- Fantoli, Annibale (2003). Galileo — For Copernicanism and the Church, 3rd English edition, tr. George V. Coyne, SJ. Vatican Observatory Publications, Notre Dame, IN. ISBN 88-209-7427-4.

- Graney, Christopher M. (2015), Setting Aside All Authority: Giovanni Battista Riccioli and the Science against Copernicus in the Age of Galileo, University of Notre Dame Press, ISBN 978-0268029883

- Heilbron, John L. (1999), The Sun in the Church: Cathedrals as Solar Observatories, Cambridge, MA: Harvard University Press, ISBN 0674005368

- Heilbron, John L. (2005). "Censorship of Astronomy in Italy after Galileo". In McMullin, Ernan. The Church and Galileo. University of Notre Dame Press, Notre Dame. ISBN 0-268-03483-4.

- Joseph, George G. (2000). The Crest of the Peacock: Non-European Roots of Mathematics, 2nd edition. Penguin Books, London. ISBN 0-691-00659-8.

- Koestler, Arthur, (1959) The Sleepwalkers: A History of Man's Changing Vision of the Universe, Penguin Books; 1986 edition: ISBN 0-14-055212-X, 1990 reprint: ISBN 0-14-019246-8

- Koyré, Alexandre (1973). The Astronomical Revolution: Copernicus – Kepler – Borelli. Ithaca, NY: Cornell University Press. ISBN 0-8014-0504-1.

- Linton, Christopher M. (2004), From Eudoxus to Einstein—A History of Mathematical Astronomy, Cambridge: Cambridge University Press, ISBN 978-0-521-82750-8

8

Interdisciplinary Aspects of Astronomy

Astronomy is one of the oldest sciences practiced and is an interdisciplinary subject. It spreads to other fields as well. Archaeoastronomy, astrochemistry along with cosmochemistry have been explained, this chapter will provide a glimpse of related fields of astronomy briefly.

Archaeoastronomy

Archaeoastronomy (also spelled archeoastronomy) is the study of how people in the past "have understood the phenomena in the sky, how they used these phenomena and what role the sky played in their cultures." Clive Ruggles argues it is misleading to consider archaeoastronomy to be the study of ancient astronomy, as modern astronomy is a scientific discipline, while archaeoastronomy considers symbolically rich cultural interpretations of phenomena in the sky by other cultures. It is often twinned with *ethnoastronomy*, the anthropological study of skywatching in contemporary societies. Archaeoastronomy is also closely associated with historical astronomy, the use of historical records of heavenly events to answer astronomical problems and the history of astronomy, which uses written records to evaluate past astronomical practice.

The rising Sun illuminates the inner chamber of Newgrange, Ireland, only at the winter solstice.

Archaeoastronomy uses a variety of methods to uncover evidence of past practices including archaeology, anthropology, astronomy, statistics and probability, and history. Because these methods are diverse and use data from such different sources, integrating them into a coherent argument has been a long-term difficulty for archaeoastronomers. Archaeoastronomy fills complementary niches in landscape archaeology and cognitive archaeology. Material evidence and its connection to the sky can reveal how a wider landscape can be integrated into beliefs about the cycles of nature, such as Mayan astronomy and its relationship with agriculture. Other examples which have brought together ideas of cognition and landscape include studies of the cosmic order embedded in the roads of settlements.

Archaeoastronomy can be applied to all cultures and all time periods. The meanings of the sky vary from culture to culture; nevertheless there are scientific methods which can be applied across cultures when examining ancient beliefs. It is perhaps the need to balance the social and scientific aspects of archaeoastronomy which led Clive Ruggles to describe it as: *"...[A] field with academic work of high quality at one end but uncontrolled speculation bordering on lunacy at the other."*

History of Archaeoastronomy

In his short history of 'Astro-archaeology' John Michell argued that the status of research into ancient astronomy had improved over the past two centuries, going 'from lunacy to heresy to interesting notion and finally to the gates of orthodoxy.' Nearly two decades later, we can still ask the question: Is archaeoastronomy still waiting at the gates of orthodoxy or has it gotten inside the gates?

— Todd Bostwick quoting John Michell

Two hundred years before Michell wrote the above, there were no archaeoastronomers and there were no professional archaeologists, but there were astronomers and antiquarians. Some of their works are considered precursors of archaeoastronomy; antiquarians interpreted the astronomical orientation of the ruins that dotted the English countryside as William Stukeley did of Stonehenge in 1740, while John Aubrey in 1678 and Henry Chauncy in 1700 sought similar astronomical principles underlying the orientation of churches. Late in the nineteenth century astronomers such as Richard Proctor and Charles Piazzi Smyth investigated the astronomical orientations of the pyramids.

The term *archaeoastronomy* was first used by Elizabeth Chesley Baity (at the suggestion of Euan MacKie) in 1973, but as a topic of study it may be much older, depending on how archaeoastronomy is defined. Clive Ruggles says that Heinrich Nissen, working in the mid-nineteenth century was arguably the first archaeoastronomer. Rolf Sinclair says that Norman Lockyer, working in the late 19th and early 20th centuries, could be called the 'father of archaeoastronomy.' Euan MacKie would place the origin even later, stating: "...the genesis and modern flowering of archaeoastronomy must surely lie in the work of Alexander Thom in Britain between the 1930s and the 1970s."

Early archaeoastronomy surveyed Megalithic constructs in the British Isles, at sites like Auglish in County Londonderry, in an attempt to find statistical patterns

In the 1960s the work of the engineer Alexander Thom and that of the astronomer Gerald Hawkins, who proposed that Stonehenge was a Neolithic computer, inspired new interest in the astronomical features of ancient sites. The claims of Hawkins were largely dismissed, but this was not the case for Alexander Thom's work, whose survey results of megalithic sites hypothesized widespread practice of accurate astronomy in the British Isles. Euan MacKie, recognizing that Thom's theories needed to be tested, excavated at the Kintraw standing stone site in Argyllshire in 1970 and 1971 to check whether the latter's prediction of an observation platform on the hill slope above the stone was correct. There was an artificial platform there and this apparent verification of Thom's long alignment hypothesis (Kintraw was diagnosed as an accurate winter solstice site) led him to check Thom's geometrical theories at the Cultoon stone circle in Islay, also with a positive result. MacKie therefore broadly accepted Thom's conclusions and published new prehistories of Britain. In contrast a re-evaluation of Thom's fieldwork by Clive Ruggles argued that Thom's claims of high accuracy astronomy were not fully supported by the evidence. Nevertheless, Thom's legacy remains strong, Krupp wrote in 1979, "Almost singlehandedly he has established the standards for archaeo-astronomical fieldwork and interpretation, and his amazing results have stirred controversy during the last three decades." His influence endures and practice of statistical testing of data remains one of the methods of archaeoastronomy.

It has been proposed that Maya sites such as Uxmal were built in accordance with astronomical alignments.

The approach in the New World, where anthropologists began to consider more fully the role of astronomy in Amerindian civilizations, was markedly different. They had access to sources that the prehistory of Europe lacks such as ethnographies and the historical records of the early colonizers. Following the pioneering example of Anthony Aveni, this allowed New World archaeoastronomers to make claims for motives which in the Old World would have been mere speculation. The concentration on historical data led to some claims of high accuracy that were comparatively weak when compared to the statistically led investigations in Europe.

This came to a head at a meeting sponsored by the International Astronomical Union (IAU) in Oxford in 1981. The methodologies and research questions of the participants were considered so different that the conference proceedings were published as two volumes. Nevertheless, the conference was considered a success in bringing researchers together and Oxford conferences have continued every four or five years at locations around the world. The subsequent conferences have resulted in a move to more interdisciplinary approaches with researchers aiming to combine the contextuality of archaeological research, which broadly describes the state of archaeoastronomy today, rather than merely establishing the existence of ancient astronomies, archaeoastronomers seek to explain why people would have an interest in the night sky.

Relations to Other Disciplines

...[O]ne of the most endearing characteristics of archaeoastronomy is its capacity to set academics in different disciplines at loggerheads with each other.

— Clive Ruggles

Archaeoastronomy has long been seen as an interdisciplinary field that uses written and unwritten evidence to study the astronomies of other cultures. As such, it can be seen as connecting other disciplinary approaches for investigating ancient astronomy: astroarchaeology (an obsolete term for studies that draw astronomical information from the alignments of ancient architecture and landscapes), history of astronomy (which deals primarily with the written textual evidence), and ethnoastronomy (which draws on the ethnohistorical record and contemporary ethnographic studies).

Reflecting Archaeoastronomy's development as an interdisciplinary subject, research in the field is conducted by investigators trained in a wide range of disciplines. Authors of recent doctoral dissertations have described their work as concerned with the fields of archaeology and cultural anthropology; with various fields of history including the history of specific regions and periods, the history of science and the history of religion; and with the relation of astronomy to art, literature and religion. Only rarely did they describe their work as astronomical, and then only as a secondary category.

Both practicing archaeoastronomers and observers of the discipline approach it from different perspectives. George Gummerman and Miranda Warburton view archaeoastronomy as part of an archaeology informed by cultural anthropology and aimed at understanding a "group's conception of themselves in relation to the heavens', in a word, its cosmology. Todd Bostwick argued that "archaeoastronomy is anthropology – the study of human behavior in the past and present." Paul Bahn has described archaeoastronomy as an area of cognitive archaeology. Other researchers relate archaeoastronomy to the history of science, either as it relates to a culture's observations of nature and the conceptual framework they devised to impose an order on those observations or as it relates to the political motives which drove particular historical actors to deploy certain astronomical concepts or techniques. Art historian Richard Poss took a more flexible approach, maintaining that the astronomical rock art of the North American Southwest should be read employing "the hermeneutic traditions of western art history and art criticism" Astronomers, however, raise different questions, seeking to provide their students with identifiable precursors of their discipline, and are especially concerned with the important question of how to confirm that specific sites are, indeed, intentionally astronomical.

The reactions of professional archaeologists to archaeoastronomy have been decidedly mixed. Some expressed incomprehension or even hostility, varying from a rejection by the archaeological mainstream of what they saw as an archaeoastronomical fringe to an incomprehension between the cultural focus of archaeologists and the quantitative focus of early archaeoastronomers. Yet archaeologists have increasingly come to incorporate many of the insights from archaeoastronomy into archaeology textbooks and, as mentioned above, some students wrote archaeology dissertations on archaeoastronomical topics.

Since archaeoastronomers disagree so widely on the characterization of the discipline, they even dispute its name. All three major international scholarly associations relate archaeoastronomy to the study of culture, using the term *Astronomy in Culture* or a translation. Michael Hoskin sees an important part of the discipline as fact-collecting, rather than theorizing, and proposed to label this aspect of the discipline *Archaeotopography*. Ruggles and Saunders proposed *Cultural Astronomy* as a unifying term for the various methods of studying folk astronomies. Others have argued that astronomy is an inaccurate term, what are being studied are cosmologies and people who object to the use of logos have suggested adopting the Spanish *cosmovisión*.

When debates polarise between techniques, the methods are often referred to by a colour code, based on the colours of the bindings of the two volumes from the first Oxford Conference, where the approaches were first distinguished. Green (Old World) archaeoastronomers rely heavily on statistics and are sometimes accused of missing the cultural context of what is a social practice. Brown (New World) archaeoastronomers in contrast have abundant ethnographic and historical evidence and have been described as 'cavalier' on matters of measurement and statistical analysis. Finding a way to integrate various approaches has been a subject of much discussion since the early 1990s.

Methodology

For a long time I have believed that such diversity requires the invention of some all-embracing theory. I think I was very naïve in thinking that such a thing was ever possible.

—Stanislaw Iwaniszewski

There is no one way to do Archaeoastronomy. The divisions between archaeoastronomers tend not to be between the physical scientists and the social scientists. Instead it tends to depend on the location of kind of data available to the researcher. In the Old World, there is little data but the sites themselves; in the New World, the sites were supplemented by ethnographic and historic data. The effects of the isolated development of archaeoastronomy in different places can still often be seen in research today. Research methods can be classified as falling into one of two approaches, though more recent projects often use techniques from both categories.

Green Archaeoastronomy

Green Archaeoastronomy is named after the cover of the book *Archaeoastronomy in the Old World*. It is based primarily on statistics and is particularly apt for prehistoric sites where the social evidence is relatively scant compared to the historic period. The basic methods were developed by Alexander Thom during his extensive surveys of British megalithic sites.

Thom wished to examine whether or not prehistoric peoples used high-accuracy astronomy. He believed that by using horizon astronomy, observers could make estimates of dates in the year to a specific day. The observation required finding a place where on a specific date the sun set into a notch on the horizon. A common theme is a mountain which blocked the Sun, but on the right day would allow the tiniest fraction to re-emerge on the other side for a 'double sunset'. The animation below shows two sunsets at a hypothetical site, one the day before the summer solstice and one at the summer solstice, which has a double sunset.

To test this idea he surveyed hundreds of stone rows and circles. Any individual alignment could indicate a direction by chance, but he planned to show that together the distribution of alignments was non-random, showing that there was an astronomical intent to the orientation of at least some of the alignments. His results indicated the existence of eight, sixteen, or perhaps even thirty-two approximately equal divisions of the year. The two solstices, the two equinoxes and four cross-quarter days, days half-way between a solstice and the equinox were associated with the medieval Celtic calendar. While not all these conclusions have been accepted, it has had an enduring influence on archaeoastronomy, especially in Europe.

Euan MacKie has supported Thom's analysis, to which he added an archaeological context by comparing Neolithic Britain to the Mayan civilization to argue for a stratified society in this period. To test his ideas he conducted a couple of excavations at proposed prehistoric observatories in Scotland. Kintraw is a site notable for its four-meter high standing stone. Thom proposed that this was a foresight to a point on the distant horizon between Beinn Shianaidh and Beinn o'Chaolias on Jura. This, Thom argued, was a notch on the horizon where a double sunset would occur at midwinter. However, from ground level, this sunset would be obscured by a ridge in the landscape, and the viewer would need to be raised by two meters: another observation platform was needed. This was identified across a gorge where a platform was formed from small stones. The lack of artifacts caused concern for some archaeologists and the petrofabric analysis was inconclusive, but further research at Maes Howe and on the Bush Barrow Lozenge led MacKie to conclude that while the term 'science' may be anachronistic, Thom was broadly correct upon the subject of high-accuracy alignments.

In contrast Clive Ruggles has argued that there are problems with the selection of data in Thom's surveys. Others have noted that the accuracy of horizon astronomy is limited by variations in refraction near the horizon. A deeper criticism of Green archaeoastronomy is that while it can answer *whether* there was likely to be an interest in astronomy in past times, its lack of a social element means that it struggles to answer *why* people would be interested, which makes it of limited use to people asking questions about the society of the past. Keith Kintigh wrote: "To put it bluntly, in many cases it doesn't matter much to the progress of anthropology whether a particular archaeoastronomical claim is right or wrong because the information doesn't inform the current interpretive questions." Nonetheless the study of alignments remains a staple of archaeoastronomical research, especially in Europe.

Brown Archaeoastronomy

In contrast to the largely alignment-oriented statistically led methods of Green archaeoastronomy, Brown archaeoastronomy has been identified as being closer to the history of astronomy or to cultural history, insofar as it draws on historical and ethnographic records to enrich its understanding of early astronomies and their relations to calendars and ritual. The many records of native customs and beliefs made by the Spanish chroniclers means that Brown archaeoastronomy is most often associated with studies of astronomy in the Americas.

One famous site where historical records have been used to interpret sites is Chichen Itza. Rather than analysing the site and seeing which targets appear popular, archaeoastronomers have instead examined the ethnographic records to see what features of the sky were important to the Mayans and then sought archaeological correlates. One example which could have been overlooked without historical records is the Mayan interest in the planet Venus. This interest is attested to by the Dresden codex which contains tables with information about the Venus's appearances in the sky. These cycles would have been of astrological and ritual significance as Venus was associated with Quetzalcoatl or Xolotl. Associations of architectural features with settings of Venus can be found in Chichen Itza.

"El Caracol" a possible observatory temple at Chichen Itza.

The Temple of the Warriors bears iconography depicting feathered serpents associated with Quetzalcoatl or Kukulcan. This means that the building's alignment towards the place on the horizon where Venus first appears in the evening sky (when it coincides with the rainy season) may be meaningful. Aveni claims that another building associated with the planet Venus in the form of Kukulcan, and the rainy season at Chichen Itza is the Caracol. This is a building with circular tower and doors facing the cardinal directions. The base faces the most northerly setting of Venus. Additionally the pillars of a stylobate on the building's upper platform were painted black and red. These are colours associated with Venus as an evening and morning star. However the windows in the tower seem to have been little more than slots, making them poor at letting light in, but providing a suitable place to view out.

Aveni states that one of the strengths of the Brown methodology is that it can explore astronomies invisible to statistical analysis and offers the astronomy of the Incas as another example. The empire of the Incas was conceptually divided using *ceques* radial routes emanating from the capital at Cusco. Thus there are alignments in all directions which would suggest there is little of astronomical significance, However, ethnohistorical records show that the various directions do have cosmological and astronomical significance with various points in the landscape being significant at different times of the year. In eastern Asia archaeoastronomy has developed from the History of Astronomy and much archaeoastronomy is searching for material correlates of the historical record. This is due to the rich historical record of astronomical phenomena which, in China, stretches back into the Han dynasty, in the second century BC.

A criticism of this method is that it can be statistically weak. Schaefer in particular has questioned how robust the claimed alignments in the Caracol are. Because of the wide variety of evidence, which can include artefacts as well as sites, there is no one way to practice archaeoastronomy.

Despite this it is accepted that archaeoastronomy is not a discipline that sits in isolation. Because archaeoastronomy is an interdisciplinary field, whatever is being investigated should make sense both archaeologically and astronomically. Studies are more likely to be considered sound if they use theoretical tools found in archaeology like analogy and homology and if they can demonstrate an understanding of accuracy and precision found in astronomy.

Source Materials

Because archaeoastronomy is about the many and various ways people interacted with the sky, there are a diverse range of sources giving information about astronomical practices.

Alignments

A common source of data for archaeoastronomy is the study of alignments. This is based on the assumption that the axis of alignment of an archaeological site is meaningfully oriented towards an astronomical target. Brown archaeoastronomers may justify this assumption through reading historical or ethnographic sources, while Green archaeoastronomers tend to prove that alignments are unlikely to be selected by chance, usually by demonstrating common patterns of alignment at multiple sites.

An alignment is calculated by measuring the azimuth, the angle from north, of the structure and the altitude of the horizon it faces The azimuth is usually measured using a theodolite or a compass. A compass is easier to use, though the deviation of the Earth's magnetic field from true north, known as its magnetic declination must be taken into account. Compasses are also unreliable in areas prone to magnetic interference, such as sites being supported by scaffolding. Additionally a compass can only measure the azimuth to a precision of a half a degree.

A theodolite can be considerably more accurate if used correctly, but it is also considerably more difficult to use correctly. There is no inherent way to align a theodolite with North and so the scale has to be calibrated using astronomical observation, usually the position of the Sun. Because the position of celestial bodies changes with the time of day due to the Earth's rotation, the time of these calibration observations must be accurately known, or else there will be a systematic error in the measurements. Horizon altitudes can be measured with a theodolite or a clinometer.

Artifacts

The Antikythera mechanism (main fragment)

For artifacts such as the Sky Disc of Nebra, alleged to be a Bronze Age artefact depicting the cosmos, the analysis would be similar to typical post-excavation analysis as used in other sub-disciplines in archaeology. An artefact is examined and attempts are made to draw analogies with historical or ethnographical records of other peoples. The more parallels that can be found, the more likely an explanation is to be accepted by other archaeologists.

A more mundane example is the presence of astrological symbols found on some shoes and sandals from the Roman Empire. The use of shoes and sandals is well known, but Carol van Driel-Murray has proposed that astrological symbols etched onto sandals gave the footwear spiritual or medicinal meanings. This is supported through citation of other known uses of astrological symbols and their connection to medical practice and with the historical records of the time.

Another well-known artefact with an astronomical use is the Antikythera mechanism. In this case analysis of the artefact, and reference to the description of similar devices described by Cicero, would indicate a plausible use for the device. The argument is bolstered by the presence of symbols on the mechanism, allowing the disc to be read.

Art and Inscriptions

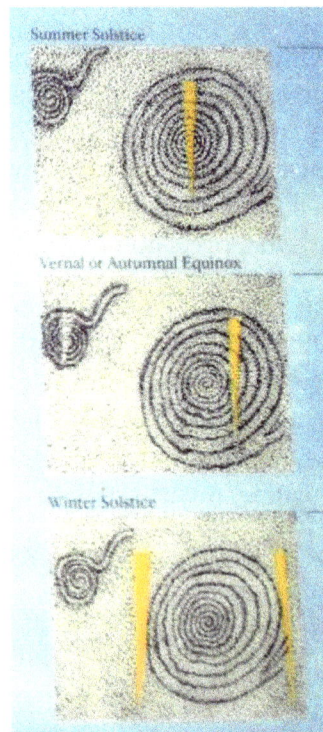

Diagram showing the location of the sun daggers on the Fajada Butte petroglyph on various days

Art and inscriptions may not be confined to artefacts, but also appear painted or inscribed on an archaeological site. Sometimes inscriptions are helpful enough to give instructions to a site's use. For example, a Greek inscription on a stele (from Itanos) has been translated as:"Patron set this up for Zeus Epopsios. Winter solstice. Should anyone wish to know: off 'the little pig' and the stele the sun turns." From Mesoamerica come Mayan and Aztec codices. These are folding books made from Amatl, processed tree bark on which are glyphs in Mayan or Aztec script. The Dresden codex contains information regarding the Venus cycle, confirming its importance to the Mayans.

More problematic are those cases where the movement of the Sun at different times and seasons causes light and shadow interactions with petroglyphs. A widely known example is the Sun Dagger of Fajada Butte at which a glint of sunlight passes over a spiral petroglyph. The location of a dagger of light on the petroglyph varies throughout the year. At the summer solstice a dagger can be seen through the heart of the spiral; at the winter solstice two daggers appear to either side of it. It is proposed that this petroglyph was created to mark these events. Recent studies have identified many similar sites in the US Southwest and Northwestern Mexico. It has been argued that the number of solstitial markers at these sites provides statistical evidence that they were intended to mark the solstices. The Sun Dagger site on Fajada Butte in Chaco Canyon, New Mexico, stands out for its explicit light markings that record all the key events of both the solar and lunar cycles: summer solstice, winter solstice, equinox, and the major and minor lunar standstills of the moon's 18.6 year cycle. In addition at two other sites on Fajada Butte, there are five light markings on petroglyphs recording the summer and winter solstices, equinox and solar noon. Numerous buildings and interbuilding alignments of the great houses of Chaco Canyon and outlying areas are oriented to the same solar and lunar directions that are marked at the Sun Dagger site.

If no ethnographic nor historical data are found which can support this assertion then acceptance of the idea relies upon whether or not there are enough petroglyph sites in North America that such a correlation could occur by chance. It is helpful when petroglyphs are associated with existing peoples. This allows ethnoastronomers to question informants as to the meaning of such symbols.

Ethnographies

As well as the materials left by peoples themselves, there are also the reports of other who have encountered them. The historical records of the Conquistadores are a rich source of information about the pre-Columbian Americans. Ethnographers also provide material about many other peoples.

Aveni uses the importance of zenith passages as an example of the importance of ethnography. For peoples living between the tropics of Cancer and Capricorn there are two days of the year when the noon Sun passes directly overhead and casts no shadow. In parts of Mesoamerica this was considered a significant day as it would herald the arrival of rains, and so play a part in the cycle of agriculture. This knowledge is still considered important amongst Mayan Indians living in Central America today. The ethnographic records suggested to archaeoastronomers that this day may have been important to the ancient Mayans. There are also shafts known as 'zenith tubes' which illuminate subterranean rooms when the sun passes overhead found at places like Monte Albán and Xochicalco. It is only through the ethnography that we can speculate that the timing of the illumination was considered important in Mayan society. Alignments to the sunrise and sunset on the day of the zenith passage have been claimed to exist at several sites. However, it has been shown that, since there are very few orientations that can be related to these phenomena, they likely have different explanations.

Ethnographies also caution against over-interpretation of sites. At a site in Chaco Canyon can be found a pictograph with a star, crescent and hand. It has been argued by some astronomers that this is a record of the 1054 Supernova. However recent reexaminations of related 'supernova petroglyphs' raises questions about such sites in general and anthropological evidence suggests other inrepretations. The Zuni people, who claim a strong ancestral affiliation with Chaco, marked their sun-watching station with a crescent, star, hand and sundisc, similar to those found at the Chaco site.

Ethnoastronomy is also an important field outside of the Americas. For example, anthropological work with Aboriginal Australians is producing much information about their Indigenous astronomies and about their interaction with the modern world.

Recreating the Ancient Sky

"...Although different ways to do science and different scientific results do arise in different cultures, this provides little support for those who would use such differences to question the scienc-es' ability to provide reliable statements about the world in which we live."

—Stephen McCluskey

Once the researcher has data to test, it is often necessary to attempt to recreate ancient sky conditions to place the data in its historical environment.

Declination

To calculate what astronomical features a structure faced a coordinate system is needed. The stars provide such a system. If you were to go outside on a clear night you would observe the stars spinning around the celestial pole. This point is +90° if you are watching the North Celestial Pole or −90° if you are observing the Southern Celestial Pole. The concentric circles the stars trace out are lines of celestial latitude, known as *declination*. The arc connecting the points on the horizon due East and due West (if the horizon is flat) and all points midway between the Celestial Poles is the Celestial Equator which has a declination of 0°. The visible declinations vary depending where you are on the globe. Only an observer on the North Pole of Earth would be unable to see any stars from the Southern Celestial Hemisphere at night. Once a declination has been found for the point on the horizon that a building faces it is then possible to say whether a specific body can be seen in that direction.

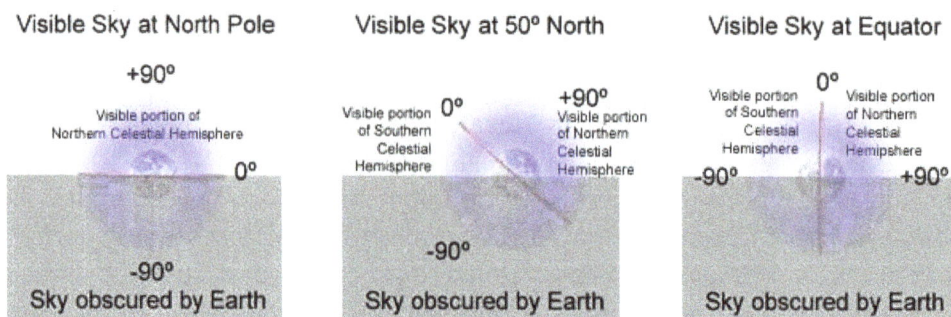

Diagram of the visible portions of sky at varying latitudes.

Solar Positioning

While the stars are fixed to their declinations the Sun is not. The rising point of the Sun varies throughout the year. It swings between two limits marked by the solstices a bit like a pendulum, slowing as it reaches the extremes, but passing rapidly through the midpoint. If an archaeoastronomer can calculate from the azimuth and horizon height that a site was built to view a declination of +23.5° then he or she need not wait until 21 June to confirm the site does indeed face the summer solstice. For more information see History of solar observation.

Lunar Positioning

The Moon's appearance is considerably more complex. Its motion, like the Sun, is between two limits — known as *luni*stices rather than *sol*stices. However, its travel between lunistices is considerably faster. It takes a sidereal month to complete its cycle rather than the year-long trek of the Sun. This is further complicated as the lunistices marking the limits of the Moon's movement move on an 18.6 year cycle. For slightly over nine years the extreme limits of the moon are outside the range of sunrise. For the remaining half of the cycle the Moon never exceeds the limits of the range of sunrise. However, much lunar observation was concerned with the *phase* of the Moon. The cycle from one New Moon to the next runs on an entirely different cycle, the Synodic month. Thus when examining sites for lunar significance the data can appear sparse due the extremely variable nature of the moon. See Moon for more details.

Stellar Positioning

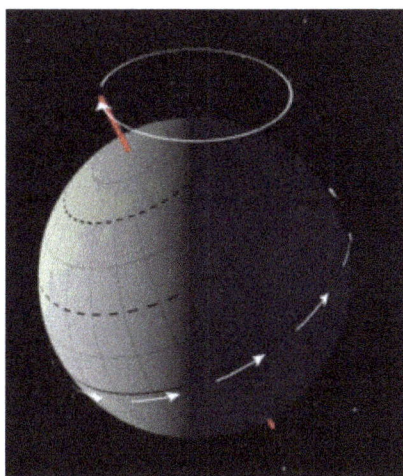

Precessional movement.

Finally there is often a need to correct for the apparent movement of the stars. On the timescale of human civilisation the stars have largely maintained the same position relative to each other. Each night they appear to rotate around the celestial poles due to the Earth's rotation about its axis. However, the Earth spins rather like a spinning top. Not only does the Earth rotate, it wobbles. The Earth's axis takes around 25,800 years to complete one full wobble. The effect to the archaeoastronomer is that stars did not rise over the horizon in the past in the same places as they do today. Nor did the stars rotate around Polaris as they do now. In the case of the Egyptian pyramids, it has been shown they were aligned towards Thuban, a faint star in the constellation of Draco. The effect can be substantial over relatively short lengths of time, historically speaking. For instance a person born on 25 December in Roman times would have been born with the sun in the constellation Capricorn. In the modern period a person born on the same date would have the sun in Sagittarius due to the precession of the equinoxes.

Transient Phenomena

Additionally there are often transient phenomena, events which do not happen on an annual cycle. Most predictable are events like eclipses. In the case of solar eclipses these can be used to date events in the past. A solar eclipse mentioned by Herodotus enables us to date a battle between the Medes and the Lydians, which following the eclipse failed to happen, to 28 May, 585 BC. Other easily calculated events are supernovae whose remains are visible to astronomers and therefore their positions and magnitude can be accurately calculated.

Halley's Comet depicted on the Bayeux tapestry

Some comets are predictable, most famously Halley's Comet. Yet as a class of object they remain unpredictable and can appear at any time. Some have extremely lengthy orbital periods which means their past appearances and returns cannot be predicted. Others may have only ever passed through the Solar System once and so are inherently unpredictable.

Meteor showers should be predictable, but some meteors are cometary debris and so require calculations of orbits which are currently impossible to complete. Other events noted by ancients include aurorae, sun dogs and rainbows all of which are as impossible to predict as the ancient weather, but nevertheless may have been considered important phenomena.

Major Topics of Archaeoastronomical Research

"What has astronomy brought into the lives of cultural groups throughout history? The answers are many and varied..."

— *Von Del Chamberlain and M. Jane Young*

The Use of Calendars

A common justification for the need for astronomy is the need to develop an accurate calendar for agricultural reasons. Ancient texts like Hesiod's Works and Days, an ancient farming manual, would appear to contradict this. Instead astronomical observations are used in combination with ecological signs, such as bird migrations to determine the seasons. Ethnoastronomical work with the Mursi of Ethiopia shows that haphazard astronomy continued until recent times in some parts of the world. All the same, calendars appear to be an almost universal phenomenon in societies as they provide tools for the regulation of communal activities.

An example of a non-agricultural calendar is the *Tzolk'in* calendar of the Maya civilization of pre-Columbian Mesoamerica, which is a cycle of 260 days. This count is based on an earlier calendar and is found throughout Mesoamerica. This formed part of a more comprehensive system of Maya calendars which combined a series of astronomical observations and ritual cycles.

Other peculiar calendars include ancient Greek calendars. These were nominally lunar, starting with the New Moon. In reality the calendar could pause or skip days with confused citizens inscribing dates by both the civic calendar and *ton theoi*, by the moon. The lack of any universal calendar for ancient Greece suggests that coordination of panhellenic events such as games or rituals could be difficult and that astronomical symbolism may have been used as a politically neutral form of timekeeping. Orientation measurements in Greek temples and Byzantine churches have been associated to deity's name day, festivities, and special events.

Myth and Cosmology

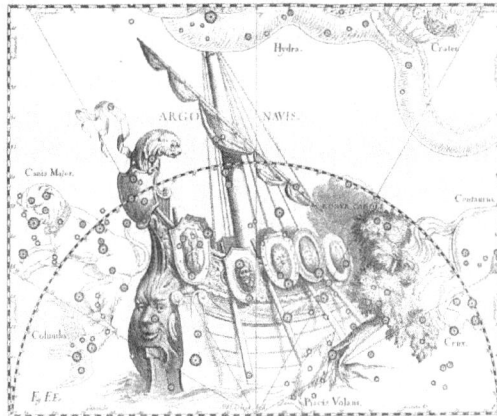

The constellation Argo Navis drawn by Johannes Hevelius in 1690.

Another motive for studying the sky is to understand and explain the universe. In these cultures myth was a tool for achieving this and the explanations, while not reflecting the standards of modern science, are cosmologies.

The Incas arranged their empire to demonstrate their cosmology. The capital, Cusco, was at the centre of the empire and connected to it by means of ceques, conceptually straight lines radiating out from the centre. These ceques connected the centre of the empire to the four *suyus*, which were regions defined by their direction from Cusco. The notion of a quartered cosmos is common across the Andes. Gary Urton, who has conducted fieldwork in the Andean villagers of Misminay, has connected this quartering with the appearance of the Milky Way in the night sky. In one season it will bisect the sky and in another bisect it in a perpendicular fashion.

The importance of observing cosmological factors is also seen on the other side of the world. The Forbidden City in Beijing is laid out to follow cosmic order though rather than observing four directions. The Chinese system was composed of five directions: North, South, East, West and Centre. The Forbidden City occupied the centre of ancient Beijing. One approaches the Emperor from the south, thus placing him in front of the circumpolar stars. This creates the situation of the heavens revolving around the person of the Emperor. The Chinese cosmology is now better known through its export as feng shui.

There is also much information about how the universe was thought to work stored in the mythology of the constellations. The Barasana of the Amazon plan part of their annual cycle based on observation of the stars. When their constellation of the Caterpillar-Jaguar (roughly equivalent to the modern Scorpius) falls they prepare to catch the pupating caterpillars of the forest as they fall from the trees. The caterpillars provide food at a season when other foods are scarce.

A more well-known source of constellation myth are the texts of the Greeks and Romans. The origin of their constellations remains a matter of vigorous and occasionally fractious debate.

The loss of one of the sisters, Merope, in some Greek myths may reflect an astronomical event wherein one of the stars in the Pleiades disappeared from view by the naked eye.

Giorgio de Santillana, professor of the History of Science in the School of Humanities at the Massachusetts Institute of Technology, along with Hertha von Dechend believed that the old mythological stories handed down from antiquity were not random fictitious tales but were accurate depictions of celestial cosmology clothed in tales to aid their oral transmission. The chaos, monsters and violence in ancient myths are representative of the forces that shape each age. They believed that ancient myths are the remains of preliterate astronomy that became lost with the rise of the Greco-Roman civilization. Santillana and von Dechend in their book *Hamlet's Mill, An Essay on Myth and the Frame of Time* (1969) clearly state that ancient myths have no historical or factual basis other than a cosmological one encoding astronomical phenomena, especially the precession of the equinoxes. Santillana and von Dechend's approach is not widely accepted.

Displays of Power

The Precinct of Amun-Re was aligned on the midwinter solstice.

By including celestial motifs in clothing it becomes possible for the wearer to make claims the power on Earth is drawn from above. It has been said that the Shield of Achilles described by Homer is also a catalogue of constellations. In North America shields depicted in Comanche petroglyphs appear to include Venus symbolism.

Solsticial alignments also can be seen as displays of power. When viewed from a ceremonial plaza on the Island of the Sun (the mythical origin place of the Sun) in Lake Titicaca, the Sun was seen to rise at the June solstice between two towers on a nearby ridge. The sacred part of the island was separated from the remainder of it by a stone wall and ethnographic records indicate that access to the sacred space was restricted to members of the Inca ruling elite. Ordinary pilgrims stood on a platform outside the ceremonial area to see the solstice Sun rise between the towers.

In Egypt the temple of Amun-Re at Karnak has been the subject of much study. Evaluation of the site, taking into account the change over time of the obliquity of the ecliptic show that the Great Temple was aligned on the rising of the midwinter sun. The length of the corridor down which sunlight would travel would have limited illumination at other times of the year.

In a later period the Serapeum in Alexandria was also said to have contained a solar alignment so that, on a specific sunrise, a shaft of light would pass across the lips of the statue of Serapis thus symbolising the Sun saluting the god.

Major Sites of Archaeoastronomical Interest

Clive Ruggles and Michel Cotte recently edited a book on heritage sites of astronomy and archaeo-astronomy that provides a list of the main sites around the world.

"At Stonehenge in England and at Carnac in France, in Egypt and Yucatán, across the whole face of the earth, are found mysterious ruins of ancient monuments, monuments with astronomical significance... They mark the same kind of commitment that transported us to the moon and our spacecraft to the surface of Mars."

—Edwin Krupp

Newgrange

The sunlight enters the tomb at Newgrange via the roofbox built above the door.

Newgrange is a passage tomb in the Republic of Ireland dating from around 3,300 to 2,900 BC For a few days around the Winter Solstice light shines along the central passageway into the heart of the tomb. What makes this notable is not that light shines in the passageway, but that it does not do so through the main entrance. Instead it enters via a hollow box above the main doorway discovered by Michael O'Kelly. It is this roofbox which strongly indicates that the tomb was built with an astronomical aspect in mind. Clive Ruggles notes:

"...Few people - archaeologists or astronomers- have doubted that a powerful astronomical symbolism was deliberately incorporated into the monument, demonstrating that a connection between astronomy and funerary ritual, at the very least, merits further investigation."

Egypt

The pyramids of Giza

Since the first modern measurements of the precise cardinal orientations of the pyramids by Flinders Petrie, various astronomical methods have been proposed for the original establishment of these orientations. It was recently proposed that this was done by observing the positions of two stars in the Plough / Big Dipper which was known to Egyptians as the thigh. It is thought that a vertical alignment between these two stars checked with a plumb bob was used to ascertain where north lay. The deviations from true north using this model reflect the accepted dates of construction.

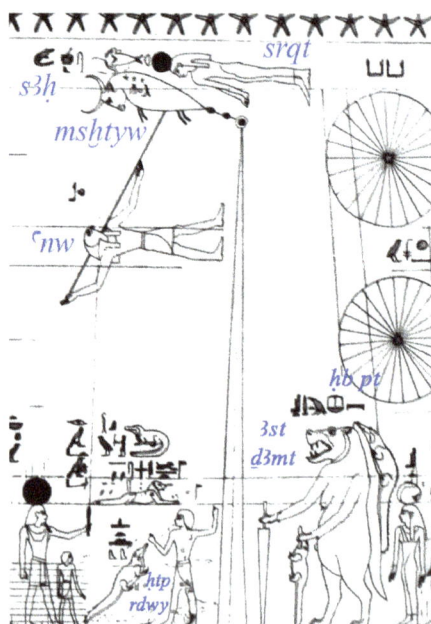

Constellations on the astronomical ceiling of Senemut Tomb

Some have argued that the pyramids were laid out as a map of the three stars in the belt of Orion, although this theory has been criticized by reputable astronomers. The site was instead probably governed by a spectacular hierophany which occurs at the summer solstice, when the sun, viewed from the Sphinx terrace, forms - together with the two giant pyramids - the symbol Akhet, which was also the name of the Great Pyramid. Further, the south east corners of all the 3 pyramids align towards the temple of Heliopolis, as first discovered by the Egyptologist Mark Lehner.

The astronomical ceiling of the tomb of Senenmut (ca 1470 BC) contains the Celestial Diagram depicting circumpolar constellations in the form of discs. Each disc is divided into 24 sections suggesting a 24-hour time period. Constellations are portrayed as sacred deities of Egypt. The observation of lunar cycles is also evident.

El Castillo

El Castillo, also known as Kukulcán's Pyramid, is a Mesoamerican step-pyramid built in the centre of Mayan center of Chichen Itza in Mexico. Several architectural features have suggested astronomical elements. Each of the stairways built into the sides of the pyramid has 91 steps. Along with the extra one for the platform at the top, this totals 365 steps, which is possibly one for each day of the year (365.25) or the number of lunar orbits in 10,000 rotations (365.01).

Plumed Serpent

A visually striking effect is seen every March and September as an unusual shadow occurs on the equinoxes. A shadow appears to descend the west balustrade of the northern stairway. The visual effect is of a serpent descending the stairway, with its head at the base in light. Additionally the western face points to sunset around 25 May, traditionally the date of transition from the dry to the rainy season.

Stonehenge

The sun rising over Stonehenge at the 2005 Summer Solstice.

Many astronomical alignments have been claimed for Stonehenge, a complex of megaliths and earthworks in the Salisbury Plain of England. The most famous of these is the midsummer alignment, where the Sun rises over the Heel Stone. However, this interpretation has been challenged by some archaeologists who argue that the midwinter alignment, where the viewer is outside Stonehenge and sees the sun setting in the henge, is the more significant alignment, and the midsummer alignment may be a coincidence due to local topography.

As well as solar alignments, there are proposed lunar alignments. The four station stones mark out a rectangle. The short sides point towards the midsummer sunrise and midwinter sunset. The long sides if viewed towards the south-east, face the most southerly rising of the moon. Aveni notes that these lunar alignments have never gained the acceptance that the solar alignments have received. The Heel Stone azimuth is one-seventh of circumference, matching the latitude of Avebury, while summer solstice sunrise azimuth is no longer equal to the construction era direction.

Maeshowe

The interior of Maeshowe chambered tomb.

This is an architecturally outstanding Neolithic chambered tomb on the Mainland of Orkney, Scotland – probably dating to the early 3rd millennium BC, and where the setting sun at midwinter shines down the entrance passage into the central chamber. In the 1990s further investigations were carried out to discover whether this was an accurate or an approximate solar alignment. Several new aspects of the site were discovered. In the first place the entrance passage faces the hills of the island Hoy, about 10 miles away. Secondly, it consists of two straight lengths, angled at a few degrees to each other. Thirdly, the outer part is aligned towards the midwinter sunset position on a level horizon just to the left of Ward Hill on Hoy. Fourthly the inner part points directly at the Barnhouse standing stone about 400m away and then to the right end of the summit of Ward Hill, just before it dips down to the notch between it at Cuilags to the right. This indicated line points to sunset on the first Sixteenths of the solar year (according to A. Thom) before and after the winter solstice and the notch at the base of the right slope of the Hill is at the same declination. Fourthly a similar 'double sunset' phenomenon is seen at the right end of Cuilags, also on Hoy; here the date is the first Eighth of the year before and after the winter solstice, at the beginning of November and February respectively – the Old Celtic festivals of Samhain and Imbolc. This alignment is not indicated by an artificial structure but gains plausibility from the other two indicated lines. Maeshowe is thus an extremely sophisticated calendar site which must have been positioned carefully in order to use the horizon foresights in the ways described.

Uxmal

Uxmal is a Mayan city in the Puuc Hills of Yucatán Peninsula, Mexico. The Governor's Palace at Uxmal is often used as an exemplar of why it is important to combine ethnographic and alignment data. The palace is aligned with an azimuth of 118° on the pyramid of Cehtzuc. This alignment corresponds approximately to the southernmost rising and, with a much greater precision, to the northernmost setting of Venus; both phenomena occur once every eight years. By itself this would not be sufficient to argue for a meaningful connection between the two events. The palace has to be aligned in one direction or another and why should the rising of Venus be any more important than the rising of the Sun, Moon, other planets, Sirius *et cetera*? The answer given is that not only does the palace point towards significant points of Venus, it is also covered in glyphs which stand for Venus and Mayan zodiacal constellations. Moreover, the great northerly extremes of Venus always occur in late April or early May, coinciding with the onset of the rainy season. The Venus glyphs placed in the cheeks of the Maya rain god Chac, most likely referring to the concomitance of these phenomena, support the west-working orientation scheme.

The Palace of the Governor at Uxmal.

Chaco Canyon

The Great Kiva at Chaco Canyon.

In Chaco Canyon, the center of the ancient Pueblo culture in the American Southwest, numerous solar and lunar light markings and architectural and road alignments have been documented. These findings date to the 1977 discovery of the Sun Dagger site by Anna Sofaer. Three large stone slabs leaning against a cliff channel light and shadow markings onto two spiral petroglyphs on the cliff wall, marking the solstices, equinoxes and the lunar standstills of the 18.6 year cycle of the moon. Subsequent research by the Solstice Project and others demonstrated that numerous building and interbuilding alignments of the great houses of Chaco Canyon are oriented to solar, lunar and cardinal directions. In addition, research shows that the Great North Road, a thirty-five mile engineered "road", was constructed not for utilitarian purposes but rather to connect the ceremonial center of Chaco Canyon with the direction north.

Lascaux Cave

According to Rappenglueck, the eyes of the bull, the bird, and the bird-man may represent the three stars Vega, Altair, and Deneb commonly known as the Summer Triangle.

In recent years, new research has suggested that the Lascaux cave paintings in France may incorporate prehistoric star charts. Michael Rappenglueck of the University of Munich argues that some of the non-figurative dot clusters and dots within some of the figurative images correlate with the constellations of Taurus, the Pleiades and the grouping known as the "Summer Triangle". Based on her own study of the astronomical significance of Bronze Age petroglyphs in the Vallée des Merveilles and her extensive survey of other prehistoric cave painting sites in the region—most of which appear to have been selected because the interiors are illuminated by the setting sun on the day of the winter solstice—French researcher Chantal Jègues-Wolkiewiez has further proposed that the gallery of figurative images in the Great Hall represents an extensive star map and that key points on major figures in the group correspond to stars in the main constellations as they appeared in the Paleolithic. Applying phylogenetics to myths of the Cosmic Hunt, Julien d'Huy suggested that the palaeolithic version of this story could be the following: there is an animal that is a horned herbivore, especially an elk. One human pursues this ungulate. The hunt locates or gets to the sky. The animal is alive when it is transformed into a constellation. It forms the Big Dipper. This story may be represented in the famous Lascaux shaft 'scene'

Fringe Archaeoastronomy

At least now we have all the archaeological facts to go along with the astronomers, the Druids, the Flat Earthers and all the rest.

—Sir Jocelyn Stephens

Archaeoastronomy owes something of this poor reputation among scholars to its occasional misuse to advance a range of pseudo-historical accounts. During the 1930s, Otto S. Reuter compiled a study entitled *Germanische Himmelskunde*, or "Teutonic Skylore". The astronomical orientations of ancient monuments claimed by Reuter and his followers would place the ancient Germanic peoples ahead of the Ancient Near East in the field of astronomy, demonstrating the intellectual superiority of the "Aryans" (Indo-Europeans) over the Semites.

Since the 19th century, numerous scholars have sought to use archaeoastronomical calculations to demonstrate the antiquity of Ancient Indian Vedic culture, computing the dates of astronomical observations ambiguously described in ancient poetry to as early as 4000 BCE. David Pingree, a historian of Indian astronomy, condemned "the scholars who perpetrate wild theories of prehistoric science and call themselves archaeoastronomers."

More recently Gallagher, Pyle, and Fell interpreted inscriptions in West Virginia as a description in Celtic Ogham alphabet of the supposed winter solstitial marker at the site. The controversial translation was supposedly validated by a problematic archaeoastronomical indication in which the winter solstice sun shone on an inscription of the sun at the site. Subsequent analyses criticized its cultural inappropriateness, as well as its linguistic and archeaoastronomical claims, to describe it as an example of "cult archaeology".

Archaeoastronomy is sometimes related to the fringe discipline of Archaeocryptography, when its followers attempt to find underlying mathematical orders beneath the proportions, size, and placement of archaeoastronomical sites such as Stonehenge and the Pyramid of Kukulcán at Chichen Itza.

Archaeoastronomical Organisations and Publications

There are currently three academic organisations for scholars of archaeoastronomy. ISAAC—the International Society for Archaeoastronomy and Astronomy in Culture—was founded in 1995 and now sponsors the Oxford conferences and *Archaeoastronomy — the Journal of Astronomy in Culture*. SEAC— La Société Européenne pour l'Astronomie dans la Culture—is slightly older; it was created in 1992. SEAC holds annual conferences in Europe and publishes refereed conference proceedings on an annual basis. There is also SIAC— La Sociedad Interamericana de Astronomía en la Cultura, primarily a Latin American organisation which was founded in 2003. Two new organisations focused on regional archaeoastronomy were founded in 2013: ASIA - the Australian Society for Indigenous Astronomy in Australia and SMART - the Society of Māori Astronomy Research and Traditions in New Zealand.

Additionally the *Journal for the History of Astronomy* publishes many archaeoastronomical papers. For twenty-seven volumes (from 1979 to 2002) it published an annual supplement *Archaeoastronomy*. The *Journal of Astronomical History and Heritage* (National Astronomical Research Institute of Thailand), *Culture & Cosmos* (University of Wales, UK) and *Mediterranean Archaeology and Archaeometry* (University of Aegean, Greece) also publish papers on archaeoastronomy.

Various national archaeoastronomical projects have been undertaken. Among them is the program at the Tata Institute of Fundamental Research named "Archaeo Astronomy in Indian Context" that has made interesting findings in this field.

Astrochemistry

Astrochemistry is the study of the abundance and reactions of chemical elements and molecules in the universe, and their interaction with radiation. The discipline is an overlap of astronomy and chemistry. The word "astrochemistry" may be applied to both the Solar System and the interstellar medium. The study of the abundance of elements and isotope ratios in Solar System objects, such as meteorites, is also called cosmochemistry, while the study of interstellar atoms and molecules and their interaction with radiation is sometimes called molecular astrophysics. The formation, atomic and chemical composition, evolution and fate of molecular gas clouds is of special interest, because it is from these clouds that solar systems form.

One particularly important experimental tool in astrochemistry is spectroscopy, the use of telescopes to measure the absorption and emission of light from molecules and atoms in various environments. By comparing astronomical observations with laboratory measurements, astrochemists can infer the elemental abundances, chemical composition, and temperatures of stars and interstellar clouds. This is possible because ions, atoms, and molecules have characteristic spectra: that is, the absorption and emission of certain wavelengths (colors) of light, often not visible to the human eye. However, these measurements have limitations, with various types of radiation (radio, infrared, visible, ultraviolet etc.) able to detect only certain types of species, depending on the chemical properties of the molecules. Interstellar formaldehyde was the first organic molecule detected in the interstellar medium.

Perhaps the most powerful technique for detection of individual chemical species is radio astronomy, which has resulted in the detection of over a hundred interstellar species, including radicals and ions, and organic (i.e. carbon-based) compounds, such as alcohols, acids, aldehydes, and ketones. One of the most abundant interstellar molecules, and among the easiest to detect with radio waves (due to its strong electric dipole moment), is CO (carbon monoxide). In fact, CO is such a common interstellar molecule that it is used to map out molecular regions. The radio observation of perhaps greatest human interest is the claim of interstellar glycine, the simplest amino acid, but with considerable accompanying controversy. One of the reasons why this detection was controversial is that although radio (and some other methods like rotational spectroscopy) are good for the identification of simple species with large dipole moments, they are less sensitive to more complex molecules, even something relatively small like amino acids.

Moreover, such methods are completely blind to molecules that have no dipole. For example, by far the most common molecule in the universe is H_2 (hydrogen gas), but it does not have a dipole moment, so it is invisible to radio telescopes. Moreover, such methods cannot detect species that are not in the gas-phase. Since dense molecular clouds are very cold (10 to 50 K [−263.1 to −223.2 °C; −441.7 to −369.7 °F]), most molecules in them (other than hydrogen) are frozen, i.e. solid. Instead, hydrogen and these other molecules are detected using other wavelengths of light. Hydrogen is easily detected in the ultraviolet (UV) and visible ranges from its absorption and emission of light (the hydrogen line). Moreover, most organic compounds absorb and emit light in the infrared (IR) so, for example, the detection of methane in the atmosphere of Mars was achieved using an IR ground-based telescope, NASA's 3-meter Infrared Telescope Facility atop Mauna Kea, Hawaii. NASA also has an airborne IR telescope called SOFIA and an IR space telescope called Spitzer. Somewhat related to the recent detection of methane in the atmosphere of Mars, scientists reported, in June 2012, that measuring the ratio of hydrogen and methane levels on Mars may help determine the likelihood of life on Mars. According to the scientists, "... low H_2/CH_4 ratios (less than approximately 40) indicate that life is likely present and active." Other scientists have recently reported methods of detecting hydrogen and methane in extraterrestrial atmospheres.

Infrared astronomy has also revealed that the interstellar medium contains a suite of complex gas-phase carbon compounds called polyaromatic hydrocarbons, often abbreviated PAHs or PACs. These molecules, composed primarily of fused rings of carbon (either neutral or in an ionized state), are said to be the most common class of carbon compound in the galaxy. They are also the most common class of carbon molecule in meteorites and in cometary and asteroidal dust (cosmic dust). These compounds, as well as the amino acids, nucleobases, and many other compounds in meteorites, carry deuterium and isotopes of carbon, nitrogen, and oxygen that are very rare on earth, attesting to their extraterrestrial origin. The PAHs are thought to form in hot circumstellar environments (around dying, carbon-rich red giant stars).

Infrared astronomy has also been used to assess the composition of solid materials in the interstellar medium, including silicates, kerogen-like carbon-rich solids, and ices. This is because unlike visible light, which is scattered or absorbed by solid particles, the IR radiation can pass through the microscopic interstellar particles, but in the process there are absorptions at certain wavelengths that are characteristic of the composition of the grains. As above with radio astronomy, there are certain limitations, e.g. N_2 is difficult to detect by either IR or radio astronomy.

Such IR observations have determined that in dense clouds (where there are enough particles to attenuate the destructive UV radiation) thin ice layers coat the microscopic particles, permitting some low-temperature chemistry to occur. Since hydrogen is by far the most abundant molecule in the universe, the initial chemistry of these ices is determined by the chemistry of the hydrogen. If the hydrogen is atomic, then the H atoms react with available O, C and N atoms, producing "reduced" species like H_2O, CH_4, and NH_3. However, if the hydrogen is molecular and thus not reactive, this permits the heavier atoms to react or remain bonded together, producing CO, CO_2, CN, etc. These mixed-molecular ices are exposed to ultraviolet radiation and cosmic rays, which results in complex radiation-driven chemistry. Lab experiments on the photochemistry of simple interstellar ices have produced amino acids. The similarity between interstellar and cometary ices (as well as comparisons of gas phase compounds) have been invoked as indicators of a connection between interstellar and cometary chemistry. This is somewhat supported by the results of the analysis of the organics from the comet samples returned by the Stardust mission but the minerals also indicated a surprising contribution from high-temperature chemistry in the solar nebula.

Research

Research is progressing on the way in which interstellar and circumstellar molecules form and interact, and this research could have a profound impact on our understanding of the suite of molecules that were present in the molecular cloud when our solar system formed, which contributed to the rich carbon chemistry of comets and asteroids and hence the meteorites and interstellar dust particles which fall to the Earth by the ton every day.

Transition from atomic to molecular gas at the border of the Orion molecular cloud.

The sparseness of interstellar and interplanetary space results in some unusual chemistry, since symmetry-forbidden reactions cannot occur except on the longest of timescales. For this reason, molecules and molecular ions which are unstable on Earth can be highly abundant in space, for example the H_3^+ ion. Astrochemistry overlaps with astrophysics and nuclear physics in characterizing the nuclear reactions which occur in stars, the consequences for stellar evolution, as well as stellar 'generations'. Indeed, the nuclear reactions in stars produce every naturally occurring chemical element. As the stellar 'generations' advance, the mass of the newly formed elements increases. A first-generation star uses elemental hydrogen (H) as a fuel source and produces helium (He). Hydrogen is the most abundant element, and it is the basic building block for all other elements as its nucleus has only one proton. Gravitational pull toward the center of a star creates massive amounts of heat and pressure, which cause nuclear fusion. Through this process of merging nuclear mass, heavier elements are formed. Carbon, oxygen and silicon are examples of elements that form in stellar fusion. After many stellar generations, very heavy elements are formed (e.g. iron and lead).

In October 2011, scientists reported that cosmic dust contains organic matter ("amorphous organic solids with a mixed aromatic-aliphatic structure") that could be created naturally, and rapidly, by stars.

On August 29, 2012, and in a world first, astronomers at Copenhagen University reported the detection of a specific sugar molecule, glycolaldehyde, in a distant star system. The molecule was found around the protostellar binary *IRAS 16293-2422*, which is located 400 light years from Earth. Glycolaldehyde is needed to form ribonucleic acid, or RNA, which is similar in function to DNA. This finding suggests that complex organic molecules may form in stellar systems prior to the formation of planets, eventually arriving on young planets early in their formation.

In September, 2012, NASA scientists reported that polycyclic aromatic hydrocarbons (PAHs), subjected to interstellar medium (ISM) conditions, are transformed, through hydrogenation, oxygenation and hydroxylation, to more complex organics - "a step along the path toward amino acids and nucleotides, the raw materials of proteins and DNA, respectively". Further, as a result of these transformations, the PAHs lose their spectroscopic signature which could be one of the reasons "for the lack of PAH detection in interstellar ice grains, particularly the outer regions of cold, dense clouds or the upper molecular layers of protoplanetary disks."

In February 2014, NASA announced the creation of an improved spectral database for tracking polycyclic aromatic hydrocarbons (PAHs) in the universe. According to scientists, more than 20% of the carbon in the universe may be associated with PAHs, possible starting materials for the formation of life. PAHs seem to have been formed shortly after the Big Bang, are widespread throughout the universe, and are associated with new stars and exoplanets.

On August 11, 2014, astronomers released studies, using the Atacama Large Millimeter/Submillimeter Array (ALMA) for the first time, that detailed the distribution of HCN, HNC, H_2CO, and dust inside the comae of comets C/2012 F6 (Lemmon) and C/2012 S1 (ISON).

For the study of the recourses of chemical elements and molecules in the universe is developed the mathematical model of the molecules composition distribution in the interstellar environment on thermodynamic potentials by professor M.Yu. Dolomatov using methods of the probability theory, the mathematical and physical statistics and the equilibrium thermodynamics. Based on this model are estimated the resources of life-related molecules, amino acids and the nitrogenous bases in the interstellar medium. The possibility of the oil hydrocarbons molecules formation is shown. The given calculations confirm Sokolov's and Hoyl's hypotheses about the possibility of the oil hydrocarbons formation in Space. Results are confirmed by data of astrophysical supervision and space researches.

In July 2015, scientists reported that upon the first touchdown of the *Philae* lander on comet 67/P's surface, measurements by the COSAC and Ptolemy instruments revealed sixteen organic compounds, four of which were seen for the first time on a comet, including acetamide, acetone, methyl isocyanate and propionaldehyde.

Cosmochemistry

Meteorites are often studied as part of cosmochemistry.

Cosmochemistry or chemical cosmology is the study of the chemical composition of matter in the universe and the processes that led to those compositions. This is done primarily through the study of the chemical composition of meteorites and other physical samples. Given that the asteroid parent bodies of meteorites were some of the first solid material to condense from the early solar nebula, cosmochemists are generally, but not exclusively, concerned with the objects contained within the Solar System.

History

In 1938, Swiss mineralogist Victor Goldschmidt and his colleagues compiled a list of what they called "cosmic abundances" based on their analysis of several terrestrial and meteorite samples. Goldschmidt justified the inclusion of meteorite composition data into his table by claiming that terrestrial rocks were subjected to a significant amount of chemical change due to the inherent processes of the Earth and the atmosphere. This meant that studying terrestrial rocks exclusively would not yield an accurate overall picture of the chemical composition of the cosmos. Therefore, Goldschmidt concluded that extraterrestrial material must also be included to produce more accurate and robust data. This research is considered to be the foundation of modern cosmochemistry.

During the 1950s and 1960s, cosmochemistry became more accepted as a science. Harold Urey, widely considered to be one of the fathers of cosmochemistry, engaged in research that eventually led to an understanding of the origin of the elements and the chemical abundance of stars. In 1956, Urey and his colleague, German scientist Hans Suess, published the first table of cosmic abundances to include isotopes based on meteorite analysis.

The continued refinement of analytical instrumentation throughout the 1960s, especially that of mass spectrometry, allowed cosmochemists to perform detailed analyses of the isotopic abundances of elements within meteorites. in 1960, John Reynolds determined, through the analysis of short-lived nuclides within meteorites, that the elements of the Solar System were formed before the Solar System itself which began to establish a timeline of the processes of the early Solar System.

Meteorites

Meteorites are one of the most important tools that cosmochemists have for studying the chemical nature of the Solar System. Many meteorites come from material that is as old as the Solar System itself, and thus provide scientists with a record from the early solar nebula. Carbonaceous chondrites are especially primitive; that is they have retained many of their chemical properties since their formation 4.56 billion years ago, and are therefore a major focus of cosmochemical investigations.

The most primitive meteorites also contain a small amount of material (< 0.1%) which is now recognized to be presolar grains that are older than the Solar System itself, and which are derived directly from the remnants of the individual supernovae that supplied the dust from which the Solar System formed. These grains are recognizable from their exotic chemistry which is alien to the Solar System (such as matrixes of graphite, diamond, or silicon carbide). They also often have isotope ratios which are not those of the rest of the Solar System (in particular, the Sun), and which differ from each other, indicating sources in a number of different explosive supernova events. Meteorites also may contain interstellar dust grains, which have collected from non-gaseous elements in the interstellar medium, as one type of composite cosmic dust ("stardust")

Recent findings by NASA, based on studies of meteorites found on Earth, suggests DNA and RNA components (adenine, guanine and related organic molecules), building blocks for life as we know it, may be formed extraterrestrially in outer space.

Comets

On 30 July 2015, scientists reported that upon the first touchdown of the *Philae* lander on comet 67/P's surface, measurements by the COSAC and Ptolemy instruments revealed sixteen organic compounds, four of which were seen for the first time on a comet, including acetamide, acetone, methyl isocyanate and propionaldehyde.

Research

In 2004, scientists reported detecting the spectral signatures of anthracene and pyrene in the ultraviolet light emitted by the Red Rectangle nebula (no other such complex molecules had ever been found before in outer space). This discovery was considered a confirmation of a hypothesis that as nebulae of the same type as the Red Rectangle approach the ends of their lives, convection currents cause carbon and hydrogen in the nebulae's core to get caught in stellar winds, and radiate outward. As they cool, the atoms supposedly bond to each other in various ways and eventually form particles of a million or more atoms. The scientists inferred that since they discovered polycyclic aromatic hydrocarbons (PAHs)—which may have been vital in the formation of early life on Earth—in a nebula, by necessity they must originate in nebulae.

In August 2009, NASA scientists identified one of the fundamental chemical building-blocks of life (the amino acid glycine) in a comet for the first time.

In 2010, fullerenes (or "buckyballs") were detected in nebulae. Fullerenes have been implicated in the origin of life; according to astronomer Letizia Stanghellini, "It's possible that buckyballs from outer space provided seeds for life on Earth."

In August 2011, findings by NASA, based on studies of meteorites found on Earth, suggests DNA and RNA components (adenine, guanine and related organic molecules), building blocks for life as we know it, may be formed extraterrestrially in outer space.

In October 2011, scientists reported that cosmic dust contains complex organic matter ("amorphous organic solids with a mixed aromatic-aliphatic structure") that could be created naturally, and rapidly, by stars.

On August 29, 2012, astronomers at Copenhagen University reported the detection of a specific sugar molecule, glycolaldehyde, in a distant star system. The molecule was found around the protostellar binary *IRAS 16293-2422*, which is located 400 light years from Earth. Glycolaldehyde is needed to form ribonucleic acid, or RNA, which is similar in function to DNA. This finding suggests that complex organic molecules may form in stellar systems prior to the formation of planets, eventually arriving on young planets early in their formation.

In September 2012, NASA scientists reported that polycyclic aromatic hydrocarbons (PAHs), subjected to interstellar medium (ISM) conditions, are transformed, through hydrogenation, oxygenation and hydroxylation, to more complex organics - "a step along the path toward amino acids and nucleotides, the raw materials of proteins and DNA, respectively". Further, as a result of these transformations, the PAHs lose their spectroscopic signature which could be one of the reasons "for the lack of PAH detection in interstellar ice grains, particularly the outer regions of cold, dense clouds or the upper molecular layers of protoplanetary disks."

In 2013, the Atacama Large Millimeter Array (ALMA Project) confirmed that researchers have discovered an important pair of prebiotic molecules in the icy particles in interstellar space (ISM). The chemicals, found in a giant cloud of gas about 25,000 light-years from Earth in ISM, may be a precursor to a key component of DNA and the other may have a role in the formation of an important amino acid. Researchers found a molecule called cyanomethanimine, which produces adenine, one of the four nucleobases that form the "rungs" in the ladder-like structure of DNA.

The other molecule, called ethanamine, is thought to play a role in forming alanine, one of the twenty amino acids in the genetic code. Previously, scientists thought such processes took place in the very tenuous gas between the stars. The new discoveries, however, suggest that the chemical formation sequences for these molecules occurred not in gas, but on the surfaces of ice grains in interstellar space. NASA ALMA scientist Anthony Remijan stated that finding these molecules in an interstellar gas cloud means that important building blocks for DNA and amino acids can 'seed' newly formed planets with the chemical precursors for life.

In January 2014, NASA reported that current studies on the planet Mars by the *Curiosity* and *Opportunity* rovers will now be searching for evidence of ancient life, including a biosphere based on autotrophic, chemotrophic and/or chemolithoautotrophic microorganisms, as well as ancient water, including fluvio-lacustrine environments (plains related to ancient rivers or lakes) that may have been habitable. The search for evidence of habitability, taphonomy (related to fossils), and organic carbon on the planet Mars is now a primary NASA objective.

In February 2014, NASA announced a greatly upgraded database for tracking polycyclic aromatic hydrocarbons (PAHs) in the universe. According to scientists, more than 20% of the carbon in the universe may be associated with PAHs, possible starting materials for the formation of life. PAHs seem to have been formed shortly after the Big Bang, are widespread throughout the universe, and are associated with new stars and exoplanets.

References

- Jordans, Frank (30 July 2015). "Philae probe finds evidence that comets can be cosmic labs". The Washington Post. Associated Press. Retrieved 30 July 2015.

- Various (24 January 2014). "Special Issue – Table of Contents – Exploring Martian Habitability". Science. 343 (6169): 345–452. Retrieved 24 January 2014.

- Hoover, Rachel (February 21, 2014). "Need to Track Organic Nano-Particles Across the Universe? NASA's Got an App for That". NASA. Retrieved February 22, 2014.

- Finley, Dave, Discoveries Suggest Icy Cosmic Start for Amino Acids and DNA Ingredients, The National Radio Astronomy Observatory, Feb. 28, 2013

- ScienceDaily Staff (26 October 2011). "Astronomers Discover Complex Organic Matter Exists Throughout the Universe". ScienceDaily. Retrieved 2011-10-27.

- Callahan, M.P.; Smith, K.E.; et al. (11 August 2011). "Carbonaceous meteorites contain a wide range of extraterrestrial nucleobases". PNAS. doi:10.1073/pnas.1106493108. Retrieved 2011-08-15.

- Steigerwald, John (8 August 2011). "NASA Researchers: DNA Building Blocks Can Be Made in Space". NASA. Retrieved 2011-08-10.

- ScienceDaily Staff (9 August 2011). "DNA Building Blocks Can Be Made in Space, NASA Evidence Suggests". ScienceDaily. Retrieved 2011-08-09.

Permissions

Index